# LONDON MATHEMATICAL SOCIETY LECTURE NOTE SERIES

Managing Editor: Professor N.J. Hitchin, Mathematical Institute,
University of Oxford, 24–29 St Giles, Oxford OX1 3LB, United Kingdom

The titles below are available from booksellers, or, in case of difficulty, from Cambridge University Press.

W0008248

London Mathematical Society Lecture Note Series. 279

# Topics in Symbolic Dynamics and Applications

Edited by

F. Blanchard,
*Institut de Mathématiques de Luminy, CNRS*

A. Maass,
*Universidad de Chile, Santiago*

A. Nogueira
*Univesidade Federal do Rio de Janeiro*

CAMBRIDGE
UNIVERSITY PRESS

CAMBRIDGE UNIVERSITY PRESS
Cambridge, New York, Melbourne, Madrid, Cape Town,
Singapore, São Paulo, Delhi, Mexico City

Cambridge University Press
The Edinburgh Building, Cambridge CB2 8RU, UK

Published in the United States of America by Cambridge University Press, New York

www.cambridge.org
Information on this title: www.cambridge.org/9780521796606

© Cambridge University Press 2000

First published 2000

*A catalogue record for this publication is available from the British Library*

ISBN 978-0-521-79660-6 Paperback

# Contents

# FOREWORD

This volume contains the courses given at the CIMPA–UNESCO Summer School on Symbolic Dynamics and its Applications held at Universidad de la Frontera, Temuco, Chile, from January 6th to 24th, 1997. This School was devised for graduate students and high level undergraduate students interested in dynamical systems emphasizing symbolic dynamics.

The scientific committee was composed of François Blanchard (IML–CNRS–Marseille), Mike Boyle (University of Maryland at College Park), Mike Keane (CWI–Amsterdam), Alejandro Maass (Universidad de Chile), Servet Martínez (Universidad de Chile) and Arnaldo Nogueira (Universidade Federal do Rio de Janeiro).

The book is divided into eight chapters, each one corresponding to a course given at Temuco and devoted to a particular, active area of symbolic dynamics or some application to ergodic theory or number theory. Each author has used his own notation, so each chapter can be considered as an independent one, even though there exist natural relations between them. Each chapter has its own bibliography; all the references in this foreword can be found in at least one of these bibliographies.

By now the reader has understood that our aim was not to make a textbook on symbolic dynamics – such textbooks exist, one is due to D. Lind and B. Marcus (Cambridge U. Press, 1997) and another to B. Kitchens (Springer-Verlag, 1998) – but rather to present the reader with a sampling of what is going on in the field and around it.

## About Symbolic Dynamics

The utilization by Hadamard, at the end of the $XIX^{th}$ century, of infinite sequences of symbols for the study of geodesics on certain surfaces marks the beginning of symbolic dynamics. From that time on symbolic sequences have been used repeatedly for coding various smooth transformations, but the interest for symbolic dynamics among mathematicians was further enhanced by the growing importance of symbolic coding in ergodic information theory; by the discovery of simple but important connections with theoretical computer science; and finally through the upsurge of applications in the fields of engineering and molecular biology, among others.

There are two points of view on symbolic dynamics that almost amount to two distinct definitions of the field. Specialists of smooth dynamics view it as a tool: whenever a smooth system has a "nice" symbolic representation, it is easy to compute its entropy, to obtain informations about the set of invariant measures and other relevant features. Here the basic question is: What symbolic representation can we find for a given smooth system and, if there is one, what are its properties? Of course, it would help to know necessary and sufficient conditions for a system to have symbolic representations at all. On the other hand, "pure" symbolic dynamists study the properties of subshifts in general or those of some class of symbolic systems.

Let us describe briefly the main objects in this theory. One starts with a finite set, say $A$, often called an alphabet. On this alphabet one construct words, that is, finite concatenations of symbols of $A$, $w = a_0...a_n$; and also infinite sequences of symbols of $A$, $(a_i)_{i \in N}$ or $(a_i)_{i \in Z}$. The set of words on the alphabet $A$ is denoted by $A^*$; the set of infinite sequences of symbols in $A$ is denoted by $A^N$ for one–sided sequences and $A^Z$ for two–sided sequences. The study of infinite symbolic sequences has been undertaken by looking directly to the sequence and producing different combinatorial objects to describe its complexity, and also by considering the "shift dynamical system" or "subshift" associated to it. Put $K = N$ or $K = Z$. We endow the set of infinite sequences of symbols $A^K$ with the natural completely discontinuous product topology and consider the action of the shift transformation $\sigma$ : $A^K \to A^K$, such that $\sigma((a_i)_{i \in K}) = (a_{i+1})_{i \in K}$. The pair $(A^K, \sigma)$ is called the full shift on $A$. A subshift is any closed (for the product topology), shift–invariant subset $X \subseteq A^K$ $(\sigma(X) = X)$.

A subshift can be described in at least two different ways. First, the closure of the orbit of one given sequence under $\sigma$ is a subshift; it is called the subshift generated by this sequence. This is the case of the so–called substitutive systems. But a subshift can also be defined as the set of all sequences such that the words appearing in such sequences do not belong to a given set of forbidden words. The most popular subshifts constructed in this way are subshifts of finite type and sofic systems. The techniques developed to analyze these systems nowadays run from pure combinatorial arguments through probability theory to abstract algebraic analysis.

Now, what is a symbolic representation? Suppose $C$ is a compact metric space (or a manifold), and $T : C \to C$ a homeomorphism (a diffeomorphism). A subshift $(X, \sigma)$ of $A^Z$ is said to be a representation of $(C, T)$ if there exists a continuous, surjective map $\pi : C \to X$ such that $\pi \circ T = \sigma \circ \pi$. Of course one wishes $\pi$ to have some suitable properties; the best situation is when $\pi$ is one–to–one except on a set of universal measure 0.

Two families of subshifts have been subject to particularly deep investigations. Both families provide symbolic representations for some natural smooth maps; for both families there exist underlying, most of the time algebraic, sometimes arithmetic, structures that provide tools for their study.

The first consists of subshifts of finite type, sofic systems and coded systems. In these subshifts the set of allowed (or, for that matter, forbidden) words is defined with the help of a matrix. A subshift of finite type (SFT) is one that can be described by local constraints, that is, by forbidding a finite set of words; a sofic system is any homomorphic image of a SFT (this definition also provides a description in terms of a finite matrix); and a coded system is defined with the help of a matrix with countable entries. Such subshifts provide natural symbolic representations of smooth transformations of tori or the unit interval, for instance Anosov diffeomorphisms. They display many chaotic properties : a dense set of periodic points, and also, except for a trivial subclass, positive entropy ("high complexity") and a wide variety of shift–invariant measures; SFTs even have equilibrium states for continuous potentials. Owing to their positive entropy and the simple structure of their word combinatorics they have been used for the construction of some error–avoiding codes.

A long–standing conjecture on subshifts of finite type was due to R. Williams. It stated that conjugacy between two subshifts of finite type was equivalent to the shift equivalence of their defining matrices. It was recently proved to be false by Kim and Roush (Annals of Math.).

On the other hand the motivations for the study of systems of "very low complexity" like Sturmian subshifts or substitution dynamical systems arise from their connections with geometry (billiards, interval exchanges, tilings), number theory (irrational rotations of the 1-torus, numeration systems), formal language theory (DOL– and TAG–systems) and $C^*$-algebras. Subshifts in this area have entropy zero and often rigid features like minimality and/or unique ergodicity. They have given rise to sophisticated numeration systems – representations of real numbers by infinite sequences of symbols – and they provide exciting interpretations of almost periodic tilings and quasi–crystal models.

Between these extreme and rather well–known classes of systems, there can be found an infinite variety of dynamical behaviours; a typical family is that of minimal subshifts, some of which have positive and some of which zero entropy, and which range from very rigid to very chaotic. One typical class is that of Toeplitz systems: they are all minimal and some have positive, some others zero entropy; some are uniquely ergodic, others have uncountably many ergodic measures; they provide an almost inexhaustible source of examples and counter–examples for symbolic dynamics and ergodic theory.

## The Contents

The chapters in this volume show different aspects of "high complexity" and "low complexity" symbolic sequences.

In Chapter 1 the author investigates symbolic sequences of "low complexity", in particular Sturmian sequences introduced by Hedlund and Morse in the thirties, and automatic sequences. Both classes of sequences have been used to model systems in crystallography and biology. The chapter begins with the definition of the so-called complexity function of a sequence, $p(n)$: for each positive integer $n$ this function counts the number of words of length $n$ appearing in the sequence. In particular, Sturmian sequences are exactly those for which $p(n) = n + 1$. Then the author describes the main techniques and results that serve to study these systems: the graph of words, the notion of special factors, and the formal power series approach.

Chapter 2 continues the study of "low complexity" symbolic systems. Here substitution dynamical systems are considered. A substitution is any map $\tau : A \to A^*$, extended to sequences of $A^{\mathbb{N}}$ in a natural way. A substitution dynamical system is any subshift generated by a fixed point of a substitution. The existence of such a fixed point is ensured by a standard condition called primitivity. In the literature, the main reference in this area is the book of Queffélec (LNM-1294-Springer-Verlag) about the spectral properties of these systems, where some general introduction is also provided. The main discussion in this chapter is focussed on the topological dynamical properties of these systems and their representation by means of Bratteli–Vershik diagrams. This recent approach derives from the remarkable work of Herman, Putnam and Skau (Int. J. of Math., 1992) about the representation of minimal Cantor systems. After reviewing the main dynamical properties of primitive substitution systems, the author describes all the steps that are necessary to construct this representation.

The two following chapters are devoted to "high complexity" systems. In particular Chapter 3, after presenting a more detailed view of these systems than the one given above, describes the main algebraic tools that have been developed in order to investigate the problem of shift equivalence between subshifts of finite type, in other words, the Williams conjecture and related questions. The main algebraic invariants that are considered are matrix invariants and dimension group type invariants. The chapter ends with a description of the most recent results about the Williams conjecture.

There is a deep connection between Chapters 2 and 3. Some Bratteli diagrams can be endowed with two dynamics, one of them describing a subshift of finite type, the other a substitution dynamical system or an odometer; this establishes a kind of duality between these two classes. It is not surprising that dimension groups play an essential rôle in the two settings, with a different topological interpretation in each (see Durand-Host-Skau, Erg. Th. Dyn. Syst. 1999).

Chapter 4 is an introduction to the study of dynamical properties of $\mathbb{Z}^d$–actions of automorphisms over a compact topological group. The topic of $\mathbb{Z}^d$–actions is a difficult one, on which fewer results are known than for $\mathbb{Z}$–symbolic actions. This corresponds to the study of multidimensional Markov shifts (or subshifts of finite type) which have a group structure, also called Markov subgroups. The chapter concentrates on the "zero–dimensional" case. In particular, a complete characterization of expansive automorphisms of zero–dimensional topological groups is provided. In the transitive and positive entropy case, they correspond to full shifts. After that, the author presents one of the main techniques used in the study of the multidimensional case: the so–called descending chain condition. The symbolic dynamics of

$(\mathbb{Z}/2\mathbb{Z})^{\mathbb{Z}^2}$ provides the main example.

In Chapter 5 the author considers symbolic systems from a probabilistic point of view. Using the concept of an asymptotic rare event to mean a sequence of measurable sets of asymptotically zero probability, the author presents some results concerning asymptotic laws for the random times of occurrence of various asymptotically rare events in the case of shifts of finite type endowed with a Hölder equilibrium state. There is also a brief review of known results in this direction for more general ergodic dynamical systems.

The last three chapters of the book deal with applications of symbolic dynamics to ergodic theory, number theory and one-dimensional dynamics.

Chapter 6 deals with the connection between dynamics, the area of number theory called diophantine problems and combinatorial Ramsey theory. Van der Waerden's theorem (for any finite partitioning of $\mathbb{N}$, one of the elements contains arbitrarily long arithmetic progression) can actually be interpreted as a statement in topological dynamics and proved in this setting; likewise Szemerédi's theorem (a subset of $\mathbb{N}$ of positive density contains arbitrarily long arithmetic progressions), first proved in a purely combinatorial way, is also an ergodic statement, with an ergodic proof due to Furstenberg. The aim of this section is to present the fundamental ideas that led to these proofs and to many further developments; one of the most important is considering symbolic sequences.

Chapter 7 is devoted to the dynamics of symbolic systems that arise when one develops numbers in some particular numeration systems. First the author studies $\beta$-expansions of numbers and the relations between the defining real number $\beta > 1$ and the associated symbolic system: the set of lexicographically maximal $\beta$-expansions happens to possess interesting symbolic properties. Here a symbolic representation of an interval transformation plays an important part. Afterwards the author considers a closely related type of integer expansions: expansions with respect to an increasing sequence of integers $\{U_0, U_1, U_2, ...\}$. The basic example is the Fibonacci sequence.

Finally, Chapter 8 gives a complete description of the symbolic dynamics of Lorenz maps. This is another instance of a symbolic representation. Lorenz maps are piecewise increasing maps of the interval. A symbolic representation helps solve the problem of their topological conjugacy: in particular, for expanding Lorenz maps the conditions of Hubbard and Sparrow solve the topological classification problem.

## Acknowledgments

We would like to express our gratitude to all the participants of the school and to the people who contributed to its organization. In particular, to Servet Martínez who promote the idea of this school and contributed to its achievement in a significant way; and to Cesar Burgueño whose help at Temuco was priceless.

We want to thank especially the authors of each chapter for their well-prepared manuscripts and the stimulating conferences they gave at Temuco.

We are also indebted to our sponsors and supporting institutions, whose interest and help was essential to organize the school: Universidad de la Frontera, Universidad de Chile, Institut de Mathématiques de Luminy (CNRS), Université de la Méditerranée, Centre International de Mathématiques Pures et Appliquées (CIMPA–UNESCO), FONDAP Program in Applied Mathematics, Fundación Andes, Conicyt, ECOS–Conicyt Program, CNRS–Conicyt Program, Regional Office for French Cooperation, Cultural Service of French Embassy in Chile, and UMALCA.

We want to particularly thank Gladys Cavallone for her huge contribution to the administrative tasks related to the summer school, for the considerable task of unifying the typography of the different chapters of this book, and for her patience with our ceaseless requests all the while.

The Editors

# Chapter 1

# SEQUENCES OF LOW COMPLEXITY: AUTOMATIC AND STURMIAN SEQUENCES

*Valérie BERTHÉ*
*Institut de Mathématiques de Luminy*
*CNRS-UPR 9016*
*Case 907, 163 avenue de Luminy*
*F-13288 Marseille Cedex 9*
*France*

The complexity function is a classical measure of disorder for sequences with values in a finite alphabet: this function counts the number of factors of given length. We introduce here two characteristic families of sequences of low complexity function: automatic sequences and Sturmian sequences. We discuss their topological and measure-theoretic properties, by introducing some classical tools in combinatorics on words and in the study of symbolic dynamical systems.

## 1.1   Introduction

The aim of this course is to introduce two characteristic families of sequences of low "complexity": automatic sequences and Sturmian sequences (complexity is defined here as the combinatorial function which counts the number of factors of given length of a sequence over a finite alphabet). These sequences

1

not only occur in many mathematical fields but also in various domains as
theoretical computer science, biology, physics, crystallography...

We first define some classical tools in combinatorics on words and in the s-
tudy of symbolic dynamical systems: the complexity function and frequencies
of factors in connection with the notions of topological and measure-theoretic
entropy (Sections 1.2 and 1.3), the graphs of words (Section 1.4), special and
bispecial factors (Section 1.5). Then we study Sturmian sequences in Section 1.6: these sequences are defined as the sequences of minimal complexity
among non-ultimately periodic sequences. This combinatorial definition has
the particularity of being equivalent to the following simple geometrical representation: a Sturmian sequence codes the orbit of a point of the unit circle
under a rotation by irrational angle $\alpha$ with respect to a partition of the unit
circle into two intervals of lengths $\alpha$ and $1 - \alpha$. Sturmian sequences have
thus remarkable combinatorial and arithmetical properties. Then we introduce automatic sequences in Section 1.7: an automatic sequence is defined
as the image by a letter-to-letter projection of a fixed point of a substitution
of constant length or equivalently as a finite-state function of the representation of the index in a given basis. We emphasize on the connections with
transcendence of formal power series with coefficients in a finite field. In
particular, we will try to answer the following question: how to recognize if
a sequence is automatic or not? We conclude this course by discussing the
connections between sequences with a linear growth order for the complexity
function, and substitutions.

## 1.2   Complexity Function

### 1.2.1   Definition

Let us introduce a combinatorial measure of disorder for sequences over a
finite alphabet: this notion is called *(symbolic) complexity*. For more information on the subject, we refer the reader to the surveys [8, 43] and to the
course [59].

In all that follows we restrict ourselves to sequences over a finite alphabet indexed by the set $\mathbb{N}$ of non-negative integers. A *factor* of the infinite
sequence $u = (u_n)_{n \in \mathbb{N}}$ is a finite block $w$ of consecutive letters of $u$, say
$w = u_{n+1} \cdots u_{n+l}$; $l$ is called the *length* of $w$, denoted by $|w|$. Let $p(n)$ denote the *complexity function* of sequence $u$ with values in a finite alphabet:
it counts the number of distinct factors of length $n$ of the sequence $u$. The
complexity function is obviously non-decreasing and for any integer $n$, one
has $1 \leq p(n) \leq d^n$, where $d$ denotes the cardinality of the alphabet.

This function can be considered to *measure the predictability* of a sequence. The first difference of the complexity function counts the number
of possible extensions in the sequence of factors of given length. We call

*right extension* (respectively *left extension*) of a factor $w$ a letter $x$ such that $wx$ (respectively $xw$) is a factor of the sequence. Let $w^+$ (respectively $w^-$) denotes the number of right (respectively left) extensions of $w$. (One may have $w^- = 0$ but always $w^+ \geq 1$.) We have

$$p(n+1) = \sum_{|w|=n} w^+ = \sum_{|w|=n} w^-,$$

and thus

$$p(n+1) - p(n) = \sum_{|w|=n} (w^+ - 1) = \sum_{|w|=n} (w^- - 1).$$

**Exercise 1.2.1** (see [31, 54]) Prove that a sequence is *ultimately periodic* (i.e., periodic from a certain index on) if and only if its complexity function satisfies

$$\exists n, \ p(n) \leq n \iff \exists C, \ \forall n \ p(n) \leq C.$$

What happens in the case of a sequence defined over $\mathbf{Z}$ ?

The complexity function is a measure of disorder connected to the topological entropy: the *topological entropy* [1] is defined as the exponential growth rate of the complexity as the length increases

$$H_{top}(u) = \lim_{n \to +\infty} \frac{\log_d(p(n))}{n}.$$

It is easy to check that this limit exists because of the subadditivity of the function $n \mapsto \log_d(p(n))$. Note that the word *entropy* is used here as a measure of randomness or disorder. For a survey on the connections between entropy and sequences, see [13].

The study of the complexity is mainly concerned with the following three questions.

- How to compute the complexity of a sequence?

- Which functions can be obtained as the complexity function of some sequence?

- Can one deduce from the complexity a geometrical representation of sequences?

We will see how to answer the first question by introducing special and bispecial factors, in some particular cases of substitutive sequences (Section 1.5). The second question is still very much in progress and far from being solved (in particular in the case of positive entropy): for a survey on the question, see [24, 43]. Although the complexity function is in general not sufficient to describe a sequence, we will see in Section 1.6 that much can be said on the geometrical properties in the case of lowest complexity, i.e., in the case of *Sturmian sequences*: these sequences are defined to have exactly $n+1$ factors of length $n$, for any integer $n$. By Exercise 1.2.1 a sequence with complexity satisfying $p(n) \leq n$ for some $n$ is ultimately periodic. Sturmian sequences have thus the minimal complexity among all sequences that are not ultimately periodic.

**Exercise 1.2.2** Deduce from Exercise 1.2.1 that every prefix of a Sturmian sequence appears at least two times in the sequence. Deduce that the factors of every Sturmian sequence appear infinitely often (such a sequence is called *recurrent*).

## 1.2.2   Frequencies and Measure-Theoretic Entropy

The purpose of this section is to introduce a more "precise" (in a sense that we will see in Section 1.2.3) measure of disorder of sequences, connected with frequencies of factors. The *frequency* $f(B)$ of a factor $B$ of a sequence (called *density* in Host's course) is defined as the limit, if it exists, of the number of occurrences of this block in the first $k$ terms of the sequence divided by $k$.

**Exercise 1.2.3** Construct a sequence for which the frequencies of letters do not exist.

Let us first introduce *the block entropies* for sequences with values in a finite alphabet in order to define the notion of *measure-theoretic entropy*. These sequences of block entropies were first introduced by Shannon in information theory, to measure the entropy of the English language (see [65]).

Let $u$ be a sequence with values in the alphabet $\mathcal{A} = \{1, \cdots, d\}$. We suppose that all the block frequencies exist for $u$. Let

$$P(x|x_1 \cdots x_n) = \frac{f(x_1 \cdots x_n x)}{f(x_1 \cdots x_n)},$$

where $x_1 \cdots x_n$ is a block of non-zero frequency and $x$ a letter. Intuitively $P(x|x_1 \cdots x_n)$ is the conditional probability that the letter $x$ follows the block $x_1 \cdots x_n$ in the sequence $u$. We are going to associate with the sequence $u$ two sequences of block entropies $(H_n)_{n \in \mathbb{N}}$ and $(V_n)_{n \in \mathbb{N}}$.

For all $n \geq 1$, let

$$V_n = \sum L(f(x_1 \cdots x_n)),$$

where the sum is over all the factors of length $n$ and $L(x) = -x \log_d(x)$, for all $x \neq 0$ and $L(0) = 0$. We put $V_0 = 0$.

For all $n \geq 1$, let

$$H_n = \sum{}' f(x_1 \cdots x_n) H(x_1 \cdots x_n), \tag{1.1}$$

where the sum is over all the blocks of length $n$ of non-zero frequency and

$$H(x_1 \cdots x_n) = \sum_{x \in \mathcal{A}} L(P(x/x_1 \cdots x_n)).$$

We put $H_0 = V_1$. The sequence $(H_n)_{n \in \mathbb{N}}$ measures in some way the properties of predictability of the initial sequence $u$.

**Exercise 1.2.4** Prove that: $\forall n \in \mathbb{N}$, $H_n = V_{n+1} - V_n$. (This classical property in information theory is called the *chain-rule*.)

Thus, $(H_n)_{n \in \mathbb{N}}$ is the discrete derivative of $(V_n)_{n \in \mathbb{N}}$. Note that $(V_n)_{n \in \mathbb{N}}$ is a non-decreasing sequence, since $H_n \geq 0$ for all $n$.

It can be shown that $(H_n)_{n \in \mathbb{N}}$ is a monotonic non-increasing sequence of $n$ (see, for instance [16]). The intuitive meaning of this is that the uncertainty about the choice of the next symbol decreases when the number of known preceding symbols increases. From the non-increasing behaviour of the positive sequence $(H_n)_{n \in \mathbb{N}}$, we deduce the existence of the limit $\lim_{n \to +\infty} H_n$. We have: $\forall n$, $H_n = V_{n+1} - V_n$ and $\sum_{k=0}^{n-1} H_k = V_n$. By taking Cesàro means, we obtain:

$$\lim_{n \to +\infty} H_n = \lim_{n \to +\infty} \frac{V_n}{n}.$$

This limit is called the *measure-theoretic entropy* of the sequence $u$, it is the limit of the entropy per symbol of the choice of a block of length $n$, when $n$ tends to infinity.

### 1.2.3 Variational Principle

What is the relation between the sequences $(H_n)_{n \in \mathbb{N}}$ and $(V_n)_{n \in \mathbb{N}}$? We have:

$$\forall n, \; nH_n \leq \sum_{k=0}^{n-1} H_k = V_n - \sum L(P(x_1 \cdots x_n)).$$

By concavity of the function $L$ we get: $\forall n \geq 1$, $V_n \leq \log_d p(n)$. Hence the following proposition:

**Proposition 1.2.5** *We have* $H_n \leq \frac{\log_d(p(n))}{n}$, *for all* $n \geq 1$.

We hence get:

$$\lim_{n \to +\infty} H_n = \lim_{n \to +\infty} \frac{V_n}{n} = H(u) \leq H_{top}(u) = \lim_{n \to +\infty} \frac{\log_d(p(n))}{n}.$$

This inequality is a particular case of a basic relationship between topological entropy and measure-theoretic entropy called the *variational principle* (for a proof see [53]).

The two limits $\lim_{n \to +\infty} H_n$ and $\lim_{n \to +\infty} \dfrac{\log_d(p(n))}{n}$ are distinct in general and the notion of measure-theoretic entropy for a sequence is more precise. But the sequences we are mostly dealing with here are *deterministic*, i.e., sequences with zero entropy. Therefore neither the metrical nor the topological entropy can distinguish between these sequences.

## 1.3 Symbolic Dynamical Systems

Recall some basic notions on symbolic dynamical systems. For a detailed introduction to the subject, see [57]. Let $\mathcal{A}$ denote a finite alphabet; here we work with the space $\mathcal{A}^\mathbb{N}$, whereas in Host's course it is $\mathcal{A}^\mathbb{Z}$.

Endow the set $\mathcal{A}^\mathbb{N}$ of all sequences with values in the finite set $\mathcal{A}$ with the product of discrete topologies on $\mathcal{A}$. This set is thus a compact space. The topology defined on $\mathcal{A}^\mathbb{N}$ is equivalent to the topology defined by the following metrics: for $x, y \in \mathcal{A}^\mathbb{N}$

$$d(x,y) = (1 + \inf\{k \geq 0;\ x_k \neq y_k\})^{-1}.$$

Two sequences are thus close to each other if their first terms coincide. The *cylinder* $[w]$, where $w = w_1 \ldots w_n$ belongs to $\mathcal{A}^n$, is the set of sequences of the form

$$[w] = \{x \in \mathcal{A}^\mathbb{N} \mid x_0 = w_1,\ x_1 = w_2, \ldots, x_{n-1} = w_n\}.$$

Cylinders are closed and open sets and span the topology.

The space $\mathcal{A}^\mathbb{N}$ is complete as a metric compact space. Let us deduce from this the existence of fixed points of substitutions. A *substitution* defined on the finite alphabet $\mathcal{A}$ is a map from $\mathcal{A}$ to the set of words defined on $\mathcal{A}$, denoted by $\mathcal{A}^*$, extended to $\mathcal{A}^*$ by concatenation, or in other words, a homomorphism of the free monoid $\mathcal{A}^*$ (see also [49] for a precise study of substitution dynamical systems).

**Exercise 1.3.1** Let $\sigma$ be a substitution and $a$ be a letter such that $\sigma(a)$ begins by $a$ and $|\sigma(a)| \geq 2$. Prove that there exists a unique sequence beginning with $a$ satisfying $\sigma(u) = u$. This sequence is called a *fixed point* of the substitution.

For instance, the Fibonacci sequence is defined as the fixed point beginning with 1 of the following substitution

$$\sigma(1) = 10, \ \sigma(0) = 1.$$

Let $T$ denote the following map defined on $\mathcal{A}^{\mathbb{N}}$, called the *one-sided shift*:

$$T((u_n)_{n\in\mathbb{N}}) = (u_{n+1})_{n\in\mathbb{N}}.$$

The map $T$ is uniformly continuous, onto but not necessarily one-to-one on $\mathcal{A}^{\mathbb{N}}$.

**Exercise 1.3.2** Recall that a sequence is said to be *recurrent* if every factor appears at least two times, or equivalently if every factor appears an infinite number of times in this sequence.

Prove that a sequence $u$ is recurrent if and only if there exists a strictly increasing sequence $(n_k)_{k\in\mathbb{N}}$ such that

$$u = \lim_{k\to+\infty} T^{n_k}u.$$

Let $u$ be a sequence with values in $\mathcal{A}$. Define $\overline{\mathcal{O}}(u)$ as the positive orbit closure of the sequence $u$ under the action of the shift $T$, i.e., the closure of the set $\mathcal{O}(u) = \{T^n(u),\ n \geq 0\}$. The set $\overline{\mathcal{O}(u)}$ is a compact metric space, and thus complete. It is also $T$-*invariant*: $T(\overline{\mathcal{O}}(u)) \subset \overline{\mathcal{O}}(u)$. In other words $T$ may be considered as acting on $\overline{\mathcal{O}}(u)$.

**Exercise 1.3.3**

1. Prove that
   $$\overline{\mathcal{O}}(u) = \{x \in \mathcal{A}^{\mathbb{N}},\ L(x) \subset L(u)\},$$
   where $L(x)$ denotes the set of factors of the sequence $x$.

2. Prove that $u$ is recurrent if and only if $T$ is onto on $\overline{\mathcal{O}}(u)$.

Let $X$ be a non-empty compact metric space and $T$ be a continuous map from $X$ to $X$. The system $(X, T)$ is called a *topological dynamical system*. For instance, $(\overline{\mathcal{O}}(u), T)$ is a topological dynamical system. A topological dynamical system is called *minimal* if every closed $T$-invariant set $E$ is either equal to the full set $X$ or to the empty set.

**Exercise 1.3.4**

- Prove that $(X, T)$ is minimal if and only if $X = \overline{\mathcal{O}(x)}$, for every element $x$ of $X$.

- A sequence is said to be *uniformly recurrent* if every factor appears infinitely often and with bounded gaps (or, equivalently, if for every integer $n$, there exists an integer $m$ such that every factor of $u$ of length $m$ contains every factor of length $n$). Prove that a sequence $u$ is uniformly recurrent if and only if $(\overline{\mathcal{O}}(u), T)$ is minimal. (If $w$ is a factor of $u$, write
  $$\overline{\mathcal{O}}(u) = \bigcup_{n\in\mathbb{N}} T^{-n}[w],$$
  and conclude by a compactness argument.)

The following special case of the Daniell-Kolmogorov consistency theorem (see for instance [73]) establishes the existence of a certain probability measure on $(\overline{\mathcal{O}(u)}, T)$. A Borel probability measure $\mu$ defined on $(\overline{\mathcal{O}(u)}, T)$ is called $T$-*invariant* if $\mu(T^{-1}(B)) = \mu(B)$, for any Borel set $B$.

**Theorem 1.3.5** *Let $u$ be a sequence on $\mathcal{A} = \{1, \ldots, d\}$. Consider a family of maps $(p_n)_{n \geq 1}$, where $p_n$ is a map from $\mathcal{A}^n$ to $\mathbb{R}$, such that*

- *for any word $w$ in $\mathcal{A}^n$, $p_n(w) \geq 0$,*

- $$\sum_{i=1}^{d} p_1(i) = 1,$$

- *for any word $w = w_1 \ldots w_n$ in $\mathcal{A}^n$, $p_n(w) = \displaystyle\sum_{i=1}^{d} p_{n+1}(w_1 \ldots w_n i)$.*

*Then there exists a unique probability measure $\mu$ on $\mathcal{A}^{\mathbb{N}}$ defined on the cylinders by $\mu([w_1 \ldots w_n]) = p_n(w_1 \ldots w_n)$.*
*Furthermore, if for any $n$ and for any word $w = w_1 \ldots w_n$ in $\mathcal{A}^n$,*

$$p_n(w) = \sum_{i=1}^{d} p_{n+1}(iw_1 \ldots w_n),$$

*then this measure is $T$-invariant.*

In particular, if all frequencies exist, then there exists a unique $T$-invariant probability measure which assigns to each cylinder the frequency of the corresponding factor. Moreover suppose the symbolic dynamical system $(\overline{\mathcal{O}(u)}, T)$ is *uniquely ergodic*, i.e., there exists a unique $T$-invariant probability measure $\mu$ on this dynamical system. Thus a precise knowledge of the frequencies allows a complete description of the measure $\mu$. For instance, a symbolic dynamical system obtained via the fixed point of a primitive substitution [49, 57], or via a Sturmian sequence is uniquely ergodic.

## 1.4   The Graph of Words

The Rauzy graph $\Gamma_n$ of words of length $n$ of a sequence on a finite alphabet $\mathcal{A}$ (of cardinality $d$) is an oriented graph (see, for instance, [58]), which is a subgraph of the de Bruijn graph of words[1] (see [32]). Its vertices are the

---

[1]The de Bruijn graph of words corresponds to the graph of words of a sequence of maximal complexity ($\forall n, p(n) = d^n$) and was introduced by de Bruijn in order to construct circular finite sequences of length $d^n$ with values in $\{0, 1, \ldots, d-1\}$ such that every factor of length $n$ appears once and only once: such a sequence corresponds to a Hamiltonian closed path in de Bruijn graph.

factors of length $n$ of the sequence and the edges are defined as follows: there is an edge from $U$ to $V$ if $V$ follows $U$ in the sequence, i.e., if there exists a word $W$ and two letters $x$ and $y$ such that $U = xW$, $V = Wy$ and $xWy$ is a factor of the sequence. There are $p(n+1)$ edges and $p(n)$ vertices, where $p(n)$ denotes the complexity function.

**Exercise 1.4.1** Prove that the graphs of words of a sequence are always connected. Prove the following equivalence (see [61]):

- the sequence $u$ is recurrent,

- every factor of $u$ appears at least twice,

- the graphs of words are strongly connected.

Let $U$ be a vertex of the graph. Denote by $U^+$ the number of edges of $\Gamma_n$ with origin $U$ and by $U^-$ the number of edges of $\Gamma_n$ with end vertex $U$. In other words, $U^+$ (respectively $U^-$) counts the number of right (respectively left) extensions of $U$. Recall that

$$p(n+1) = \sum_{|U|=n} U^+ = \sum_{|U|=n} U^-,$$

and thus

$$p(n+1) - p(n) = \sum_{|U|=n} (U^+ - 1) = \sum_{|U|=n} (U^- - 1).$$

**Exercise 1.4.2** Recall that a Sturmian sequence is defined as a sequence of complexity function $p(n) = n + 1$, for every positive integer $n$, and that it is recurrent (Exercise 1.2.2).

- For any positive integer $n$, prove that there exists a unique factor of length $n$ having two right (respectively left) extensions: such a factor is called a *right* (respectively *left*) *special factor* (or also *expansive* factor) and is denoted from now on by $R_n$ (respectively $L_n$).

- Prove that the graph of words $\Gamma_n$ of a Sturmian sequence has the two possible forms given in figure 1.1.

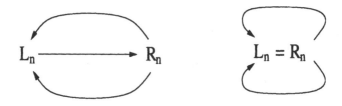

Figure 1.1:

- Deduce from the morphology of the graph of words $\Gamma_n$ that every Sturmian sequence is uniformly recurrent. One can first prove that every factor of a Sturmian sequence is a subfactor of a factor of the form $R_n$ and then deduce from the morphology of the graph $\Gamma_n$ that $R_n$ appears with bounded gaps.

**Exercise 1.4.3**

- Prove that if the sequence $u$ is uniformly recurrent and non-constant, then the graph $\Gamma_n$ has not edge of the form $U \to U$, for $n$ large enough.

- Suppose that the sequence $u$ is uniformly recurrent. Prove that if the graph of words $\Gamma_{n+1}$ is Hamiltonian (i.e., there exists a closed oriented path passing exactly once through every vertex), then the graph $\Gamma_n$ is Eulerian (there exists a closed path passing exactly once through every edge) and that $U^+ = U^-$, for every vertex of $\Gamma_n$. Is the converse true?

## 1.4.1   The Line Graph

The *line graph* $D(\Gamma_n)$ of the graph of words $\Gamma_n$ is defined as follows: its vertices are the edges of $\Gamma_n$ (i.e., the factors of length $n+1$); given two vertices $u$ and $v$ in $D(\Gamma_n)$, there is an edge from $u$ to $v$ if the end point of the edge labelled $u$ in $\Gamma_n$ is the origin of the edge labelled $v$. It is easily seen that the edges of the line graph correspond to words of length $n+2$ such that their prefix and their suffix of length $n+1$ are factors of the sequence $u$. The line graph of $\Gamma_n$ is thus a subgraph of $\Gamma_{n+1}$.

**Exercise 1.4.4** Study the evolution of the graph of words from $\Gamma_n$ to $\Gamma_{n+1}$ for a Sturmian sequence by using the line graph. (Distinguish between the two possible forms of the graph).

**Remark 1.4.5** In [61] Rote uses the graph of words and the line graph for the study of sequences of complexity $p(n) = 2n$, for every $n$ (see also [41]).

The study of the evolution of the graph of words for any Sturmian sequence is a very powerful method and contains all the information concerning the sequence: Arnoux and Rauzy have thus proved that every Sturmian sequence is generated by the composition of two substitutions (see [11]); one can also study the frequencies of factors of given length (see Section 1.8.2 and [14]) or covering numbers for rotations (see [15, 26]).

### 1.4.2   Graph and Frequencies

Let us see how to deduce from the morphology of the graphs of words results concerning the frequencies of factors. This follows an idea of Dekking who expressed the block frequencies for the Fibonacci sequence, by using the graph of words (see [34]).

In this section we restrict ourselves to sequences for which the frequencies exist. Observe that the function which associates to an edge labelled by $xWy$ the frequency of the factor $xWy$ is a *flow*. Indeed, it satisfies Kirchhoff's current law: the total current flowing into each vertex is equal to the total current leaving the vertex. This common value is equal to the frequency of the word corresponding to this vertex.

**Lemma 1.4.6** *Let $U$ and $V$ be two vertices linked by an edge such that $U^+ = 1$ and $V^- = 1$. Then the two factors $U$ and $V$ have the same frequency.*

**Proof.** Write $U = xW$ and $V = Wy$, where $x$ and $y$ are letters. As $U^+ = 1$, $U$ has a unique right extension $y$. Similarly, $V$ has a unique left extension $x$. Thus $f(U) = f(Uy) = f(xWy) = f(xV) = f(V)$, where $f$ denotes the frequency. ∎

A *branch* of the graph $\Gamma_n$ is a sequence of maximal length $(U_1, \ldots, U_m)$ of connected edges of $\Gamma_n$, possibly empty, satisfying

$$U_i^+ = 1, \text{ for } i < m, \quad U_i^- = 1, \text{ for } i > 1.$$

Therefore, the edges of a branch have the same frequency and the number of frequencies of factors of given length is bounded by the number of branches of the corresponding graph, as expressed below (see [18]).

**Theorem 1.4.7** *For a recurrent sequence of complexity function $p(n)$, the frequencies of factors of given length, say $n$, take at most $3(p(n+1) - p(n))$ values.*

**Proof.** Let $V_1$ denote the set of factors of length $n$ having more than one extension. In other words $V_1$ is the subset of vertices of the graph $\Gamma_n$ defined as follows: $U \in V_1$ if and only if $U^+ \geq 2$. The cardinality of $V_1$ satisfies

$$\mathrm{card}(V_1) = \sum_{|U|=n,\ U^+ \geq 2} 1 \leq \sum_{|U|=n} (U^+ - 1) = p(n+1) - p(n).$$

Let $V_2$ denote the subset of vertices of the graph $\Gamma_n$ defined as follows: $U \in V_2$ if and only if $U^+ = 1$ and if $V$ denotes the unique vertex such that there is an edge from $U$ to $V$ in $\Gamma_n$, then $V^- \geq 2$. In other words, $U$ belongs to $V_2$ if and only if $U = xW$, where $x$ is a letter and where the factor $W$ of the sequence $u$ has a unique right extension but at least two left extensions. The cardinality of $V_2$ satisfies:

$$\text{card}(V_2) \leq \sum_{V^- \geq 2} V^- = \sum_{V^- \geq 2} (V^- - 1) + \sum_{V^- \geq 2} 1 \leq 2(p(n+1) - p(n)).$$

Thus there are at most $3(p(n+1)) - p(n))$ factors in $V_1 \cup V_2$.

Let $U$ be a factor of length $n$ belonging neither to $V_1$ nor to $V_2$: $U^+ = 1$ and the unique word $V$ such that there is an edge from $U$ to $V$ in $\Gamma_n$ satisfies $V^- = 1$. The two factors $U$ and $V$ thus have the same frequency. Now consider the path of the graph beginning at $U$ and consisting of vertices which do not belong to $V_1$ nor to $V_2$. The last vertex of this path belongs to either $V_1$ or to $V_2$, and has the same frequency as $U$. ∎

**Remark 1.4.8** In fact we have proved that the frequencies of factors of length $n$ take at most $p(n+1) - p(n) + r_n + l_n$ values, where $r_n$ (respectively $l_n$) denotes the number of factors having more than one right (respectively left) extension.

We deduce from this result that if $p(n+1) - p(n)$ is uniformly bounded with $n$, the frequencies of factors of given length take a finite number of values. Indeed, using a theorem of Cassaigne quoted below (see [23]), we can easily state the following corollary.

**Theorem 1.4.9** *If the complexity $p(n)$ of a sequence on a finite alphabet is sub-affine, i.e.,*

$$\exists (a,b), \ \forall n, \ p(n) \leq an + b,$$

*then $p(n+1) - p(n)$ is bounded.*

**Corollary 1.4.10** *If a sequence has a sub-affine complexity then the frequencies of its factors of given length take a finite number of values.*

We will see in Section 1.7.6 (Theorem 1.7.26) that fixed points of uniform substitutions (i.e., substitutions such that the images of the letters have the same length) or fixed point of primitive substitutions have sub-affine complexities. In particular, in the Sturmian case ($\forall n, p(n) = n+1$), Theorem 1.4.7 implies that the frequencies of factors of given length of a Sturmian sequence take at most three values. We will come back to this in Section 1.6.2.

Note that Theorem 1.4.9 above does not hold anymore for the second-difference of the complexity $p(n+2) + p(n) - 2p(n+1)$ in the case of a sub-quadratic complexity (see the counterexample in [42]).

## 1.5   Special factors

The aim of this section is to introduce the notions of special and bispecial factors in order to evaluate the second-difference of the complexity, which is often easier to compute than the complexity itself. Indeed, in the case of a low complexity, the number of special factors and bispecial factors, which is low, is quite easy to evaluate. For a more detailed exposition, see [24].

Let $u$ be a sequence with values in a two-letter alphabet $\mathcal{A}$. Let $w^+$ (respectively $w^-$) denote the number of right (respectively left) extensions of $w$. Recall that

$$p(n+1) - p(n) = \sum_{|w|=n} (w^+ - 1) = \sum_{|w|=n} (w^- - 1).$$

A factor is said to be a *left special factor* (respectively *right special factor*) if it has more than one left (respectively *right*) extension. We use the notation of [24], in which the case of a bigger-sized alphabet is also considered. A factor is said to be *bispecial* if it is both a right and a left special factor. More precisely, we distinguish three cases according to the cardinality $c(w)$ of $L(u) \cup \mathcal{A}w\mathcal{A}$, where $L(u)$ denotes the set of factors of the sequence $u$ and $w$ is a bispecial factor, the operation considered here being the concatenation. We have obviously $2 \leq c(w) \leq 4$.

- If $c(w) = 2$, then $w$ is called *a weak bispecial factor*,

- if $c(w) = 3$, then $w$ is called *an ordinary bispecial factor*,

- if $c(w) = 4$, then $w$ is called *a strict bispecial factor*.

**Exercise 1.5.1** Let $b_w(n)$ (respectively $b_s(n)$) denote the number of weak (respectively strict) bispecial factors of the sequence $u$ of size $n$. Prove that the second-difference of the complexity is given by

$$p(n+2) + p(n) - 2p(n+1) = s(n+1) - s(n) = b_s(n) - b_w(n),$$

where $s(n) = p(n+1) - p(n)$.

**Exercise 1.5.2** Consider the Fibonacci sequence. Prove that every factor $w$ can be uniquely written as follows: $w = r_1 \sigma(x) r_2$, where $x$ is a factor, $r_1 \in \{\varepsilon, 0\}$, and $r_2 = 1$ if the last letter of $w$ is 1, and $r_2 = 0$, otherwise. Prove by induction that the bispecial factors of the Fibonacci sequence are all ordinary. Deduce that this sequence is Sturmian.

**Exercise 1.5.3** Let $u$ be the Thue-Morse sequence defined as the fixed point beginning by 0 of the following substitution: $\sigma(0) = 01$ and $\sigma(1) = 10$. We will compute the complexity function of the Thue-Morse sequence in two ways.

1. Prove that every factor $w$ can be written as follows: $w = r_1\sigma(x)r_2$, where $x$ is a factor and $r_i \in \{\varepsilon, a, b\}$. If $|w| \geq 5$, then this decomposition is unique.

2. Prove that $p(2n) = p(n) + p(n+1)$ and that $p(2n+1) = 2p(n+1)$, for $n \geq 1$. Give an expression for the complexity function (see for instance [20]).

3. Find by induction the expressions of $b_s(n)$, $b_o(n)$ and $b_w(n)$ by studying the small length cases. Deduce an expression for the complexity.

## 1.6 Sturmian sequences

Sturmian sequences have received considerable attention in the literature. We refer the reader to the impressive bibliography of [21]. A recent account on the subject can also be found in [12].

### 1.6.1 A Particular Coding of Rotations

We introduce a large family of Sturmian sequences obtained by coding the orbit of a point of the unit circle under an irrational rotation.

Let $\{x\}$ denote as usual the fractional part of $x$ (i.e., if $\lfloor x \rfloor$ denotes the largest integer not exceeding $x$, then $\{x\} = x - \lfloor x \rfloor$). Let $\alpha$ be an irrational number in $]0, 1[$ and consider the rotation $R_\alpha$ of angle $\alpha$ defined on the unit circle (identified with $[0, 1[$ or with the unidimensional torus $\mathbb{R}/\mathbb{Z}$): we have $R_\alpha(x) = x + \alpha \pmod 1$. The *positive orbit* of a point $x$ of the unit circle under the rotation by angle $\alpha$ is the set of points $\{\{\alpha n + x\}, n \geq 0\}$. We code the information concerning the orbit of $x$ by a binary sequence. The *coding* of the orbit of $x$ under the rotation by angle $\alpha$ with respect to the partition $\mathcal{P} = \{[0, 1-\alpha[, [1-\alpha, 1[\}$ is the sequence $u = (u_n)_{n \in \mathbb{N}}$ defined on the alphabet $\{0, 1\}$ as follows:

$$u_n = 1 \Leftrightarrow \{x + n\alpha\} \in [0, 1-\alpha[.$$

We could also choose to code the orbit of the rotation with respect to the partition $\mathcal{P}' = \{]0, 1-\alpha], ]1-\alpha, 1]\}$. As $\alpha$ is irrational, the two sequences obtained by coding with respect to $\mathcal{P}$ or to $\mathcal{P}'$ are ultimately equal. The results stated below on coding with respect to $\mathcal{P}$ are obviously true for $\mathcal{P}'$.

Let $I_0 = [0, 1-\alpha[$ and $I_1 = [1-\alpha, 1[$. A finite word $w_1 \cdots w_n$ defined on the alphabet $\{0, 1\}$ is a factor of the sequence $u$ if and only if there exists an integer $k$ such that

$$\{x + k\alpha\} \in I(w_1, \ldots, w_n) = \bigcap_{j=0}^{n-1} R_\alpha^{-j}(I_{w_{j+1}}).$$

As $\alpha$ is irrational, the sequence $(\{x+n\alpha\})_{n\in\mathbb{N}}$ is dense in the unit circle, which implies that $w_1 w_2 \ldots w_n$ is a factor of $u$ if and only if $I(w_1, \ldots, w_n) \neq \emptyset$. In particular, the set of factors does not depend on the initial point $x$ of this coding. Furthermore, one can check that the sets $I(w_1, \ldots, w_n)$ are connected and are bounded by the points $\{k(1 - \alpha)\}$, for $0 \leq k \leq n - 1$. There are $n + 1$ such intervals ($\alpha$ is irrational) and thus $n + 1$ factors of length $n$: the sequence $u$ is therefore Sturmian.

**Remark 1.6.1** In the general case of a coding of an irrational rotation with respect to a partition of the unit circle in $l$ intervals, the complexity has the form $p(n) = an + b$, for $n$ large enough (see [2]). Conversely, every sequence of ultimately affine complexity is not necessarily obtained as a coding of a rotation. See for instance [61], where Rote studies the case of sequences of complexity $p(n) = 2n$, for every $n$. However, if the complexity of a sequence $u$ has the form $p(n) = n + k$, for $n$ large enough, then $u$ is the image of a Sturmian sequence by a morphism, up to a prefix of finite length (see for instance [2, 25, 36]).

**Exercise 1.6.2** Let $u$ be a coding of the rotation by angle $\alpha$ defined as above. Prove that the orbit closure $\overline{\mathcal{O}(u)}$ is the set of sequences obtained by coding every point of the unit circle with respect to the partition $\mathcal{P}$ or $\mathcal{P}'$ under the rotation by angle $\alpha$. (Use the fact that the set of factors of a coding depends neither on the initial point $x$ nor on the choice of the partition but only on the angle of the rotation). Deduce the minimality of $(\overline{\mathcal{O}(u)}, T)$.

Now a natural question is whether all Sturmian sequences are obtained by coding a rotation as defined above. The answer is yes and is due to Morse and Hedlund (see [55]): this shows that in this case of low disorder, one can give a geometrical description of sequences defined up to their complexity function. When the complexity grows, this becomes much more difficult (see for instance [11] for a geometrical representation of a particular class of sequences of complexity $2n + 1$).

**Theorem 1.6.3** *[Hedlund and Morse] A sequence $u$ is Sturmian if and only if there exist an irrational $\alpha$ in $]0, 1[$ and $x$ on the unit circle such that $u$ is the coding of the orbit of $x$ under the rotation by angle $\alpha$ with respect to one of the partitions $\{[0, 1 - \alpha[, [1 - \alpha, 1[\}$ or $\{]0, 1 - \alpha], ]1 - \alpha, 1]\}$.*

Sturmian sequences are also characterized by the following properties.

- Sturmian sequences are exactly the non-ultimately periodic balanced sequences over a two-letter alphabet. A sequence is balanced if the difference between the number of occurrences of a letter in any two factors of the same length is bounded by one in absolute value.

- Sturmian sequences are codings of trajectories of irrational initial slope in a square billiard obtained by coding horizontal sides by the letter 0 and vertical sides by the letter 1.

- One can also consider Sturmian sequences as approximations of a line of irrational slope in the upper half-plane.

The last three properties can be easily deduced from the representation by a rotation: they are just geometrical reformulations; the first characterization of Sturmian sequences in terms of the balance property is much more difficult to establish and is an important step in the proof of Theorem 1.6.3.

## 1.6.2   Frequencies of Factors of Sturmian Sequences

We now consider properties of frequencies of factors of Sturmian sequences.

The frequency of the factor $w_1 \ldots w_n$ exists and is equal to the density of the set

$$\{k \mid \{x + k\alpha\} \in I(w_1, \ldots, w_n)\},$$

which is equal to the length of $I(w_1, \ldots, w_n)$, by uniform distribution of the sequence $(\{x + n\alpha\})_{n \in \mathbb{N}}$. The lengths of these intervals are equal to the frequencies of factors of length $n$.

But we deduce from Theorem 1.4.7 the following result.

**Theorem 1.6.4** *The frequencies of factors of given length of a Sturmian sequence take at most three values.*

Theorem 1.6.4 implies that the lengths of the intervals $I(w_1, \ldots, w_n)$, and thus the lengths of the intervals obtained by placing the points $0, \{1 - \alpha\}, \ldots, \{n(1-\alpha)\}$ on the unit circle, take at most three values. We thus have proved the following classical result in Diophantine approximation, called the *three-distance theorem* (see the survey [3]). In fact, this point of view and more precisely, the study of the evolution of the graphs of words with respect to the length $n$ of the factors, allows us to give a proof of the most complete version of the three distance theorem, i.e., to express the exact number of factors having each of the three frequencies and the frequencies themselves (for more details, the reader is referred to [14]).

The three distance theorem was initially conjectured by Steinhaus and proved by V. T. Sós (see [67, 68, 69, 70]).

**Theorem 1.6.5** *Let $0 < \alpha < 1$ be an irrational number and $n$ a positive integer. The points $\{i\alpha\}$, for $0 \le i \le n$, partition the unit circle into $n + 1$ intervals, the lengths of which take at most three values, one being the sum of the other two.*

*More precisely, let* $(\frac{p_k}{q_k})_{k \in \mathbb{N}}$ *and* $(c_k)_{k \in \mathbb{N}}$ *be the sequences of the convergents and partial quotients associated to* $\alpha$ *in its continued fraction expansion (if* $\alpha = [0, c_1, c_2, \ldots]$, *then* $\frac{p_n}{q_n} = [0, c_1, \ldots, c_n]$). *Let* $\eta_k = (-1)^k(q_k \alpha - p_k)$. *Let* $n$ *be a positive integer. There exists a unique expression for* $n$ *of the form*

$$n = mq_k + q_{k-1} + r,$$

*with* $1 \le m \le c_{k+1}$ *and* $0 \le r < q_k$. *Then, the circle is divided by the points* $0, \{\alpha\}, \{2\alpha\}, \ldots, \{n\alpha\}$ *into* $n+1$ *intervals which satisfy:*

- $n + 1 - q_k$ *of them have length* $\eta_k$ *(which is the largest of the three lengths),*

- $r + 1$ *have length* $\eta_{k-1} - m\eta_k$,

- $q_k - (r + 1)$ *have length* $\eta_{k-1} - (m-1)\eta_k$.

## 1.7 Automatic Sequences

The aim of this section is to introduce automatic sequences and to study them in connection with properties of algebraicity of formal power series over a finite field.

Let $k$ be an integer greater than or equal to 2. Recall the definition of a finite complete deterministic $k$-automaton (also called *2-tape automaton or transducer*). For more details, see the courses of Frougny, see [30] or see the surveys [5, 9]. A $k$-automaton is represented by a directed graph defined by:

- a finite set of *states* $S = \{a_1, a_2, \cdots, a_d\}$, one of these states $i = a_1$ is called the *initial state*;

- $k$ *transition maps* (or "edges") from the set of states $S$ into itself, denoted by $0, 1, \cdots, k-1$;

- a set $Y$ and a map $\varphi$ from $S$ into a set $Y$, called *output function* or *exit map*.

A sequence $(u(n))_{n \in \mathbb{N}}$ with values in $Y$ is called *k-automatic* (it is also called a *k-uniform tag sequence* or a *k-recognizable* sequence) if it is generated by a $k$-automaton as follows: let $\omega(n)$ be the base $k$ expansion of the integer $n$; starting from the initial state one feeds the automaton the sequence $\omega(n)$, the digits being read in growing order of powers; after doing this the automaton is in the state $a(n)$ (the automaton is said to be fed in *reverse* reading.) Then put $u(n) = \varphi(a_f)$. One can similarly give another definition of $k$-automaticity by reading the digits in the reverse order, i.e., by starting with the most significant digit, but these two notions are easily seen to be equivalent. (The automaton is said to be fed in *direct* reading.)

## 1.7.1   Automata and Transcendence

Let $\mathbb{F}_q$ be the finite field with $q$ elements, and let $p$ denote its characteristic. We thus have $p$ prime and $q = p^s$, where $s$ is a nonzero natural integer. The field $\mathbb{F}_q((1/X))$ of *Laurent formal power series* with coefficients in $\mathbb{F}_q$ is the field of formal power series series of the form

$$u(-d)X^d + \cdots + u(0) + u(1)X^{-1} + \cdots,$$

where the coefficients $u(i)$ belong to $\mathbb{F}_q$. Similarly, we denote by $\mathbb{F}_q((X))$ the field of formal power series of the form

$$u(-d)X^{-d} + \cdots + u(0) + u(1)X + \cdots$$

A formal power series $F$ is called *algebraic over* $\mathbb{F}_q(X)$ if there exists a nontrivial polynomial $P$ with coefficients in $\mathbb{F}_q(X)$ such that $P(F) = 0$. In the converse case, $F$ is called *transcendental* over $\mathbb{F}_q(X)$.

The following theorem due to Christol, Kamae, Mendès France and Rauzy (see [27, 28] and also [5]) gives a necessary and sufficient condition of algebraicity for a formal power series with coefficients in a finite field.

**Theorem 1.7.1 (Christol, Kamae, Mendès France and Rauzy)**
  *Let $u = (u(n))_{n \in \mathbb{N}}$ be a sequence with values in $\mathbb{F}_q$. The following conditions are equivalent:*

1.  *the formal power series $\sum u(n)X^n$ is algebraic over the field $\mathbb{F}_q(X)$,*

2.  *the $q$-kernel $N_q(u)$ of the sequence $u$ is finite, where $N_q(u)$ is the set of subsequences of the sequence $(u(n))_{n \in \mathbb{N}}$ defined by*

$$N_q(u) = \{(u(q^k n + r))_{n \in \mathbb{N}};\ k \geq 0;\ 0 \leq r \leq q^k - 1\},$$

3.  *the sequence $u$ is $q$-automatic,*

4.  *the sequence $u$ is the image by a letter-to-letter projection of a fixed point of a substitution of constant length.*

The last equivalence is due to Cobham (see [30]) and the equivalence between 2 and 3 dates back to Eilenberg in [39].

**Remark 1.7.2**

- The same theorem holds by considering a series $\sum u(n)X^{-n}$ in $\mathbb{F}_q((1/X))$ with the same definition for the $q$-kernel. Indeed, $\sum u(n)X^{-n}$ is transcendental over $\mathbb{F}_q(X)$ if and only if $\sum u(n)X^n$ is transcendental over $\mathbb{F}_q(X)$.

- The proof of the equivalence between 2 and 3, and between 2 and 4, is constructive.

- The equivalences between 2, 3 and 4 is true for any sequence taking its values in a set of cardinality $q$, where $q$ is not necessarily a power of a prime.

- The notion of $p$-automaticity can also be expressed as follows in terms of first-order logic: a sequence is generated by a $p$-substitution if and only if it is $p$-definable (it can be defined in the theory $(\mathbb{N}, +, V_p)$, where $V_p$ is the function "valuation" that associates to $x$ the highest power of $p$ that divides $x$ (or 1 if $x = 0$)). For more details, the reader is referred to the survey [22].

- Durand has extended in [37] this characterization of automatic sequences to the case of uniformly recurrent sequences generated by substitutions of non-constant length, by introducing the notion of return words.

**Exercise 1.7.3**

1. Show that a sequence is $p$-automatic if and only if it is $p^k$-automatic for any non-zero power of the prime $p$.

2. Consider the Thue-Mose sequence $(S_2(n))_{n \in \mathbb{N}}$ defined over $\mathbb{F}_2$, where $S_2(n)$ is the sum modulo 2 of the coefficients of the digits in the base 2 expansion of the integer $n$. Build a 2-automaton generating the Thue-Morse sequence.

3. Build a $d$-automaton generating the characteristic sequence of the set of powers of a fixed positive integer $d$.

4. Build a $d$-automaton generating the characteristic sequence of the set of integers divisible by $d$.

5. Build a 2-automaton generating the characteristic sequence of the set of integers of base 2-expansion of the form $1^n 0^m 1$, for $n, m > 0$ and $n + m$ odd.

6. Prove that the characteristic sequence of the set of integers of base 2-expansion of the form $1^n 0^{n+1} 1$, for $n > 0$, is not 2-automatic.

## 1.7.2   Applications

In this section we give some applications of Theorem 1.7.1. The first two are easy consequences of the theorem.

- Let $\sum u(n)X^n$ be an algebraic formal power series. Let $a$ and $b$ be two natural integers. The series $\sum u(an+b)X^n$ is algebraic.

- Let $p \geq 2$ be an integer. Let $S_p(n)$ be the sum modulo $p$ of the digits of $n$ in base $p$. The series $\sum S_p(n)X^n$ is algebraic.

**Remark 1.7.4** The series $\sum S_p(n^2)X^n$ is transcendental. More precisely, let $R$ be a polynomial with coefficients in $\mathbb{Q}$ such that $R(\mathbb{N}) \subset \mathbb{N}$. The formal power series $\sum S_p(R(n))X^n$ is algebraic over $\mathbb{F}_p$ if and only if the degree of $R$ is less than or equal to 1 (see [4]).

The **Hadamard product** of two series $\sum u(n)X^n$ and $\sum v(n)X^n$ is defined as the series $\sum u(n)v(n)X^n$. By considering the notion of $q$-kernel, we easily deduce the following.

**Theorem 1.7.5** *The Hadamard product of two algebraic formal power series with coefficients in a finite field is algebraic.*

Note that the following theorem, due to Cobbham [29], produces more examples of transcendental series. We will not give here the proof of this theorem, which is rather difficult.

**Theorem 1.7.6** *Let $u$ be a sequence which is both $k$-automatic and $k'$-automatic. If $k$ and $k'$ are multiplicatively independent (i.e., if $\frac{\log(k)}{\log(k')}$ is irrational), then the sequence $u$ is ultimately periodic.*

We deduce from this theorem the following result of transcendence, which answers, in the case of formal power series with values in a finite field, an analogous question attributed to Mahler and still open: given $(u(n))_{n\in\mathbb{N}}$ a binary sequence such that the series $\sum u(n)2^{-n}$ and $\sum u(n)3^{-n}$ are algebraic over $\mathbb{Q}$, is this sequence ultimately periodic?

**Theorem 1.7.7** *Let $(u(n))_{n\in\mathbb{N}}$ be a binary sequence such that $\sum u(n)X^n$ considered as an element of $\mathbb{F}_2((X))$ and $\sum u(n)X^n$ considered as an element of $\mathbb{F}_3((X))$ are algebraic. Then, this sequence is ultimately periodic, i.e., both series are rational.*

Another application of Cobham's Theorem is the following (see for instance [7]).

**Theorem 1.7.8** *Let $r$ be an integer greater than or equal to 2. The series $\sum_{k=0}^{+\infty} X^{r^k}$ is algebraic over $\mathbb{F}_q(X)$ if and only if $r$ is a power of $p$.*

**Proof.** Write

$$\sum_{k=0}^{+\infty} X^{r^k} = \sum_{n\geq 1} u(n)X^n,$$

where $u = (u(n))_{n\in\mathbb{N}}$ is the characteristic sequence of the set of powers of $r$. The series $\sum_{k=0}^{+\infty} X^{r^k}$ is algebraic over $\mathbb{F}_q(X)$ if and only if the sequence $u$ is $p$-automatic. But it is easily seen that the sequence $u$ is $r$-automatic (Exercise 1.7.3) and not ultimately periodic. Hence the series $\sum_{k=0}^{+\infty} X^{r^k}$ is algebraic over $\mathbb{F}_q(X)$ if and only if $r$ is a power of $p$. ∎

**Remark 1.7.9** The formal power series $\sum u(n)X^n$ belongs to $\mathbb{F}_q(X)$ if and only if the sequence $(u(n))_{n\in\mathbb{N}}$ is ultimately periodic. Note that in the real case we just have the following implication: if the sequence $(u(n))_{n\in\mathbb{N}}$ is ultimately periodic, then the series $\sum u(n)X^n$ belongs to $\mathbb{Q}(X)$. The rational series $\sum nX^n$ shows that the converse is not true.

It is natural to consider the connections between transcendence in the real case and in positive characteristic. Indeed, a formal power series is algebraic in positive characteristic if the sequence of its coefficients has some kind of order, whereas irrational algebraic real numbers cannot have a too regular expansion. Loxton and van der Poorten [51] have conjectured the following (this conjecture is often quoted as a theorem, but there seems to be a gap in the proof).

**Conjecture 1.7.10** *If the sequence of the coefficients in the base $q$-expansion of a real number is automatic, then this number is either rational or transcendental.*

This conjecture illustrates, like Cobhams's Theorem, the fact that transcendence deeply depends on the frame in which it is considered.

## 1.7.3   The Multidimensional Case

The Christol, Kamae, Mendès France and Rauzy theorem can be generalized to the multidimensional case. In particular, Salon has generalized this theorem to the case of a formal power series with a finite number of indeterminates and with coordinates in a finite field, say $\sum_{n_i\geq 0} u(n_1, n_2, \cdots, n_d)X_1^{n_1} \cdots X_d^{n_d}$ (see for instance [62] and [63]). The generalization of the $q$-kernel is given in this case by:

$$N_q(u(n_1, n_2, \cdots, n_d)) = \{u(q^k n_1 + r_1, q^k n_2 + r_2, \cdots, q^k n_d + r_d),$$

$$k \geq 0, \ 0 \leq n_i \leq q^k - 1, \text{ for } i = 1 \text{ to } d\}.$$

Recall that a formal power series $F = \sum_{n_i \geq 0} u(n_1, n_2, \cdots, n_d) X_1^{n_1} \cdots X_d^{n_d}$ is said to be algebraic over $\mathbb{F}_q(X_1, X_2, \cdots, X_d)$ if there exists a nontrivial polynomial $P$ with coefficients in $\mathbb{F}_q(X_1, X_2, \cdots, X_d)$ such that $P(F) = 0$.

The notions of automaton and substitution can also be generalized in two dimensions. A two-dimensional substitution of constant length $l$ associates to each letter a square array of letters of size $(l, l)$. A two-dimensional $k$-automaton is defined similarly as a one-dimensional $k$-automaton but in this case the edges are labelled by pairs of integers in $[0, k-1]^2$. A sequence $(u(m, n))_{(m,n) \in \mathbb{Z}^2}$ is generated by the automaton $\mathcal{A}$ by reading simultaneously the digits of the base $k$ expansions of $m$ and $n$, the shortest expansion being completed with leading zeroes to get two strings of symbols of the length of the longest expansion (without leading zeroes).

We thus have the following theorem due to Salon (see [62] and [63]).

**Theorem 1.7.11** *The series $\sum u(n_1, n_2, \cdots, n_d) X_1^{n_1} \cdots X_d^{n_d}$ is algebraic over $\mathbb{F}_q(X_1, X_2, \cdots, X_d)$ if and only if the $q$-kernel of the sequence $u$ is finite.*

The following results are easy applications of this theorem.

- Let $u$ be an algebraic formal power series. The double formal power series $\sum u(m+n) X^m Y^n$ is algebraic.

- Let $\sum u(m, n) X^m Y^n$ be algebraic. Let $a, b, c, d$ be four integers. The series $\sum u(am+b, cn+d) X^m Y^n$ is algebraic.

**Exercise 1.7.12** Consider the substitution $\sigma : \{0, 1\} \to \{0, 1\} \times \{0, 1\}$ defined by

$$\sigma(0) = \begin{matrix} 00 \\ 00 \end{matrix},$$

$$\sigma(1) = \begin{matrix} 11 \\ 10 \end{matrix}.$$

Prove that the double-sequence fixed point of this substitution generated by the successive images of 1 is equal to Pascal's triangle reduced modulo 2. Find the substitution generating Pascal's triangle modulo a prime $p$.

**Remark 1.7.13** The double-sequence corresponding to Pascal's triangle modulo an integer $d$ is automatic if and only if $d$ is a power of a prime (see [10]).

## 1.7.4 Application to Diagonals

Another interesting consequence of this generalization to the multidimensional case is given by the following results. The *diagonal* of a double formal power series $\sum u(m, n) X^m Y^n$ is defined as the series $\sum u(n, n) X^n$.

**Theorem 1.7.14** *The diagonal of an algebraic formal power series with co-efficients in a finite field is algebraic.*

**Proof.** Consider either the notion of $q$-kernel or the one-dimensional substitution defined by associating to each letter the "diagonal" of the square array of letters associated by the initial substitution. ∎

Theorem 1.7.14 was first proved by Furstenberg in [46] and can be compared to the following theorem, also due to Furstenberg.

**Theorem 1.7.15** *A series with coefficients in a finite field is algebraic if and only if there exists a rational double formal power series such that the initial series is the diagonal of this double series.*

Observe that this result still holds on $\mathbb{C}$ when considering two-indeterminate series but is false for series involving more indeterminates. For a survey on the subject, the reader is referred to [6].

**Exercise 1.7.16**

1. Consider the Thue-Morse sequence $(S_2(n))_{n\in\mathbb{N}}$. Prove that the series $\sum S_2(n)X^n$ is the diagonal of the rational function in $\mathbb{F}_2(X,Y)$ defined by $Y(1 + Y(1 + XY) + X(1 + XY)^{-2})^{-1}$.

2. Let $(u(n))_{n\in\mathbb{N}}$ be a sequence with values in the finite set $X$. Prove that if $\sum u(n)X^n$ is algebraic, then, for any $x \in X$, $\sum_{v(n)=x} X^n$ is algebraic.

3. Let $(v(n))_{n\in\mathbb{N}}$ be the characteristic sequence of the set of powers of the prime $p$. Prove that the series $\sum v(n)X^n$ is the diagonal of the rational fraction of $\mathbb{F}_p(X,Y)$ defined by $X/(1 - (X^{p-1} + Y))$.

4. Let $(w(n))_{n\in\mathbb{N}}$ be the characteristic sequence of the set of integers of base 2-expansion of the form $1^n0^m1$, for $n, m > 0$ and $n + m$ odd. Let $(x(n))_{n\in\mathbb{N}}$ be the characteristic sequence of the set of integers of base 2-expansion of the form $1^n0^{n+1}1$, for $n > 0$. Let $(y(n))_{n\in\mathbb{N}}$ be the characteristic sequence of the set of squares. Prove that the Hadamard product of the series $\sum_{n\geq0} w(n)X^n$ and $\sum_{n\geq0} y(n)X^n$ is equal to $\sum_{n\geq0} x(n)X^n$. Deduce from Exercise 1.7.3 that the sequence $(y(n))_{n\in\mathbb{N}}$ is not 2-automatic (see [60] and the survey [71]).

**Remark 1.7.17** Christol, Kamae, Mendès France and Rauzy's theorem can also be extended to a general field of positive characteristic, which is not necessarily finite. Such a generalization is due to Sharif and Woodcock (see [66]) and Harase (see [48]). The results on the Hadamard poduct and on the diagonal still hold in this context. We can deduce namely the following corollary proved first by Deligne in [35].

**Corollary 1.7.18** *The Hadamard product of two algebraic formal power series with coefficients in a field of positive characteristic is algebraic. The diagonal of an algebraic formal power series with coefficients in a field of positive characteristic is algebraic.*

**Remark 1.7.19** Fresnel, Koskas and de Mathan have also generalized effectively Christol, Kamae, Mendès France and Rauzy's Theorem to the case of an infinite ground field [44].

### 1.7.5   Transcendence of the Bracket Series

The purpose of this section is to prove the following result, which gives us an example of application of the Christol, Kamae, Mendès France and Rauzy theorem. This result was first proved by Wade in [72]; the proof below is due to Allouche (see [7] and also [52]).

**Theorem 1.7.20** *The series* $\displaystyle\sum_{k=1}^{+\infty} \frac{1}{[k]}$ *is transcendental over* $\mathbb{F}_q(X)$.

This proof makes use of the following consequence of the Christol, Kamae, Mendès France and Rauzy theorem.

**Proposition 1.7.21** *Let* $(u(n))_{n\in\mathbb{N}}$ *be a sequence with values in* $\mathbb{F}_q$. *If the series* $\sum u(n)X^{-n}$ *is algebraic over* $\mathbb{F}_q(X)$, *then the sequence* $(u(q^n - 1))_{n\in\mathbb{N}}$ *is ultimately periodic.*

**Proof.** Suppose that the series $\sum u_n X^{-n}$ is algebraic; the sequence $u = (u(n))_{n\in\mathbb{N}}$ is thus $q$-automatic. Let $\mathcal{A}$ denote a finite $q$-automaton which generates the sequence $u$. The subsequence $(u(q^n - 1))_{n\in\mathbb{N}}$ is obtained by reading in the automaton $\mathcal{A}$ strings of ones. As the number of states of $\mathcal{A}$ is finite, a sufficiently long string of ones meets twice the same state. The sequence of states met is thus ultimately periodic, which implies that $(u(q^n - 1))_{n\in\mathbb{N}}$ is also ultimately periodic. ∎

**Remark 1.7.22** Let $\overline{UT^nV}$ denote, for any natural integer $n$, the integer of base-$q$ expansion $UT^nV$, where $U, T, V$ are words defined over $\{0, 1, \cdots, q - 1\}$. We can similarly prove that if the series $\sum u_n X^{-n}$ is algebraic, then the sequence $(u(\overline{UT^nV}))_{n\in\mathbb{N}}$ is ultimately periodic. This result corresponds to the classical *pumping lemma* in automata theory.

**Exercise 1.7.23** Give another proof of Proposition 1.7.21, by using the notion of $q$-kernel.

**Proof.** Let us prove Theorem 1.7.20. We have:

$$\sum_{k\geq1}\frac{1}{[k]} = \sum_{k\geq1}\frac{1}{(X^{q^k}-X)} = \sum_{k\geq1}\frac{1}{X^{q^k}(1-(\frac{1}{X})^{q^k-1})}$$

$$= \sum_{k\geq1}\frac{1}{X^{q^k}}\sum_{j\geq0}(\frac{1}{X})^{j(q^k-1)} = \frac{1}{X}\sum_{k\geq1,\,j\geq0}(\frac{1}{X})^{(j+1)(q^k-1)}$$

$$= \frac{1}{X}\sum_{k\geq1,\,j\geq1}\frac{1}{X^{j(q^k-1)}} = \frac{1}{X}\sum_{n\geq1}a(n)X^{-n},$$

where $a(n)$ is the number (modulo the characteristic $p$) of decompositions of the integer $n$ as $n = j(q^k-1)$, with $k\geq1$ and $j\geq1$, i.e.,

$$a(n) = \sum_{k\geq1,\,(q^k-1)|n}1.$$

Clearly the series $\sum_{k\geq1}\frac{1}{[k]}$ is transcendental over $\mathbb{F}_q(X)$ if and only if the series $X\sum_{k\geq1}\frac{1}{[k]}$ is transcendental. Suppose that the series $\sum_{n\geq1}a(n)X^{-n}$ is algebraic over $\mathbb{F}_q(X)$. This implies that the sequence $(a(n))_{n\in\mathbb{N}}$ is $q$-automatic and in particular that the subsequence $a((q^n-1))_{n\in\mathbb{N}}$ is ultimately periodic. This assertion leads to a contradiction.

Indeed, it is easily seen that $q^k-1$ divides $q^n-1$ if and only if $k$ divides $n$. We thus have

$$a(q^n-1) = \sum_{k\geq1,\,(q^k-1)|(q^n-1)}1 = \sum_{k\geq1,\,k|n}1.$$

The subsequence $a((q^n-1))_{n\in\mathbb{N}}$ is supposed to be ultimately periodic. Thus there exist $n_0\geq1$ and $T\geq1$ such that:

$$\forall n\geq n_0,\ \sum_{k\geq1,\,k|n}1 = \sum_{k\geq1,\,k|n+T}1 \mod p.$$

This implies

$$\forall n\geq n_0,\ \forall\mu\in\mathbb{N},\ \sum_{k\geq1,\,k|n}1 = \sum_{k\geq1,\,k|n(1+\mu T)}1 \mod p.$$

By the primes in arithmetic progression theorem, there exists a prime number $\omega>n_0$ and such that $\omega=1+\mu T$ for some integer $\mu$. Then for $n=\omega$

$$\sum_{k\geq1,\,k|\omega}1 = \sum_{k\geq1,\,k|\omega^2}1 \mod p,$$

i.e.,

$$2 = 3 \mod p,$$

which is the desired contradiction. ∎

**Exercise 1.7.24** Let us give another proof of this result which does not involve the primes in arithmetic progression theorem. This very nice proof is due to Mendès France and Yao ([52]).

1. Prove that for any positive integers $u, v, w$, the number $q^w - 1$ divides $q^u(q^v - 2) + 1$ if and only if $w$ divides the greatest common divisor of $u$ and $v$.

2. Define the sequence $a_u = (a(q^u n + 1))_{n \in \mathbb{N}}$, for a fixed positive integer $u$. Let $u, v$ be two distinct positive integers. Let $h$ be the smallest integer such that $h$ divides $u$ and $h$ does not divide $v$. Prove that $a_u(q^h - 2) - a_v(q^h - 2) \equiv 1$.

3. Deduce from this the transcendence of $\sum_{n \geq 1} a(n) X^{-n}$.

4. Prove similarly the following theorem [52].

**Theorem 1.7.25** *Let $(n_k)_{k \in \mathbb{N}}$ be a sequence of elements of $\mathbb{F}_q$ which is not ultimately equal to zero. The formal power series $\sum_{k \geq 1} \frac{n_k}{[k]}$ is transcendental over $\mathbb{F}_q(X)$.*

## 1.7.6    Complexity and Frequencies

Recall that automatic sequences (and more generally substitutive sequences) have a strong underlying structure with respect to the complexity, as expressed by the following properties.

**Theorem 1.7.26**

1. *The complexity of a fixed point of a primitive substitution [57] or of a fixed point of a substitution of constant length [30] satisfies*

$$\forall n, \ p(n) \leq Cn.$$

2. *The complexity of a substitution fixed point satisfies [40, 56]*

$$\forall n, \ p(n) \leq Cn^2.$$

This disproves the automaticity of a high complexity sequence. Note that automatic sequences have rather similar complexity functions but very dissimilar spectral properties (see for instance [57]).

Let us review some results on the frequencies of factors of a fixed point of a substitution of constant length. For general results in the case of primitive substitutions, see [49, 57]. Recall in particular that for such sequences the frequencies exist and are strictly positive.

The following equirepartition result holds for frequencies of factors of primitive substitutions of constant length (see [35]).

**Theorem 1.7.27** *Let $\sigma$ be a primitive substitution of constant length. There exist $C_2 > C_1 > 0$ such that*

$$C_1 \leq nf(w) \leq C_2,$$

*for all $n \geq 1$ and all factors $w$ of length $n$.*

One proves the following result by applying the properties of stochastic matrices to the $n$-th power of the matrix of the substitution divided by $l^n$, where $l$ denotes the length of the substitution (see [30, 49, 57]).

**Theorem 1.7.28** *If the frequencies of the factors of an automatic sequence exist, they are rational. Furthermore, if the corresponding substitution is primitive, then the frequencies exist.*

In particular, a Sturmian sequence cannot be automatic.

Cobham gives more precise results for automatic sequences with a letter of 0 frequency [30]. In particular we have the following result which can be considered as a criterion for testing the automaticity.

**Theorem 1.7.29** *Let $u$ be an automatic sequence and let $a$ be a letter which occurs in $u$ infinitely often with 0 frequency. Then the gaps between successive occurrences of $a$ in $u$ satisfy*

$$\limsup_{n \to +\infty} \frac{\alpha_{n+1}}{\alpha_n} > 1,$$

*where $\alpha_n$ denotes the index of the $n$-th occurrence of $a$ in the sequence $u$.*

**Exercise 1.7.30** Apply Theorem 1.7.29 to the characteristic sequence of the set of squares (see also Exercise 1.7.16).

**Remark 1.7.31** Yao produces in [74] non-automaticity criteria motivated by a transcendence criterion due to de Mathan (see [33]). Note that Koskas gives a proof using automata of this last criterion in [50].

# 1.8   Conclusion

Let us conclude this lecture by discussing the connections between Sturmian sequences and automatic sequences (Section 1.8.1), and more generally
between sequences of sub-affine complexity and substitutions (Section 1.8.2).

## 1.8.1   Automaticity and Sturmian sequences

Shallit introduces in [64] a *measure of automaticity* of a sequence $u$ over a
finite alphabet: the $k$-automaticity of $u$, $A_u^k(n)$, is defined as the smallest
possible number of states in any deterministic finite automaton which generates the prefix of size $n$ of this sequence. (By Christol, Kamae, Mendès
France and Rauzy's theorem a sequence has a finite measure of automaticity
if and only if this sequence is automatic.) This measure tells quantitatively
how "close" a sequence is to being $k$-automatic.

**Remark 1.8.1** The automaton is fed with the digits $i$, starting from the
*least significant digit*: there are languages of low automaticity whose mirror
image has high automaticity (see [47]).

A sequence can fail to be $k$-automatic if all the sequences in the $k$-kernel
are distinct. A sequence is said to be *maximally diverse* if the subsequences
$\{(u(kn + r))_{n \in \mathbb{N}} : k \geq 1, 0 \leq r \leq k - 1\}$ are all distinct. Shallit proves
in [64] that Sturmian sequences are maximally diverse, which shows that
they are very far from being automatic, even when they are fixed points of
substitution. More precisely, he deduces from the three distance theorem a
measure of automaticity for some Sturmian sequences [64].

**Theorem 1.8.2** *Let $0 < \alpha < 1$ be an irrational number with bounded partial
quotients. Let $u_n = \lfloor (n + 1)\alpha \rfloor - \lfloor n\alpha \rfloor$, for $n \geq 1$. The automaticity of the
sequence $(u_n)_{n \geq 1}$ has the same order of magnitude as $n^{1/5}$.*

## 1.8.2   Sub-affine Complexity

Numerous combinatorial, ergodic or arithmetical properties hold in the case
of a sub-affine complexity function. Consider a dynamical system generated
by a minimal sequence with sub-affine complexity. Ferenczi proves in [42]
the absence of strong mixing. Boshernitzan produces in [17, 19] an explicit
upper bound for the number of ergodic measures. He proves furthermore
that the following conditions imply the unique ergodicity [17, 19]:

$$\liminf_{n \to +\infty} \frac{p(n)}{n} < 2, \text{ or } \limsup_{n \to +\infty} \frac{p(n)}{n} < 3.$$

Furthermore Ferenczi deduces from Theorem 1.4.9 the following result [42]: a symbolic dynamical system generated by a minimal sequence of sub-affine complexity is generated by a finite number of substitutions (such a system is called *S-adic* following Vershik's terminology). One thus explicitely knows *S*-adic expansions of Sturmian sequences or of Arnoux-Rauzy sequences [11]. The reciprocal of this result is false. Consider indeed a substitution of quadratic complexity: it thus provides a counter-example.

The question of finding a characterization of sequences of sub-affine complexity in terms of *S*-adic expansions remains open. Note that Durand gives in [38] a sufficient (but not necessary) condition for a *S*-adic sequence to have sub-affine complexity.

# Bibliography

[1] R. L. Adler, A. G. Konheim, M. H. McAndrew, *Topological Entropy*, Trans. Amer. Math. Soc. **114**, pp. 309–319 (1965).

[2] P. Alessandri, *Codages de Rotations et Basses Complexités*, Université Aix-Marseille II, Thèse (1996).

[3] P. Alessandri, V. Berthé, *Three Distance Theorems and Combinatorics on Words*, Enseig. Math. **44**, pp. 103–132 (1998).

[4] J.-P. Allouche, *Somme des Chiffres et Transcendence*, Bull. Soc. Math. France **110**, pp. 279–285 (1982).

[5] J.-P. Allouche, *Automates Finis en Théorie des Nombres*, Expo. Math. **5**, pp. 239–266 (1987).

[6] J.-P. Allouche, *Note sur un Article de Sharif et Woodcock*, Séminaire de théorie des Nombres de Bordeaux, Série II **1**, pp. 631–633 (1989).

[7] J.-P. Allouche, *Sur la Transcendance de la Série Formelle Π*, Séminaire de Théorie des Nombres de Bordeaux **2**, pp. 103–117 (1990).

[8] J.-P. Allouche, *Sur la Complexité des Suites Infinies*, Bull. Belg. Math. Soc. **1**, pp. 133–143 (1994).

[9] J.-P. Allouche, M. Mendès France, *Automata and Automatic Sequences*, Actes de l'École de Physique Théorique des Houches: "Beyond Quasicrystals", Les Éditions de Physique, Springer, pp. 293–367 (1995).

[10] J.-P. Allouche, F. von Haeseler, H.-O. Peitgen, G. Skordev, *Linear Cellular Automata, Finite Automata and Pascal's Triangle*, Discrete Applied Mathematics **66**, pp. 1–22 (1996).

[11] P. Arnoux, G. Rauzy, *Représentation Géométrique de Suites de Complexité* $2n + 1$, Bull. Soc. Math. France **199**, pp. 199–215 (1991).

[12] J. Berstel, *Recent Results in Sturmian Words*, Developments in Language Theory II (DLT'95) (Dassow, Rozenberg, Salomaa eds) World Scientific, pp. 13–24 (1996).

[13] V. Berthé, *Entropy in Deterministic and Random Systems*, Actes de l'École de Physique Théorique des Houches: "Beyond quasicrystals", Les Éditions de Physique, Springer, pp. 441–463 (1995).

[14] V. Berthé, *Fréquences des Facteurs des Suites Sturmiennes*, Theoret. Comput. Sci. **165**, pp. 295–309 (1996).

[15] V. Berthé, N. Chekhova, S. Ferenczi, *Covering Numbers: Arithmetics and Dynamics for Rotations and Interval Exchanges*, J. Anal. Math., to appear.

[16] P. Billingsley, *Ergodic Theory and Information*, John Wiley and Sons, New York (1965).

[17] M. Boshernitzan, *A Unique Ergodicity of Minimal Symbolic Flows with Linear Block Growth*, J. Anal. Math. **44**, pp. 77–96 (1984).

[18] M. Boshernitzan, *A Condition for Minimal Interval Exchange Maps to be Uniquely Ergodic*, Duke Math. J. **52**, pp. 723–752 (1985).

[19] M. Boshernitzan, *A Condition for Unique Ergodicity of Minimal Symbolic Flows*, Ergodic Theory Dynam. Sys. **12**, pp. 425–428 (1992).

[20] S. Brlek, *Enumeration of Factors in the Thue-Morse Word*, Discr. Appl. Math. **24**, pp. 83–96 (1989).

[21] T. C. Brown, *Descriptions of the Characteristic Sequence of an Irrational*, Canad. Math. Bull **36**, pp. 15–21 (1993).

[22] V. Bruyère, G. Hansel, C. Michaux, *Logic and p-Recognizable Sets of Integers*, Bull. Belg. Math. Soc. **1**, pp. 191–238 (1994).

[23] J. Cassaigne, *Special Factors of Sequences with Linear Subword Complexity*, Developments in Language Theory II (DLT'95) (Dassow, Rozenberg, Salomaa eds) World Scientific, pp. 25–34 (1996).

[24] J. Cassaigne, *Complexité et Facteurs Spéciaux*, Bull. Belg. Math. Soc. **4**, pp. 67–88 (1997).

[25] J. Cassaigne, *Sequences with Grouped Factors*, Developments in Language Theory III (DLT'97), Publications of the Aristotle University of Thessaloniki, pp. 211–222 (1998).

[26] N. Chekhova, *Covering Numbers of Rotations*, Theoret. Comput. Sci., to appear.

[27] G. Christol, *Ensembles presque Périodiques k-Reconnaissables*, Theoret. Comp. Sci. **9**, pp. 141–145 (1979).

[28] G. Christol, T. Kamae, M. Mendès France, G. Rauzy, *Suites Algébriques, Automates et Substitutions*, Bull. Soc. Math. France **108**, pp. 401-419 (1980).

[29] A. Cobham, *On the Base Dependence of Sets of Numbers Recognisable by Finite Automata*, Math. Systems Theory **3**, pp. 186–192 (1969).

[30] A. Cobham, *Uniform Tag Sequences*, Math. Systems Theory **6**, pp. 164–192 (1972).

[31] E. M. Coven, G. A. Hedlund, *Sequences with Minimal Block Growth*, Math. Systems Theory **7**, pp. 138–153 (1973).

[32] N. G. de Bruijn, *A Combinatorial Problem*, Rominklijke Netherland Academic Van Wetenschepen Proc. **49** Part. 20, pp. 758–764 (1946).

[33] B. de Mathan, *Irrationality Measures and Transcendence in Positive Characteristic*, J. Number Theory **54**, pp. 93–112 (1995).

[34] F. M. Dekking, *On the Prouhet-Thue-Morse Measure*, Acta Universitatis Carolinae, Mathematica et Physica **33**, pp. 35–40 (1992).

[35] P. Deligne, *Intégration sur un Cycle Évanescent*, Invent. Math. **76**, pp. 129–143 (1984).

[36] G. Didier, *Caractérisation des N-Écritures et Applications à l'Étude des Suites de Complexité Ultimement $n + c^{ste}$*, Theoret. Comput. Sci. **215**, pp. 31–49 (1999).

[37] F. Durand, *A Characterization of Substitutive Sequences using Return Words*, Discrete Math. **179**, pp. 89–101 (1998).

[38] F. Durand, *Linearly Recurrent Subshifts*, Ergod. Theory Dynam. Sys., to appear.

[39] S. Eilenberg, *Automata, Languages, and Machines*, Academic Press, Vol. A (1974).

[40]  A. Ehrenfeucht, K. P. Lee, G. Rozenberg, *Subwords of Various Classes of Deterministic Developmental Languages without Intercations*, Theoret. Comput. Sci. 1, pp. 59–75 (1975).

[41]  S. Ferenczi, *Les Transformations de Chacon : Combinatoire, Structure Géométrique, Lien avec les Systèmes de Complexité* $2n + 1$, Bull. Soc. Math. France **123**, pp. 271–292 (1995).

[42]  S. Ferenczi, *Rank and Symbolic Complexity*, Erg. Theory Dynam. Sys. **16**, pp. 663–682 (1996).

[43]  S. Ferenczi, *Complexity of Sequences and Dynamical Systems*, Discrete Math., to appear.

[44]  J. Fresnel, M. Koskas, B. de Mathan, *Automata and Transcendence in Positive Characteristic*, J. Number Theory, to appear.

[45]  C. Frougny, *Number Representation and Finite Automata*, this volume.

[46]  H. Furstenberg, *Algebraic Functions over Finite Fields*, J. Algebra **7**, pp. 271–277 (1967).

[47]  I. Glaister, J. Shallit, *Automaticity III: Polynomial Automaticity and Context-Free Languages (extended abstract)*, Mathematical Foundations of Computer Science 1996 (Cracow), Lecture Notes in Comput. Sci. **1113**, pp. 382–393 (1996).

[48]  T. Harase, *Algebraic Elements in Formal Power Series Rings*, Israel J. Math. **63**, pp. 281–288 (1988).

[49]  B. Host, *Substitutions and Bratelli Diagrams*, this volume.

[50]  M. Koskas, *Complexité de Suites, Fonctions de Carlitz*, University of Bordeaux I, Thesis (1995).

[51]  J. H. Loxton, A. J. van der Poorten, *Arithmetic Properties of Automata: Regular Sequences*, J. Reine Angew. Math. **392**, pp. 57–69 (1988).

[52]  M. Mendès France, J. Y. Yao, *Transcendence and the Carlitz-Goss Gamma Function*, J. Number Theory **63**, pp. 396–402 (1997).

[53]  M. Misiurewicz, *A Short Proof of the Variational Principle for a* $\mathbb{Z}_+^n$ *Action on a Compact Space*, Int. Conf. Dyn. Systems in Math. Physics, Société Mathématique de France, Astérisque **40**, pp. 147–158 (1976).

[54]  M. Morse, G. A. Hedlund, *Symbolic Dynamics*, Amer. J. Math. **60**, pp. 815–866 (1938).

[55] M. Morse, G. A. Hedlund, *Symbolic Dynamics II: Sturmian Trajectories*, Amer. J. Math. **62**, pp. 1–42 (1940).

[56] J.-J. Pansiot, *Complexité des Facteurs des Mots Infinis Engendrés par Morphismes Itérés*, Lecture Notes in Comput. Sci. **172**, pp. 380–389 (1984).

[57] M. Queffélec, *Substitution Dynamical Systems. Spectral Analysis*, Lecture Notes in Math. **1294**, Springer-Verlag (1987).

[58] G. Rauzy, *Suites à Termes dans un Alphabet Fini*, Sém. de Théorie des Nombres de Bordeaux, 25-01–25-16 (1983).

[59] G. Rauzy, *Low Complexity and Geometry*, Dynamics of Complex Interacting Systems (Santiago, 1994), pp. 147–177, Nonlinear Phenom. Complex Systems, 2, Kluwer Acad. Publ., Dordrecht (1996).

[60] R. Ritchie, *Finite automata and the set of squares*, J. Assoc. Comput. Mach. **10**, pp. 528–531 (1963).

[61] G. Rote, *Sequences with Subword Complexity 2n*, J. Number Theory **46**, pp. 196–213 (1994).

[62] O. Salon, *Suites Automatiques à Multi-indices et Algébricité*, C. R. Acad. Sci. Paris, Série I **305**, pp. 501–504 (1987).

[63] O. Salon, *Suites Automatiques à Multi-indices*, Séminaire de Théorie des Nombres, Exposé 4, 1986-1987.

[64] J. Shallit, *Automaticity IV: Sequences, Sets, and Diversity*, J. Théor. Nombres Bordeaux **8**, pp. 347–367 (1996).

[65] C. E. Shannon, *A Mathematical Theory of Communication*, Bell System Tech. J. **27**, pp. 379–423, 623–656 (1948).

[66] H. Sharif, C. F. Woodcock, *Algebraic Functions over a Field of Positive Characteristic and Hadamard Products*, J. Lond. Math. Soc. **37**, pp. 395–403 (1988).

[67] V. T. Sós, *On the Theory of Diophantine Approximations I*, Acta Math. Acad. Sci. Hungar. **8**, pp. 461–472 (1957).

[68] V. T. Sós, *On the Distribution mod 1 of the Sequence nα*, Ann. Univ. Sci. Budapest, Eötvös Sect. Math. **1**, pp. 127–134 (1958).

[69] J. Surányi, *Über die Anordnung der Vielfachen einer reellen Zahl mod 1*, Ann. Univ. Sci. Budapest, Eötvös Sect. Math. **1**, pp. 107–111 (1958).

[70] S. Świerczkowski, *On Successive Settings of an Arc on the Circumference of a Circle*, Fundamenta Math. **46**, pp. 187–189 (1958).

[71] D. S. Thakur, *Automata and Transcendence*, Contemporary Mathematics, **210**, pp. 387–399 (1998).

[72] L. J. Wade, *Certain Quantities Transcendental over $GF(p^n, x)$*, Duke Math. J. **8**, pp. 701-720 (1941).

[73] P. Walters, *An Introduction to Ergodic Theory*, Graduate Texts in Mathematics **79**.

[74] J.-y. Yao, *Critères de Non-Automaticité et leurs Applications*, Acta Arith. **80**, pp. 237–248 (1997).

# Chapter 2

# SUBSTITUTION SUBSHIFTS AND BRATTELI DIAGRAMS

*Bernard HOST*
*Équipe d'Analyse et de Mathématiques Appliquées*
*Université de Marne la Vallée*
*2 rue de la Butte Verte*
*93166 Noisy le Grand cedex*
*France*

The goal of these lectures is to introduce, through the example of substitution dynamical systems, some of the basic concepts of topological dynamics: minimality, unique ergodicity, Kakutani-Rohlin partitions, ...

Moreover, we shall present here the less classical notions of Bratteli diagrams and Bratteli–Vershik systems. It became clear in the last years that these objects provide a very fruitful link between topological dynamics and the theory of $C^*$-algebras. In particular, an algebraic invariant of $C^*$-algebras, the so–called "dimension group", has been proved to have a dynamical interpretation, and can now be used for the classification of dynamical systems.

Since we cannot completely develop this theory in the available space, we shall only explain how to construct a Bratteli–Vershik model in the case of substitution dynamical systems.

## 2.1  Subshifts

### 2.1.1  Notation: Words, Sequences, Morphisms

**Words.** An alphabet is a finite set of symbols called letters. If $A$ is an alphabet, a word of $A$ is a finite (non–empty) sequence of letters, and $A^+$ denotes the set of words. For $u = u_1 u_2 \ldots u_n \in A^+$, $|u| = n$ is called the length of $u$. For each letter $a$ we denote by $|u|_a$ the number of occurrences of $a$ in $u$, and the vector $(|u|_a \, ; \, a \in A)$ is sometimes called the composition vector of $u$. $A^*$ consists of $A^+$ and the empty word $\emptyset$ of length 0.

Given a word $u = u_1 \ldots u_m$ and an interval $J = \{i, \ldots, j\}$ contained in $\{1, \ldots, m\}$, $u_J$ denotes the word $u_i u_{i+1} \ldots u_j$. A factor of $u$ is a word $v$ such that $v = u_J$ for some interval $J \subset \{1, \ldots, m\}$.

**Sequences.** Elements of $A^{\mathbb{Z}}$ are called sequences over the alphabet $A$. For a sequence $x$ we use the notation $x_J$ and the term 'factor' in exactly the same way they are used for words.

Let $x$ be a sequence, $n$ an integer and $u, v$ words. We say that $n$ is an occurrence of $u$ in $x$ if $x_{[n,n+|u|)} = u$; we say that $n$ is an occurrence of $u.v$ in $x$ if $x_{[n-|u|,n+|v|)} = uv$. The language $\mathcal{L}(x)$ of the sequence $x$ is the set of all words that are factors of $x$.

**The Shift Map.** Unless this produces some ambiguity, $T$ will denote the shift map on $A^{\mathbb{Z}}$, $T$ is defined as

$$(Tx)_n = x_{n+1} \text{ for every } x \in A^{\mathbb{Z}} \text{ and every } n \in \mathbb{Z}$$

and it is a bijection of $A^{\mathbb{Z}}$ onto itself.

The orbit $\mathcal{O}_x$ of the sequence $x \in A^{\mathbb{Z}}$ is

$$\mathcal{O}_x = \{T^n x \, ; \, n \in \mathbb{Z}\} \, .$$

**Morphisms.** Let $A, B$ be two alphabets, and let $\phi \colon A \to B^*$ be any map. By concatenation, $\phi$ can be extended to a map $A^+ \to B^*$, which will also be denoted by $\phi$. This map is a morphism for the concatenation, a morphism for short, which means that $\phi(uv) = \phi(u)\phi(v)$ for every words $u, v$.

If $\phi$ maps $A$ to $B^+$, it also defines a map from $A^{\mathbb{Z}}$ to $B^{\mathbb{Z}}$, which will be written as $\phi$ too:

$$x \quad = \quad \ldots x_{-2} x_{-1} \quad | \quad x_0 x_1 x_2 \ldots \in A^{\mathbb{Z}},$$

we set: $\quad \phi(x) \quad = \quad \ldots \phi(x_2)\phi(x_{-1}) \quad | \quad \phi(x_0)\phi(x_1)\phi(x_2) \ldots$

where a vertical bar separates the negative and nonnegative halves of a sequence. The map $\phi \colon A^{\mathbb{Z}} \to B^{\mathbb{Z}}$ satisfies

$$\forall x \in A^{\mathbb{Z}}, \ \phi(Tx) = T^{|\phi(x_0)|}\phi(x) \text{ and } \phi(T^{-1}x) = T^{-|\phi(x_{-1})|}\phi(x) \, .$$

## 2.1.2   Subshifts

**The Full Shift.** Let $A$ be endowed with the discrete topology and $A^{\mathbf{Z}}$ with the product topology. This topology on $A^{\mathbf{Z}}$ is associated to the following distance:

$$\forall x, y \in A^{\mathbf{Z}}, \ d(x,y) = \Big(1 + \inf\{n \geq 0 \,;\, x_n \neq y_n \text{ or } x_{-n} \neq y_{-n}\}\Big)^{-1}.$$

$A^{\mathbf{Z}}$ is a non–empty and compact metric space. More precisely, it is a Cantor space, which means that

i) The sets which are both closed and open (in short, the clopen sets) span the topology;

ii) the space has no isolated points.

Moreover, $T \colon A^{\mathbf{Z}} \to A^{\mathbf{Z}}$ is obviously a homeomorphism.

Let $A, B$ be alphabets, and $\phi : A \to B^+$ be a map; then the associated map $\phi \colon A^{\mathbf{Z}} \to B^{\mathbf{Z}}$ is continuous.

**Definition 2.1.1** *Dynamical Systems. A (topological) dynamical system $(X, T)$ is a non–empty compact metric space $X$ endowed with an homeomorphism $T \colon X \to X$.*

Thus $(A^{\mathbf{Z}}, T)$ is a dynamical system, called the full shift on the alphabet $A$. When $(X, T)$ is a dynamical system and $Y$ a subset of $X$, $Y$ is said to be invariant if $T^{-1}Y = Y$.

**Definition 2.1.2** *Subshifts. A subshift on the alphabet $A$ is the dynamical system consisting of a non–empty closed $T$–invariant subset $X$ of $A^{\mathbf{Z}}$, endowed with the restriction of $T$ to $X$.*

The restriction of $T$ to $X$ is often written $T_X$, $T$ for short when this cannot lead to some ambiguity.

Given a subshift $(X, T)$ on the alphabet $A$, and two words $u, v$ on $A$, put

$$[u] = \{x \in X \,;\, x_{[0,|u|)} = u\} \text{ and } [u.v] = \{x \in X \,;\, x_{[-|u|,|v|)} = uv\}.$$

Subsets of $X$ of this kind are called cylinder sets. Notice that these notations depend on the given subshift, which is supposed to be determined by the context. Cylinder sets are clopen and span the topology of $X$.

The language $\mathcal{L}(X)$ of the subshift $(X, T)$ is the set of words $u$ that are factors of at least one element of $X$, i.e., such that the cylinder set $[u]$ is not empty. The language of $X$ characterizes $X$, because

$$X = \{x \in A^{\mathbf{Z}} \,;\, \mathcal{L}(x) \subset \mathcal{L}(X)\}.$$

Conversely, let $\mathcal{L}$ be a language, i.e., a subset of $A^*$, and $X$ the set of sequences $x \in A^{\mathbb{Z}}$ every factor of which belongs to $\mathcal{L}$. Then $X$ is a subshift, called the subshift associated to $\mathcal{L}$. One has $\mathcal{L} = \mathcal{L}(X)$ for some subshift $X$ if and only if:

i) $u \in \mathcal{L}$ and ($v$ is a factor of $u \implies v \in \mathcal{L}$) and

ii) $\forall u \in \mathcal{L}\ \exists a, b \in A$ such that $aub \in \mathcal{L}$.

The subshift spanned by a sequence $x \in A^{\mathbb{Z}}$ is the closure $\overline{\mathcal{O}}_x$ of the $T$–orbit of $x$. For every $y \in A^{\mathbb{Z}}$ one has:

$$y \in \overline{\mathcal{O}}_x \iff \mathcal{L}(y) \subset \mathcal{L}(x) ,$$

thus $\mathcal{L}(\overline{\mathcal{O}}_x) = \mathcal{L}(x)$.

### 2.1.3 Minimal Systems

**Proposition 2.1.3** *and Definition: Minimal Systems. For a dynamical system $(X, T)$ the following properties are equivalent:*

i) *The only invariant open subsets of $X$ are $\emptyset$ and $X$.*

ii) *The only invariant closed subsets of $X$ are $\emptyset$ and $X$.*

iii) *Every $x \in X$ has a dense orbit.*

iv) *For every non–empty open subset $U$ of $X$ and every $x \in X$ there exists $n \in \mathbb{Z}$ with $T^n x \in U$.*

v) *For every non–empty open subset $U$ of $X$ there exists an integer $n > 0$ such that:*

$$\forall x \in X\ \exists k \text{ with } 0 < k \leq n \text{ and } T^k x \in U .$$

*A system $(X, T)$ that satisfies any of these properties is said to be minimal.*

**Proof of iv) $\Rightarrow$ v).** Suppose that v) fails; let $U$ be a non–empty open set and, for every $n > 0$, let $x_n$ be a point of $X$ with $T^k x_n \notin U$ for all $k \in \{1, \ldots, 2n\}$. For each $n$, let $y_n = T^n x_n$, and let $(y_{n_j})$ be a subsequence of the sequence $(y_n)$, converging to some $y \in X$. Then $T^k y \notin U$ for any $k$, and iv) fails. ∎

Given any dynamical system $(X, T)$ there exists a non–empty closed invariant subset $Y$ of $X$ such that $(Y, T)$ is minimal.

**Proposition 2.1.4** *Minimal Subshifts. For a subshift* $(X, T)$ *the following properties are equivalent:*

  *i)* $(X, T)$ *is minimal.*

  *ii)* $\mathcal{L}(x) = \mathcal{L}(X)$ *for all* $x \in X$.

  *iii) for every* $u \in \mathcal{L}(X)$ *there exists an integer* $L > 0$ *such that* $u$ *is a factor of every* $v \in \mathcal{L}(X)$ *of length* $\geq L$.

**Definition 2.1.5** *Uniformly Recurrent Languages and Sequences. A language* $\mathcal{L}$ *is uniformly recurrent if for every* $u \in \mathcal{L}$ *there exists an integer* $L > 0$ *such that every* $v \in \mathcal{L}$ *of length* $\geq L$ *contains* $u$ *as a factor.*

  *A sequence* $x$ *is uniformly recurrent if its language* $\mathcal{L}(x)$ *is uniformly recurrent.*

The sequence $x$ is uniformly recurrent if and only if for every factor $u$ of $x$ there exists an integer $L > 0$ such that every interval of length $L$ in $\mathbb{Z}$ contains at least one occurrence of $u$ in $x$. It means that every factor $u$ of $x$ has infinitely many positive and infinitely many negative occurrences in $x$, and that the difference between two consecutive occurrences is bounded, that is, $u$ occurs in $x$ 'with bounded gaps'.

**Proposition 2.1.6** *The subshift spanned by a sequence is minimal if and only if the sequence is uniformly recurrent.*

## 2.2 Substitutions

### 2.2.1 Substitution Subshifts

Let $A$ be an alphabet consisting of at least 2 letters. A substitution on the alphabet $A$ is a map $\sigma \colon A \to A^+$.

Using the extension to words by concatenation, $\sigma$ can be iterated; for each integer $k > 0$, $\sigma^k \colon A \to A^+$ is again a substitution.

Here we consider only primitive substitutions, i.e. substitutions $\sigma$ on $A$ such that

$$\exists k > 0 \; \forall a, b \in A, \; b \text{ appears in } \sigma^k(a) \ . \tag{2.1}$$

It follows that

$$\forall a, b \in A, b \text{ appears in } \sigma^m(a) \text{ for every } m \text{ large enough.} \tag{2.2}$$

Moreover, for every letter $a$ one has $|\sigma^{km}(a)| \geq \mathrm{Card}(A)^m$ for all $m \geq 0$, thus $|\sigma^n(a)|$ tends to $+\infty$ when $n \to +\infty$.

**Definition 2.2.1** *The language $\mathcal{L}_\sigma$ of $\sigma$ is the set of words that are factors of $\sigma^n(a)$ for some $a \in A$ and some $n \geq 1$.*

*The subshift $X_\sigma$ associated to this language is called the substitution subshift defined by $\sigma$. Denote by $T_\sigma$ the restriction of the shift map $T$ to $X_\sigma$.*

**Remark 2.2.2** As seen in the preceding section, $\mathcal{L}(X_\sigma) = \mathcal{L}_\sigma$. Clearly, one has

$$\sigma(X_\sigma) \subset X_\sigma .$$

Let $u \in \mathcal{L}_\sigma$. By (2.2), $u$ is a factor of $\sigma^n(a)$ for every letter $a$ and every $n$ large enough. It follows that for every $n > 0$ the substitutions $\sigma^n$ and $\sigma$ define the same language, hence the same subshift.

**Proposition 2.2.3** *Every substitution subshift is minimal.*

**Proof.** Let $k$ be as in (2.1), in the definition of primitivity above.

Let $u \in \mathcal{L}_\sigma$. There exists a letter $b$ and an integer $n \geq 1$ such that $u$ is a factor of $\sigma^n(b)$. Put

$$L = 3 \sup_{a \in A} |\sigma^{n+k}(a)| ,$$

and let $v \in \mathcal{L}_\sigma$ be of length $\geq L$.

There exists a letter $c$ and an integer $p \geq 0$ such that $v$ is a factor of $\sigma^{n+k+p}(c)$. Writing $\sigma^p(c) = c_1 \ldots c_j$ we get

$$\sigma^{n+k+p}(c) = \sigma^{n+k}(c_1) \ldots \sigma^{n+k}(c_j) ,$$

and the words $\sigma^{n+k}(c_i)$ are of length $\leq L/3$. As $|v| \geq L$, it follows that at least one of these words, say $\sigma^{n+k}(c_i)$, is a factor of $v$. By definition of $k$, $b$ occurs in $\sigma^k(c_i)$, thus $\sigma^n(b)$ is a factor of $\sigma^{n+k}(c_i)$, thus a factor of $v$. As $u$ is a factor of $\sigma^n(b)$ we get that $u$ is a factor of $v$. We have proved that:

For every $u \in \mathcal{L}_\sigma$ there exists an integer $L > 0$ such that $u$ is a factor of every $v \in \mathcal{L}_\sigma$ of length $\geq L$, which implies minimality. ∎

## 2.2.2    Fixed Points

A fixed point of the substitution $\sigma$ is a sequence $x \in A^{\mathbf{Z}}$ such that $\sigma(x) = x$. When $x$ is a fixed point, the following property holds: $x_{-1}$ is the last letter of $\sigma(x_{-1})$ and $x_0$ is the first letter of $\sigma(x_0)$.

Conversely, if $r$ and $\ell$ are letters such that $r$ is the last letter of $\sigma(r)$ and $\ell$ the first letter of $\sigma(\ell)$ there exists a unique fixed point $x$ such that $x_{-1} = r$ and $x_0 = \ell$.

**Proof.** Uniqueness: Let $x$ be such a fixed point. By induction, $\sigma^n(\ell)$ is a prefix of $x_{[0,\infty)}$ for all $n$. Now, let $\jmath$ be a nonnegative integer. We have

$x_j = \left(\sigma^n(x_0)\right)_{j+1}$, where $n$ is any integer with $|\sigma^n(x_0)| > j$. For $j < 0$ we have $x_j = \left(\sigma^n(r)\right)_{j+1+|\sigma^n(r)|}$, where $n$ is any integer with $|\sigma^n(r)| > -j$.

Existence: Let $x$ be defined by the formulas above; it is easy to check that it satisfies the requested properties. ∎

The following proposition can be considered as a definition.

**Proposition 2.2.4** *For a fixed point $x$ of the substitution $\sigma$, the following properties are equivalent:*

   *i) the word $x_{-1}x_0$ has at least two occurrences in $x$;*

   *ii) the word $x_{-1}x_0$ belongs to $\mathcal{L}_\sigma$;*

   *iii) $x \in X_\sigma$;*

   *iv) $x$ is uniformly recurrent.*

   *A fixed point is called admissible when it satisfies any of these properties.*

**Remark 2.2.5** For any admissible fixed point $x$ of $\sigma$ we have $X_\sigma = \overline{\mathcal{O}}_x$. In the literature, this property is often taken as the definition of a substitution subshift.

**Proof.** i) $\Rightarrow$ ii). If $x_{-1}x_0$ has an occurrence in $x_{[0,+\infty)}$, then it is a factor of $\sigma^n(x_0)$ for $n$ large enough, thus belonging to $\mathcal{L}_\sigma$. The same is true if it has an occurrence in $x_{(-\infty,-1]}$.

ii) $\Rightarrow$ iii). Every factor of $x$ is a factor of $\sigma^n(x_{-1}x_0) \in \mathcal{L}_\sigma$ for some $n$, and all these words belong to $\mathcal{L}_\sigma$.

iii) $\Rightarrow$ iv) and iv) $\Rightarrow$ i) are obvious. ∎

For every word $r\ell$ of length 2, let $\phi(r\ell)$ be the word of length 2 consisting of the last letter of $\sigma(r)$ followed by the first letter of $\sigma(\ell)$. If $r\ell \in \mathcal{L}_\sigma$, $\phi(r\ell)$ belongs to $\mathcal{L}_\sigma$ too, because it is a factor of $\sigma(r\ell)$. Thus $\phi$ maps the set of words of length 2 belonging to $\mathcal{L}_\sigma$ to itself, and since this set is finite, there exists an element $r\ell$ of this set and an integer $n \geq 1$ such that $\phi^{(n)}(r\ell) = r\ell$. This means that $r$ is the last letter of $\sigma^n(r)$ and $\ell$ the first letter of $\sigma^n(\ell)$. We have proved that there exists $n \geq 1$ such that $\sigma^n$ admits at least one admissible fixed point.

Since $\sigma$ and $\sigma^n$ define the same system, we can assume without loss of generality that all the substitutions we consider have at least one admissible fixed point, i.e., that any substitution subshift is the orbit closure of some admissible fixed point.

In the study of substitution subshifts, some technical difficulties arise from the fact that a substitution can have several fixed points. We introduce a class of substitutions which are easier to study.

**Definition 2.2.6** A substitution $\sigma$ on the alphabet $A$ is called proper if there exist letters $r, \ell \in A$ such that:$\forall a \in A$, $r$ is the last letter and $\ell$ the first letter of $\sigma(a)$.

If $\sigma$ is proper, then it has a unique fixed point, which is also the unique fixed point of $\sigma^n$ for every $n$. This fixed point is admissible.

## 2.3  Unique Ergodicity

Given a dynamical system $(X, T)$, an invariant measure is a Borel probability measure $\mu$ on $X$ such that $\mu(T^{-1}E) = \mu(E)$ for every Borel subset $E$ of $X$. The existence of at least one invariant measure is a general result. We say that a system is uniquely ergodic if it admits only one invariant measure. The goal of this section is to prove:

**Theorem 2.3.1** *Every substitution subshift is uniquely ergodic.*

(Remember that we consider primitive substitutions only.)

The theorem follows from the next proposition:

**Proposition 2.3.2** *Let $\sigma$ be a primitive substitution. For every $u \in \mathcal{L}_\sigma$ there exist a constant $p_u > 0$ and a sequence $\epsilon_u(n)$ tending to 0 at infinity such that, for every $v \in \mathcal{L}_\sigma$ such that $|v| \geq |u|$ one has:*

$$\left| p_u - \frac{1}{|v| - |u| + 1} \text{(Number of occurrences of } u \text{ in } v) \right| \leq \epsilon_u(|v|) \, .$$

**Proof of Theorem 2.3.1.** Let $\mu$ be an invariant measure of $(X_\sigma, T_\sigma)$, and $u \in \mathcal{L}_\sigma$ be a word of length $k$. By invariance we obtain that for every $n > 0$

$$\mu([u]) = \int \frac{1}{n} \sum_{j=1}^{n} 1_{[u]}(T^j x) \, d\mu(x)$$

$$= \int \frac{1}{n} \text{(Number of occurrences of } u \text{ in } x_{[1, n+k-1]}) \, d\mu(x) \, .$$

Thus $\left| \mu([u]) - p_u \right| \leq \epsilon_u(n = k - 1)$, and the result follows. ∎

### 2.3.1  Perron-Frobenius Theorem

**Definition 2.3.3** A square matrix $M$ with non–negative entries is said to be primitive when there exists an integer $n > 0$ such that all entries of $M^n$ are positive.

**Theorem 2.3.4** *Let $M$ be a primitive $d \times d$ matrix.*

*i) $M$ has a unique eigenvalue $\theta$ of maximum modulus, and this eigenvalue is positive.*

*ii) The maximal eigenvalue $\theta$ is algebraically simple.*

*iii) To this eigenvalue corresponds an eigenvector with positive entries.*

Let $M$ and $\theta$ be as in the theorem. For the eigenvalue $\theta$, $M$ admits a unique right eigenvector $(p_1, \ldots, p_d)$ and a unique left eigenvector $(q_1, \ldots, q_d)$ such that:

$$\sum_{i=1}^{d} p_i = 1 \text{ and } \sum_{i=1}^{d} p_i q_i = 1 \ ,$$

and these vectors have positive entries. We get:

$$\forall i, j \in \{1, \ldots, d\}, \ \theta^{-n} M_{i,j}^n \to p_i q_j \text{ as } n \to \infty \ . \tag{2.3}$$

More precisely, there exist $\rho \in (0, \theta)$ and $C > 0$ such that

$$\forall i, j \in \{1, \ldots, d\}, \ \forall n \geq 0, \ \left| M_{i,j}^n - \theta^n p_i q_j \right| \leq C \rho^n \ . \tag{2.4}$$

## 2.3.2 Density of Letters

**Notation.** Let $u$ be a word on the alphabet $A$. For every letter $a$ denote by $|u|_a$ the number of occurrences of $a$ in $u$.

Henceforth, $\sigma$ will be a primitive substitution on the alphabet $A$. The matrix $M$ of the substitution $\sigma$ is the square matrix with rows and columns indexed by $A$ defined as follows:

$$\forall a, b \in A, \ M_{a,b} = |\sigma(b)|_a \ .$$

Notice that for all $n > 0$ the matrix of the substitution $\sigma^n$ is $M^n$. It follows that the matrix $M$ is primitive. Let $\theta$, $(p_a ; a \in A)$, $(q_a ; a \in A)$, $C$ and $\rho$ be defined as above. Remark that $\theta > 1$.

By (2.4), for every $a, b \in A$ and every $n \geq 0$,

$$\left| |\sigma^n(b)|_a - \theta^n p_a q_b \right| \leq C \rho^n$$

thus

$$\left| |\sigma^n(b)| - \theta^n q_b \right| \leq C' \rho^n \tag{2.5}$$

and

$$\left| |\sigma^n(b)|_a - p_a |\sigma^n(b)| \right| \leq C'' \rho^n \ . \tag{2.6}$$

for some constants $C', C'' > 0$.

**Lemma 2.3.5** *There exist constants $c > 0$ and $0 < \alpha < 1$ such that for every non–empty word $v \in \mathcal{L}_\sigma$ and every letter $a \in A$ one has*

$$\Big| |v|_a - p_a|v| \Big| \leq c|v|^\alpha .$$

**Proof.** Let $v \in \mathcal{L}_\sigma$ be non–empty. From the definition of $\mathcal{L}_\sigma$ we immediately obtain that there exist an integer $n \geq 0$ and words $v_i$ $(0 \leq i \leq n)$ and $w_i$ $(0 \leq i < n)$ such that

i) $|v_i| \leq L$ for $0 \leq i \leq n$ and $|w_i| < L$ for $0 \leq i < n$, where $L = \max_{a \in A} |\sigma(a)|$.

ii) $v_n$ is not empty.

iii) $v = v_0\, \sigma(v_1) \ldots \sigma^{n-1}(v_{n-1})\, \sigma^n(v_n)\, \sigma^{n-1}(w_{n-1}) \ldots \sigma(w_1)\, w_0$.

By (2.6)

$$\text{for } 0 \leq k \leq n, \ \Big| |\sigma^n(v_k)|_a - p_a|\sigma^n(v_k)| \Big| \leq LC''\rho^k,$$

and a similar inequality holds for the $w_k$'s. Thus

$$\Big| |v|_a - p_a|v| \Big| \leq 2LC'' \sum_{k=0}^{n} \rho^k \leq c'\rho^n$$

for some constant $c' > 0$. But by (2.5) $|v| \geq |\sigma^n(v_n)| \geq c''\theta^n$ for some constant $c'' > 0$, and the result follows. ∎

### 2.3.3   The Substitution $\sigma_k$ on the Words of Length $k$

Let $k \geq 1$ be an integer, and $A_k$ be the set of words of length $k$ belonging to $\mathcal{L}_\sigma$. We consider $A_k$ as an alphabet, and define a substitution $\sigma_k$ on it in the following way.

Let $u = u_1, ..., u_k \in A_k$, $\sigma(u) = v = v_1 \ldots v_m$ and $p = |\sigma(u_1)|$. Set

$$\sigma_k(u) = v_{[1,k]}\, v_{[2,k+1]} \cdots v_{[p,p+k-1]} .$$

In other words, $\sigma_k(u)$ consists of the ordered list of the first $|\sigma(u_1)|$ factors of length $k$ of $\sigma(u)$. Given $n > 1$, remark that $\sigma_k^n$ is associated to $\sigma^n$ in the same way as $\sigma_k$ is associated to $\sigma$: $\sigma_k^n(u)$ consists of the ordered list of the first $|\sigma^n(u_1)|$ factors of length $k$ of $\sigma^n(u)$. If $n$ is large enough, every $v \in A_k$ is a factor of $\sigma^n(a)$ for every $a \in A$, thus occurs in $\sigma_k^n(u)$ for every $u \in A_k$. We have proved that the substitution $\sigma_k$ is primitive.

Let $w$ be a word of length $n > 0$ over the alphabet $A_k$. From the definitions of $\mathcal{L}_\sigma$ and $\mathcal{L}_{\sigma_k}$ it can be checked that: $w \in \mathcal{L}_{\sigma_k}$ if and only if there exists a word $v \in \mathcal{L}_\sigma$ of length $n + k - 1$ such that

$$w = v_{[1,k]} v_{[2,k+1]} \cdots v_{[n,n+k-1]} .$$

Clearly, given $u \in \mathcal{A}_k$, the number of occurrences of the symbol $u$ in $w$ is equal to the number of occurrences of $u-$ considered as a word of length $k$ on the alphabet $A-$ in $v$. Applying Lemma 2.3.5 to the substitution $\sigma_k$, we obtain that for every $u \in \mathcal{L}_\sigma$ there exist constants $p_u > 0$, $c > 0$ and $\alpha < 1$ such that: for every $v \in \mathcal{L}_\sigma$ of length $\geq |u|$,

$$\left| (\text{Number of occurrences of } u \text{ in } v) - p_u \left( |v| - |u| + 1 \right) \right| \leq c(|v| - |u| + 1)^\alpha .$$

This proves Proposition 2.3.2. ∎

## 2.4  Structure of Substitution Subshifts

### 2.4.1  Structure of Substitution Dynamical Systems

In the sequel, $(X_\sigma, T_\sigma)$ is the system associated to the primitive substitution $\sigma$ on the alphabet $A$, $\omega$ is an admissible fixed point of $\sigma$, $r = \omega_{-1}$ and $\ell = \omega_0$.

There exist substitutions $\sigma$ such that the system $X_\sigma$ is finite, i.e., such that every $x$ in $X_\sigma$ is periodic, or equivalently that $\omega$ is periodic. As these substitutions are of little interest from a dynamical point of view, we henceforth consider only aperiodic substitutions, i.e., substitutions giving rise to infinite systems. Notice that there is an algorithm [9], [8] which decides when a given substitution is aperiodic or not.

The systems arising from (primitive, aperiodic) substitutions have a simple self–similar structure which we now describe. Remark first that by the first definition of $X_\sigma$ one has $\sigma(X_\sigma) \subset X_\sigma$. The result below states that, although neither is $\sigma : A \to A^+$ assumed to be one-to-one, nor is $\{\sigma(a) \,;\, a \in A\}$ assumed to be a code, the restriction of $\sigma$ to $X_\sigma$ behaves almost as if these properties were true. Let us begin with some notation.

Since $\omega = \sigma(\omega)$, it can be written

$$\omega = \ldots \sigma(\omega_{-2})\sigma(\omega_{-1}) \mid \sigma(\omega_0)\sigma(\omega_1) \ldots,$$

where the vertical bar separates $\omega_{(-\infty,-1]}$ and $\omega_{[0,+\infty)}$. Let $E = \{e_j \,;\, j \in \mathbb{Z}\}$ be the set of 'natural cutting points' arising from this decomposition, defined by

$$e_j = \begin{cases} -\left| \sigma\left( \omega_{[j,0)} \right) \right| & \text{if} \quad j < 0 \\ 0 & \text{if} \quad j = 0 \\ \left| \sigma\left( \omega_{[0,j)} \right) \right| & \text{if} \quad j > 0 \end{cases}$$

**Theorem 2.4.1** *Let $\sigma$ be an aperiodic primitive substitution on the alphabet A.*

i) [6] *There exists $L > 0$ such that*

$$n \in E, \ m \in \mathbf{Z}, \ \omega_{[n-L,n+L)} = \omega_{[m-L,m+L)} \ \Rightarrow m \in E \ .$$

ii) [7] *There exists $M > 0$ such that*

$$i, j \in E, \ \omega_{[e_i - M, e_i + M)} = \omega_{[e_j - M, e_j + M)} \ \Rightarrow \omega_i = \omega_j \ .$$

**Remark 2.4.2** Earlier proofs of the same (or apparently the same) result do exist [11] and [12], but they are false. Until Mossé's papers, people had to make extra hypotheses ('recognizability', injectivity) on the substitution they were dealing with.

The theorem has the following topological interpretation (for proofs, see [2] or [3]):

**Corollary 2.4.3**

i) *The map $\sigma\colon X_\sigma \to X_\sigma$ is one–to–one and open.*

ii) *Every $x \in X_\sigma$ can be written in a unique way as follows:*

$$x = T_\sigma{}^k\big(\sigma(y)\big) \ \text{with } y \in X_\sigma \text{ and } 0 \le k < |\sigma(y_0)| \ .$$

iii) *For every $x \in X_\sigma$, $|\sigma(x_0)|$ is the first return time of $\sigma(x)$ to $\sigma(X_\sigma)$, i.e., the smallest positive integer $k$ such that $T^k\sigma(x) \in \sigma(X_\sigma)$.*

iv) *The map $\sigma\colon X_\sigma \to \sigma(X_\sigma)$ defines an isomorphism of the system $(X_\sigma, T_\sigma)$ with the system induced by $T_\sigma$ on $\sigma(X_\sigma)$.*

The same results hold for $\sigma^n$ for each $n > 0$. For every letter $a \in A$, write $[a] = \{x \in X_\sigma \ ; \ x_0 = a\}$. From the preceding theorem we get immediately:

**Corollary 2.4.4** *For every $n > 0$,*

$$\mathcal{P}_n = \big\{T^k\big(\sigma^n([a])\big) \ ; \ a \in A, \ 0 \le k < |\sigma^n(a)|\big\}$$

*is a clopen partition of $X_\sigma$.*

The partition $\mathcal{P}_n$ is a Kakutani–Rohlin partition (see [1]). The base of the partition $\mathcal{P}_n$ is

$$\bigcup_{a \in A} \sigma^n([a]) = \sigma^n(X_\sigma) \ .$$

**Proposition 2.4.5** *The sequence of partitions* $(\mathcal{P}_n)$ *is nested, i.e.:*

*i) The sequence of bases* $(\sigma^n(X_\sigma))$ *is decreasing.*

*ii) For every* $n$, $\mathcal{P}_{n+1} \succeq \mathcal{P}_n$ *as partitions.*

*Moreover, if the substitution* $\sigma$ *is proper, then*

*iii) The intersection of the bases consists of one point only (which is the unique fixed point of* $\sigma$ *).*

*iv) The sequence of partitions spans the topology of* $X_\sigma$.

**Proof.** i) is obvious. Remember that, for every $x \in X_\sigma$,

$$\sigma^n(Tx) = T^{|\sigma^n(x_0)|}\sigma^n(x) .$$

Let $a$ be a letter and $k$ be an integer with $0 \le k < |\sigma^{n+1}(a)|$. Set $b_1 \ldots b_m = \sigma(a)$, so that

$$|\sigma^n(b_1 \ldots b_m)| = |\sigma^{n+1}(a)| > k .$$

Let $j \in \{0, \ldots m-1\}$ be the integer defined by

$$|\sigma^n(b_1 \ldots b_j)| \le k < |\sigma^n(b_1 \ldots b_{j+1})| .$$

From the remark above it follows immediately that

$$T^k\sigma^{n+1}([a]) \subset T^\ell\sigma^n([b_j])$$

where $\ell = k - |\sigma^n(b_1 \ldots b_j)|$, and ii) is proved.
iii) is obvious.

Now we prove iv). Assume that $\sigma$ is proper. Let $p, r, \ell$ be as in the definition of a proper substitution.

Given an integer $m > 0$, we claim that for $n$ large enough $x_{[-m,m]}$ is constant on each element of $\mathcal{P}_n$.

For $n \ge p$, write $R_n = |\sigma^{n-p}(r)|$ and $L_n = |\sigma^{n-p}(\ell)|$. Choose $n$ large enough for $R_n$ and $L_n$ to be greater than $m$. Fix $a \in A$ and $0 \le k < |\sigma^n(a)|$.

For each $x \in T^k\sigma^n([a])$, there exists $y \in X_\sigma$ with $y_0 = a$ and $x = T^k\sigma^n(y)$. The word $\sigma^n(a)\sigma^n(y_1)$ is a prefix of $(\sigma^n(y))_{[0,\infty)}$, and $\sigma^{n-p}(\ell)$ is a prefix of $\sigma^n(y_1)$, thus $\sigma^n(a)\sigma^{n-p}(\ell)$ is a prefix of $(\sigma^n(y))_{[0,\infty)}$. Analogously, $\sigma^{n-p}(r)$ is a suffix of $(\sigma^n(y)_{(-\infty,-1]}$, and

$$\left(\sigma^n(y)\right)_{[-R_n,\,K_n+L_n)} = \sigma^{n-p}(r)\sigma^n(a)\sigma^{n-p}(\ell)$$

where $K_n = |\sigma^n(a)|$, and we get:

$$x_{[-m,m]} = \left(\sigma^{n-p}(r)\sigma^n(a)\sigma^{n-p}(\ell)\right)_{[R_n + k - m,\, R_n + k + m]};$$

which does not depend on $x$, but only on $a$ and $k$: our claim is proved, and iv) follows. ∎

## 2.5   Substitutions and Bratteli Diagrams

### 2.5.1   Bratteli Diagram Associated to a Substitution

**Definition 2.5.1** Let $\sigma$ be a primitive substitution on an alphabet $A$. To this substitution we associate an infinite directed graph $\mathcal{B} = (V, E)$ called its Bratteli diagram:

1) The set of vertices is

$$V = \{v_0\} \bigcup \{(a, n) \; ; \; a \in A, \; n \geq 1\} \, .$$

Thus $V$ is given with a partition in levels $(V_n \; ; \; n \geq 0)$ where $V_0$ consists of the unique vertex $v_0$ called the origin; for $n \geq 1$, $V_n$ is in a one–to–one correspondence with $A$: the vertex $(a, n)$ is called the vertex of label $a$ at level $n$.

2) The set $E$ of edges is given with a partition in levels

$$E = \bigcup_{n \geq 1} E_n$$

such that, for all $n$, $s(E_n) \subset V_{n-1}$ and $r(E_n) \subset V_n$ (denote by $r(e)$ the range of the edge $e$, and by $s(e)$ its source). For each $a \in A$, $E_1$ contains exactly one edge from $v_0$ to $(a, 1)$. For $n > 1$ and $a, b \in A$, the number of edges from $(a, n-1)$ to $(b, n)$ is equal to the number of occurrences of $a$ in $\sigma(b)$.

This Bratteli diagram is stationary: this means that all levels except the first are identical.

Let $m, n$ be integers with $m < n$ and $a, b$ letters. The number of paths in $\mathcal{B}$ from $(a, m)$ to $(b, n)$ is equal to the number of occurrences of $a$ in $\sigma^{n-m}(b)$. Since $\sigma$ is primitive, there exists an integer $k \geq 1$ such that:

> For all $n \geq 1$ and all $a, b \in A$, there exists at least one path      (2.7)
> from the vertex $(a, n)$ to the vertex $(b, n + k)$.

**The Path Space.** An infinite path in $\mathcal{B}$ is a sequence $x = (x_n \; ; \; n \geq 1)$ of edges such that $x_n \in E_n$ and $r(x_n) = s(x_{n+1})$ for all $n$. Thus the set $X_{\mathcal{B}}$ of infinite paths is a subset of $\prod E_n$. Let this product be endowed with the product topology, and $X_B$ with the induced topology. Then $X_B$ is a compact metric space.

Let $\sim$ be the cofinal relation in $X_B$: when $x$ and $y$ are infinite paths, $x \sim y$ if $x_n = y_n$ for all $n$ large enough.

From (2.7) it follows that

$$\text{Every equivalence class for } \sim \text{ is dense in } X_B. \qquad (2.8)$$

**Ordering the Bratteli Diagram.** For every $n \geq 1$, we define a partial order on $E_n$:

1) Two edges $e, e' \in E_n$ are comparable if and only if they have the same range. In particular, two different edges in $E_1$ are not comparable.

2) Let $b \in A$ and $n > 1$, and write $\sigma(b) = a_1 \ldots a_k$. To the set $r^{-1}(b, n)$ of edges which range equal to $(b, n)$, we give the following linear order: the first edge sources at $(a_1, n-1)$; the second edge sources at $(a_2, n-1)$ ... the last edge sources at $(a_k, n-1)$.

A Bratteli diagram with an order of this type is called a stationary ordered Bratteli diagram.

Let $m, n$ be two integers with $m < n$. The set of paths from level $m$ to level $n$ can be ordered lexicographically: two such paths $x, y$ are comparable if they end at the same vertex of $V_n$; in this case, $x > y$ if there exists $k \in (m, n]$ such that $x_k > y_k$ and the paths $x$ and $y$ agree at all levels from $k$ through $n$.

In the same way, to the order of the edges corresponds an order on the set $X_B$: given two infinite paths $x, y$, we say that $x > y$ if there exists $n$ such that

i) $x_k = y_k$ for all $k > n$; and

ii) $x_n > y_n$.

Notice that two infinite paths $x, y$ are comparable if and only if $x \sim y$: each equivalence class for $\sim$ is linearly ordered.

**Telescoping.** Let $k$ and $p$ be positive integers. Let $\mathcal{A} = (V', E')$ be the graph defined by

$$V' = \{v_0\} \cup \bigcup_{n \geq 0} V_{k+pn} \text{ and } E' = \bigcup_{n \geq 1} E'_n$$

where $E'_1$ is the set of paths that source at the origin and end in $V_k$, and for $n > 1$, $E'_n$ is the set of paths that source in $V_{k+(n-2)p}$ and end in $V_{k+(n-1)p}$.

With this order, the graph $\mathcal{A}$ shares almost all the properties of an ordered stationary Bratteli diagram, except that in general there are several edges from the origin to a given vertex in $V_1' = V_k$. This cannot occur when $k = 1$: in that case $\mathcal{B}'$ is the Bratteli diagram associated to the substitution $\sigma^p$.

It is easy to check that the set $X_{\mathcal{A}}$ of infinite paths in $\mathcal{A}$ can be identified to $X_{\mathcal{B}}$, in a manner which is compatible with the topology and the order.

**Forrest's Lemma.** [10] Let $\mathcal{A} = (V', E')$ be a stationary ordered Bratteli diagram, except that we allow multiple edges between the origin and each vertex of the first level. Then there exist an ordinary stationary ordered Bratteli diagram $\mathcal{B} = (V, E)$, and a homeomorphism $\phi\colon X_{\mathcal{A}} \to X_{\mathcal{B}}$ which is also an order isomorphism.

## 2.5.2   The Vershik Map

**Definition 2.5.2** Let $\mathcal{B} = (V, E)$ be the stationary ordered Bratteli diagram associated to the substitution $\sigma$ on $A$ as above. Remember that the order on $X_{\mathcal{B}}$ is linear when restricted to each equivalence class for $\sim$.

Let $X_{\min}$ and $X_{\max}$ be the set of minimal (resp. maximal) elements for this order: $X_{\min}$ (resp. $X_{\max}$) consists of the paths containing only minimal (resp. maximal) edges. By compactness, these sets are not empty.

It is easy to check that every infinite path $x \notin X_{\max}$ has a successor, i.e., that the set $\{y \in X_{\mathcal{B}} \; ; \; y > x\}$ has a smallest element. Also, each infinite path $x \notin X_{\min}$ has a predecessor. Let

$$V_{\mathcal{B}} : X_{\mathcal{B}} \setminus X_{\max} \longrightarrow X_{\mathcal{B}} \setminus X_{\min}$$

be the successor map. This map is called the Versik map of the ordered Bratteli diagram $\mathcal{B}$. It is one-to-one, onto, and in fact it is an homeomorphism. In general it cannot be extended to a continuous map from $X_{\mathcal{B}}$ to itself.

**Properly Ordered Stationary Bratteli Diagrams.** We say that the stationary ordered Bratteli diagram $\mathcal{B}$ is properly ordered if $X_{\max}$ and $X_{\min}$ both contain a unique infinite path.

These paths are denoted by $x_{\max}$ and $x_{\min}$ respectively. We get:

**Proposition 2.5.3** *The stationary ordered Bratteli diagram associated to a substitution $\sigma$ is properly ordered if and only if some power of $\sigma$ is a proper substitution.*

Suppose that $\mathcal{B}$ is properly ordered. We extend the Vershik map to the whole set $X_{\mathcal{B}}$ by defining $V_{\mathcal{B}}(x_{\max}) = x_{\min}$. This extended map is a homeomorphism of $X_{\mathcal{B}}$ to itself, thus $(X_{\mathcal{B}}, V_{\mathcal{B}})$ is a topological dynamical system, called the Bratteli–Vershik system associated to the ordered Bratteli diagram.

Remark that the $V_B$-orbits are exactly the equivalence classes for $\sim$, except that the equivalence classes of $x_{\min}$ and of $x_{\max}$ form a unique orbit. From (2.8) it follows that

$$\text{The Bratteli–Vershik system } (X_B, V_B) \text{ is minimal} \qquad (2.9)$$

### 2.5.3 An Isomorphism

**Theorem 2.5.4** *Let $B = (V, E)$ be the stationary ordered Bratteli diagram associated to a proper primitive substitution $\sigma$ on the alphabet $A$.*

*1) If $\sigma$ is aperiodic, the substitution dynamical system $(X_\sigma, T_\sigma)$ associated to $\sigma$ is isomorphic to the Bratteli–Vershik system $(X_B, V_B)$.*

*2) If $\sigma$ is periodic, the Bratteli–Vershik system $(X_B, V_B)$ is isomorphic to an odometer with a stationary base.*

The definition of an odometer with a stationary base will be given in the proof.

**Proof.** We begin with some remarks which will be used in both proofs.

Fix $n \geq 0$. Let $K_n$ be the set of paths from the origin to the level $n + 1$. For $y \in K_n$, let $[y]$ be the set of infinite paths which agree with $y$ until the level $n + 1$; $[y]$ is a clopen set in $X_B$. Clearly, $\mathcal{Q}_n = ([z] \; ; z \in K_n)$ is a clopen partition of $X_B$. For each $a \in A$, the subset $K_{n,a}$ of $K_n$ consisting in paths with end at $V(n+1, a)$ contains $|\sigma^n(a)|$ elements and is linearly ordered. Let $y_{n,a}$ be its first element: it is the unique path from the origin to $V(n+1, a)$ using only minimal edges.

By definition of the Vershik map, for every $y \in K_{n,a}$ there exists a unique integer $k$ with $0 \leq k < |\sigma^n(a)|$ and $[y] = V_B^k[y_{n,a}]$. Thus the partition $\mathcal{Q}_n$ can be written in the form of a Kakutani–Rohlin partition:

$$\mathcal{Q}_n = \left( V_B^k([y_{n,a}]) \; ; a \subset A, \; 0 \leq k < |\sigma^n(a)| \right) .$$

**Proof of 1).** The base of this partition is $\bigcup_{a \in A}[y_{n,a}]$; the towers are indexed by $A$, and the $a$-tower is of height $|\sigma^n(a)|$. Thus this partition has the same structure as the partition $\mathcal{P}_n$ of $X_\sigma$ introduced in Section 2.4.

We look now at the relations between the partitions $\mathcal{Q}_{n+1}$ and $\mathcal{Q}_n$. Fix $a \in A$ and put $\sigma(a) = b_1 \ldots b_m$. The infinite paths belonging to $[y_{n+1,a}]$ follow minimal edges between the origin and the level $n + 1$, thus they pass through $V(n, b_1)$, and belong to $[y_{n,b_1}]$: we get $[y_{n+1,a}] \subset [y_{n,b_1}]$ and

$$V_B^j \left( [y_{n+1,a}] \right) \subset [y_{n,b_1}] \text{ for } 0 \leq j < |\sigma^n(b_1)| .$$

Since there are $|\sigma^n(b_1)|$ paths between the origin and $V(n, b_1)$, each infinite path in $V_B^{|\sigma^n(b_1)|}([y_{n+1,a}])$ passes through the second vertex which ends at $V(n+2, a)$, thus through $V(n+1, b_2)$, and from the origin to this vertex it consists of minimal edges, thus agrees with $y_{n,b_2}$. We get $V_B^{|\sigma^n(b_1)|}([y_{n+1,a}]) \subset [y_{n,b_2}]$, and

$$V_B^{j+|\sigma^n(b_1)|}([y_{n+1,a}]) \subset V_B^j([y_{n,b_2}]) \text{ for } 0 \le j < |\sigma^n(b_2)| \ .$$

By induction, for $1 \le k < m$,

$$V_B^{j+|\sigma^n(b_1...b_k)|}([y_{n+1,a}]) \subset V_B^j([y_{n,b_{k+1}}]) \text{ for } 0 \le j < |\sigma^n(b_{k+1})| \ .$$

Finally, $\mathcal{Q}_{n+1}$ is finer than $\mathcal{Q}_n$, and the inclusions between elements of these partitions are exactly the same as between the corresponding elements of $\mathcal{P}_{n+1}$ and $\mathcal{P}_n$. We know that the sequence of partitions $(\mathcal{P}_n)$ converges to the point partition of $X_\sigma$ (see Proposition 2.4.5). Moreover, for each $n$, $\mathcal{Q}_n$ separates infinite paths which differ before the level $n+1$. Thus the sequence $(\mathcal{Q}_n)$ converges to the point partition of $X_B$. Therefore, we can define a homeomorphism $\phi\colon X_B \to X_\sigma$ by mapping each element of $\mathcal{Q}_n$ to the corresponding element of $\mathcal{P}_n$ for each $n$. Fix $x \ne x_{\max}$. For $n$ large enough, the element of $\mathcal{Q}_n$ in which $x$ lies is not the top level of a tower, and the element of $\mathcal{P}_n$ containing $\phi(x)$ is not the top of a tower of this partition. By construction of $\phi$, we get $\phi(V_B(x)) = T_\sigma(\phi(x))$. The same result also holds for $x = x_{\max}$ by density. We get $\phi \circ V_B = T_\sigma \circ \phi$, and $\phi$ is an isomorphism of $(X_B, V_B)$ to $(X_\sigma, T_\sigma)$.

Proof of 2). Choose a stationary label (in an alphabet $B$) for the edges belonging to $E_n$ for $n > 1$. Each infinite path $x = x_1 x_2 x_3 \ldots$ is characterized by the labels of $x_2, x_3, \ldots$: it is not necessary to specify $x_1$, because this edge is the unique one from the origin to the source of $x_2$. Let $b_2 b_3 \ldots$ be associated to $x$ in this way. Let $a$ be the source of $x_2$, and $b$ be the label of the minimal edge which ends in $V(n, a)$ for any $n > 1$. Then $bb_2 b_3 \ldots$ is the label of a well–defined infinite path, which we denote by $\Sigma(x)$. This map $\Sigma\colon X_B \to X_B$ is continuous. Notice that $\Sigma(x_{\min}) = x_{\min}$.

We define a map $\phi\colon X_B \to A$ by $\phi(x) = a$ if $x$ passes through $V(1, a)$, and a map $\Phi\colon X_B \to A^{\mathbf{Z}}$ by $\Phi(x)_k = \phi(V_B^k(x))$ for all $k \in \mathbf{Z}$. $\Phi$ is continuous, and by construction $\Phi \circ V_B = T \circ \Phi$, where $T$ is the shift on $A^{\mathbf{Z}}$. Looking carefully the definitions of $\Sigma$, $\Phi$ and $V_B$ we find that $\Phi \circ \Sigma = \sigma \circ \Phi$, where $\sigma\colon X_\sigma \to X_\sigma$ is defined in Section 2.2. Since $x_{\min}$ is invariant under $\Sigma$, $\Phi(x_{\min})$ is invariant under $\sigma$; this implies it is the unique fixed point of the proper substitution $\sigma$, and it is periodic. Let $p$ be its period and, for each letter $a$, $p_a$ be the number of occurrences of $a$ in a period. By density, $\phi(V_B^p(x)) = \phi(x)$ for all $x \in X_B$.

For $n \geq 1$, we can define in the same way a map $\phi_n \colon X_{\mathcal{B}} \to A$ by $\phi_n(x) = a$ if $x$ passes through $V(n+1, a)$, and a map $\Phi_n \colon X_{\mathcal{B}} \to A^{\mathbb{Z}}$ by $\Phi_n(x)_k = \phi_n(V_{\mathcal{B}}^k(x))$ for each $k \in \mathbb{Z}$. It can be checked easily that $\Phi_n(\Sigma^n(x))$ is the image of $\Phi(x)$ by the morphism $a \mapsto aa \ldots a$ ($|\sigma^n(a)|$ times) from $A$ to $A^*$. In particular, $\Phi_n(x_{\min})$ is the image of $\Phi(x_{\min})$ by this morphism. It follows that $\Phi_n(x_{\min})$ is periodic of period $p_n = \sum_a p_a |\sigma^n(a)|$, and, by density, the same is true for every $x \in X_{\mathcal{B}}$. Therefore the partition $\mathcal{Q}_n$ is periodic of period $p_n$: for each $x \in X_{\mathcal{B}}$, $x$ and $V_{\mathcal{B}}^{p_n}(x)$ belong to the same element of this partition. The paths $x$ and $V_{\mathcal{B}}^{p_n}(x)$ agree up to the level $n+1$, and it follows that $p_n$ is a period of $\Phi(x)$. Thus $p_n$ can be written $p_n = k_n p$ for some integer $k_n$. Moreover, the word $\phi(V_{\mathcal{B}}^k(x))_{[1,p_n)}$ consists in a concatenation of words of the kind $\sigma^n(a)$, and the word $\sigma^n(a)$ occurs $p_a$ times. We get:

$$\sum_a M_{b,a}^n p_a = k_n p_b \text{ for all } b \in B,$$

where $M$ is the matrix of $\sigma$. In other words, the vector $(p_a)$ is an eigenvector of $M^n$ for the eigenvalue $k_n$. It follows that $k_n = k^n$ for all $n$ and some integer $k$. Finally, the partition $\mathcal{Q}_n$ is periodic of period $pk^n$.

Let us give the definition of an odometer. Let $Y$ be the set of sequences $t = (t_0, t_1, t_2, \ldots)$, where the $t_n$ are integers, $0 \leq t_0 < p$ and $0 \leq t_n < pk^n$ for $n \geq 1$. $Y$ endowed with the product topology is a compact metric space. Let $S \colon Y \to Y$ be defined by: $S(p-1, k-1, k-1, \ldots) = (0, 0, 0, \ldots)$; for any other $t$, $S(t)$ is the successor of $t$ in $Y$ endowed with the alphabetic order (we can describe $S$ as the addition of 1 with carry). Then $S$ is a homeomorphism. The dynamical system $(Y, S)$ is called the odometer with base $(p, k, k, k, \ldots)$, and we call it a stationary odometer.

There is a more algebraic definition of this system. $Y$ can be considered as the inverse limit of the sequence of groups $(\mathbb{Z}/pk^n\mathbb{Z})$ with the natural homomorphisms from $\mathbb{Z}/pk^{n+1}\mathbb{Z}$ to $\mathbb{Z}/pk^n\mathbb{Z}$, endowed with the inverse limit topology. Then $S \colon Y \to Y$ is the addition of 1. Also, $Y$ is the dual group of the countable subgroup $\{jp^{-1}k^{-n} \bmod 1 \; ; \; j \in \mathbb{Z}, n \in \mathbb{N}\}$ of $\mathbb{T}$.

End of the proof. Let $\mathcal{R}_n$ be the partition of $Y$ given by the first $n+1$ digits. It is periodic of period $pk^n$; clearly the sequence $(\mathcal{R}_n)$ separates the points of $Y$.

Thus both $(X_{\mathcal{B}}, V_{\mathcal{B}})$ and $(Y, S)$ have a sequence of partitions converging to the point partition and with the same structure. As in the proof of 1), it can be proved that these systems are isomorphic. ∎

## 2.5.4    The General Case

The systems associated to non–proper substitutions can also be represented as stationary Bratteli–Vershik systems, as stated in the next theorem, given without proof.

**Theorem 2.5.5** *Let $\sigma$ be an aperiodic primitive substitution on the alphabet A. Then there exists a properly ordered stationary Bratteli diagram $\mathcal{B}$ such that the systems $(X_\sigma, T_\sigma)$ and $(X_\mathcal{B}, V_\mathcal{B})$ are isomorphic.*

*Equivalently, there exists a primitive proper substitution $\tau$ such that the systems $(X_\sigma, T_\sigma)$ and $(X_\tau, T_\tau)$ are isomorphic.*

# Bibliography

A large part of the contents of this course can be found in the following article:

[1] F. Durand, B.Host, C.F. Skau, *Substitution Dynamical Systems, Bratteli Diagrams and Dimension Groups*, Erg. Th. Dyn. Sys., to appear.

For a general introduction to substitution dynamical systems, see:

[2] M. Queffélec, *Substitution Dynamical Systems*, Lectures Notes in Math. 1294 (1987).

[3] B. Host, *Valeurs Propres des Systèmes Dynamiques Définis par des Substitutions de Longueur non Constante*, Erg. Th. Dyn. Sys. **6**, pp. 529–540 (1986).

The Vershik map was introduced by Anatoly Vershik, but in a measure-theoretical (not topological) context. The fundamental papers on the recent developments of the theory of dimension groups, Bratteli diagrams, and of their use in topological dynamics are:

[4] T. Giordano, I. Putnam, C.F. Skau, *Topological Orbit Equivalence and $C^*$-Crossed Products*, J. Reine Angew. Math. **469**, pp. 51–111 (1995).

[5] R.H. Herman, I. Putnam, C.F. Skau, *Ordered Bratteli Diagrams, Dimension Groups and Topological Dynamics*, Internat. J. Math. **3**, pp. 827–864 (1992).

For a more combinatorial point of view and proofs of the 'recognizability' property, see:

[6] B. Mossé, *Puissances de Mots et Reconnaissabilité des Points Fixes d'une Substitution*, Theor. Computer Sci. **99**, pp. 327–334 (1992).

[7] B. Mossé, *Reconnaissabilité des Substitutions et Complexité des Suites Automatiques*, Bull. Soc. Math. France **124**, pp. 101–118 (1996).

[8] T. Harju, M. Linna, *On the Periodicity of Morphisms of the Free Monoid*, Informatique théorique et applications / Theoretical Informatics and Applications **20(1)**, pp. 47–54 (1986).

[9] J.J. Pansiot, *Decidability of Periodicity of Infinite Words*, Informatique Théorique et Applications / Theoretical Informatics and Applications **20(1)**, pp. 43–46 (1986).

[10] A.H. Forrest, *K-Groups Associated with Substitution Minimal Systems*, Israel J. of Mathematics **98**, pp. 101–139 (1997).

[11] J.C. Martin, *Substitutions Minimal Flows*, Amer. J. Math. **93**, pp. 503–526 (1971).

[12] J.C. Martin, *Substitutions Minimal Flows Arising from Substitutions of non Constant Lenths*, Math. System Theory **7**, pp. 73–82 (1973).

# Chapter 3

# ALGEBRAIC ASPECTS OF SYMBOLIC DYNAMICS

*Mike BOYLE*
*Department of Mathematics*
*University of Maryland*
*College Park, MD 20742-4015 U.S.A.*

This is an introduction to some algebraic aspects of symbolic dynamics, with emphasis on shifts of finite type, based on four lectures at the Temuco School.

## 3.1   Introduction

These four lectures are an introduction to some algebraic aspects of symbolic dynamics, with no assumption of previous background. The intent is much more to describe ideas and invariants than to give proofs, and this tendency becomes more pronounced in the last two lectures.

Lecture I (sections 3.2 - 3.6) is elementary background on subshifts. The sections 3.2 - 3.3 are elementary (but cover a lot of ground) and the proofs left to the reader are feasible.

Lecture II (section 3.7) describes and mostly explains the standard matrix invariants for shifts of finite type.

Lecture III (section 3.8) explains the basics of dimension groups and shift equivalence, with some hints at further developments. There is a greater emphasis on the order structure of dimension groups than we need for the application in this lecture. This reflects the fundamental importance of the order structure of these groups in other situations.

In Lecture IV (sections 3.9 - 3.11), we throw proof to the winds to describe one of the deepest results on SFT's, the Kim-Roush-Wagoner Factorization Theorem, and the Kim-Roush counterexamples to Williams' Conjecture.

We cover a lot, but still we provide only a narrow algebraic slice of symbolic dynamics. (For example we don't even mention systems of algebraic origin! See the lectures of Kitchens and the book [8] of Schmidt.) For a general introduction to the symbolic dynamics around SFT's, see the books of Lind and Marcus [7] and Kitchens [6].

In these few lectures, I give few references and don't mention many of the people who made contributions to what I describe. The Lind-Marcus book [7] does cover the historical development, and contains a long appendix surveying advanced topics, which I strongly recommend as an antidote to my omissions. The Kitchens book [6] also contains historical information, and further coverage of some advanced topics. For something shorter, which in addition covers more on the polynomial matrices of Section 8.4, see my survey [1]. The references [1], [6] and [7] have extensive bibliographies.

Since the original delivery of these lectures, some errata were corrected and small changes made; and a final paragraph was added on the Kim-Roush counterexample [4] to the irreducible Williams' Conjecture.

# 3.2    General Subshifts

## 3.2.1    Dynamical Systems

For the purposes of these lectures, a topological dynamical system will be a continuous map $T$ from a compact metric space $X$ into itself. We can represent this as $(X, T)$ or just $T$. Apart from occasional remarks, $T$ will be a homeomorphism.

## 3.2.2    Full Shifts

The system which is the full shift on $n$ symbols (also called the $n$-shift) is defined as follows. We give a finite set of $n$ elements — say, $\{0, 1, ..., n-1\}$ — the discrete topology. (This finite set is often called the *alphabet*.) We let $X$ be the product of countably many copies of this set, with the copies indexed by $\mathbb{Z}$. We think of an element $x$ of $X$ as a doubly infinite sequence

$$x = ...x_{-1}x_0x_1...$$

where each $x_i$ is one of the $n$ elements. $X$ is given the product topology and thus becomes a compact metrizable space. A metric compatible with the topology is given by defining, when $x$ is not equal to $y$,

$$\text{dist}(x, y) = 2^{-k}, \quad \text{where } k = \min\{|i| : x_i \neq y_i\}.$$

That is, two sequences are close if they agree in a large stretch of coordinates around the zero coordinate.

A finite sequence of elements of the alphabet is called a word. If $W$ is a word of length $j - i + 1$, then the set of sequences $x$ such that $x_i...x_j = W$ is called a cylinder set. The cylinder sets are closed and open, and they give a basis for the product topology on $X$. Thus $X$ is zero dimensional.

There is a natural shift map $S$ sending $X$ into $X$, defined by shifting the index set by one: $(Sx)_i = x_{i+1}$. (This is the "dynamics" in symbolic dynamics.) It is easy to see that $S$ is bijective, $S$ sends cylinders to cylinders, and thus $S$ is a homeomorphism. The full shift on $n$ symbols is the system $(X, S)$.

### 3.2.3 Subshifts

A subshift is a subsystem of some full shift $(X, T)$ on $n$ symbols. This means that it is a homeomorphism obtained by restriction of $T$ to some compact subset $Y$ invariant under the shift and its inverse. The complement of $Y$ is open and is thus a union of cylinder sets. Because $Y$ is shift invariant, it follows that there is a (countable) list of words such that $Y$ is precisely the set of all sequences $y$ such that for every word $W$ on the list, for every $i \le j$, $W$ is not equal to $y_i...y_j$. That is, $Y$ is the subset of all sequences which avoid the forbidden words.

Concisely: any subshift may defined by excluding a countable collection of words.

If $Y$ is a set which may be obtained by forbidding a finite list of words, then the subshift is called a subshift of finite type, or just a shift of finite type (SFT). Of course, if $S$ is SFT, then $S$ could also be defined by excluding an infinite set of words (for example, excluding 00 in the 2-shift is equivalent to excluding all words which contain 00). (But any list of forbidden words defining an SFT will by a compactness argument contain a finite list of forbidden words which defines the same SFT.)

### 3.2.4 Examples

Let $S$ be the subshift defined by restricting the two-shift to the set $Y$ of sequences in which the word 00 never occurs. (That is: $S$ is the subshift of the twoshift defined by excluding the word 00.) Then $S$ is SFT.

Let $T$ be the subshift defined by excluding all words 100...01 where the number of zeros is odd. Then $T$ is a subshift, and we will see below it is not SFT. ($T$ is the "even system" of B. Weiss.)

## 3.3   Block Codes

### 3.3.1   Codes

Suppose $(X, S)$ and $(Y, T)$ are subshifts. A map $f$ from $X$ to $Y$ is called a code, if it is continuous and intertwines the shifts, i.e. $fS = Tf$ . We think of a code as a homomorphism of these dynamical systems.

If a code $f$ is surjective, then it is called a quotient map or factor map or epimorphism of subshifts. If it is injective, then it is called an embedding or monomorphism of subshifts. If it is injective and surjective, then it is an isomorphism or conjugacy of subshifts. This notion of isomorphism is our fundamental equivalence relation.

### 3.3.2   Block Codes

Now suppose $F$ is a function from words of length $2n + 1$ which occur in $S$-sequences into some finite set $A$. This function $F$ gives a rule for taking an input sequence $x$ of $S$ and producing an output sequence $fx$. The sequence $fx$ is determined by defining each of its coordinate symbols $(fx)_i$ by the rule

$$(fx)_i = F(x_{i-n}...x_{i+n}).$$

It is easy to see that the map $f$ produced in this way will be continuous and commute with the shift. That is, such a map $f$ defines a code (called a block code) from $S$ into the full shift on the alphabet $A$ (or into any subshift which contains the image of $f$).

Example. The shift map itself is given by the code $(fx)_i = x_{i+1}$.

Example. Let $S$ be the twoshift and $F(ijk) = j + k$ (mod 2), that is $(fx)_i = x_i + x_{i+1}$ (mod 2). Then at a sample point $x$ we see

$$
\begin{aligned}
x &= \ldots x_0 x_1 \ldots & &= \ldots 01101011... \\
fx &= \ldots (fx)_0 (fx)_1 \ldots & &= \ldots 1011110 \ldots
\end{aligned}
$$

Notational caveat: sometimes "block code" is used to refer only to $f$ such that $(fx)_i = F(x_i...x_{i+n})$ for some function $F$.

### 3.3.3   Curtis-Hedlund-Lyndon

The "Curtis-Hedlund-Lyndon Theorem" asserts that every code is a block code.

**Proof.** Suppose $f$ is a code from $S$ to $T$. By continuity, for each symbol $a$ in the alphabet of $T$, the inverse image of the closed open set $\{x : x_0 = a\}$ is closed open, i.e. compact open.

Any open set is a union of cylinders and any compact open set is a union of finitely many cylinders. If a cylinder $C$ is defined on coordinates $[j, k]$ and $j' < j < k < k'$, then $C$ is a union of cylinders defined on coordinates $[j', k']$. Because the alphabet of $T$ is finite, it follows that we may choose $N$ sufficiently large that for every $a$, the inverse image of $\{x : x_0 = a\}$ is a union of cylinders defined on coordinates $[-N, N]$.

Now define $F(i_{-N} \ldots i_N) = a$ if $\{x : x_{-N} \ldots x_N\}$ is contained in the inverse image of $\{x : x_0 = a\}$. Because $f$ commutes with the shift, we have for any $i$ that

$$
\begin{aligned}
(fx)_i &= (T^{-i} f S^i x)_i = (f S^i x)_0 \\
&= F( [S^i x]_{-N} \ldots [S^i x]_N ) \\
&= F( x_{i-N} \ldots x_N ) \quad \blacksquare
\end{aligned}
$$

It is often convenient to abuse notation and use the same symbol for the maps we call $f$ and $F$ above.

### 3.3.4 Higher Block Presentations

Let $S$ be a subshift. Suppose $n$ is a positive integer and $j, k$ are nonnegative integers such that $j + k + 1 = n$. Given this we will define a code $f$ with domain the subshift $S$ by the rule $(fx)_i = [x_{i-j} \ldots x_{i+k}]$. So, for the output sequence $fx$, the symbol $(fx)_i$ is the word of length $n$ in $x$ in the coordinates $i - j$ through $i + k$. For example, with $n = 2$ and $[j, k] = [0, 1]$, we would see

$$
\begin{aligned}
x &= \ldots x_0 x_1 \ldots &&= \ldots 012330 \ldots \\
fx &= \ldots (fx)_0 (fx)_1 \ldots &&= \ldots [01][12][23][33][30] \ldots
\end{aligned}
$$

The shift map on the set of all output sequences is a subshift $T$, and $f$ is an isomorphism from $S$ to $T$. ($f$ is a block code and it is clearly one-to-one.) Given $n$, the map for the chosen $[j, k]$ is the map for $[0, n-1]$ followed by the $j$th power of the shift. So, given $n$, the system $T$ doesn't depend on the choice of $j$ (although the map does). $T$ is called the $n$-block presentation of $S$.

Exercise. Construct a one-block isomorphism between the $n$-block presentation of $S$ and the subshift obtained by passing to the 2-block presentation $n - 1$ times.

### 3.3.5 One-Block Codes

A one-block code $f : S \to T$ is a block code for which the symbol $(fx)_0$ is determined by the symbol $x_0$. So, f is given by a map from the alphabet of $S$ to the alphabet of $T$.

Suppose $h : S \to T$ is a code, where $(hx)_0$ is determined by say $x_{-n} \ldots x_n$. Let $S'$ be the $(2n + 1)$-block presentation of $S$. Let $g : S \to S'$ be the isomorphism given by $(gx)_i = [x_{i-n} \ldots x_{i+n}]$. Let $h' : S' \to T$ be the one-block map given by

$$h' : [x_{-n} \ldots x_n] \mapsto h(x_{-n} \ldots x_n) .$$

Then $h = h'g$, and we have presented the map $h$ as a one-block map on a higher block presentation. This is often convenient.

## 3.4   Shifts of Finite Type

### 3.4.1   Vertex Shifts

We now define vertex shifts, which are examples of shifts of finite type. Notation: throughout these lectures, "graph" means "directed graph".

For some $n$, let $A$ be an $n \times n$ zero-one matrix. We think of $A$ as the adjacency matrix of a graph with $n$ vertices; the vertices index the rows and the columns, and $A(i,j)$ is the number of edges from vertex $i$ to vertex $j$. Let $Y$ be the space of doubly infinite sequences $y$ such that for every $k$ in $Z$, $A(y_k, y_{k+1}) = 1$. We think of $Y$ as the space of doubly infinite walks through the graph, where the walks/itineraries are presented by recording the vertices traversed. The restriction of the shift to $Y$ is a shift of finite type: a sufficient list of forbidden words is the set of words $ij$ such that there is no arc from $i$ to $j$.

Claim. Suppose $(Y, S)$ is SFT, then it is isomorphic to a vertex shift.

**Proof.** Let $L$ be a finite list of words such that $Y$ is the set of all sequences (from some full shift) in which no word of $L$ occurs. Now, if a word $W$ has length less than $n$, then excluding $W$ is equivalent to excluding all words of length $n$ which contain $W$. So without loss of generality, we can assume that all the words in the list $L$ have the same length, say $n + 1$. (Then we say $S$ is an "$n$-step" SFT.)

It follows that the $n$-block presentation of $S$ is a "1-step" SFT, definable by excluding some words of length two. (For example, a forbidden word 12345 in $S$ corresponds to a forbidden word [1234][2345] in the 4-block presentation of $S$.) So after passing to the higher block presentation, WLOG we may suppose $n = 1$. We claim that $S$ is then a vertex shift.

Let the alphabet of $S$ be the vertex set of a graph. In this graph, there is an edge from $i$ to $j$ iff the word $ij$ is not forbidden in $S$. Now the vertex shift for this graph has the same alphabet as $S$, and the same list of forbidden words of length 2, so it equals $S$. ∎

## 3.4.2    Edge Shifts

Again let $A$ be an adjacency matrix for a directed graph, but now allow
multiple edges: so, the entries of $A$ are nonnegative integers. Let the set of
edges be the alphabet. Let $Y$ be the set of sequences $y$ such that for all $k$,
the terminal vertex of $y_k$ is the initial vertex of $y_{k+1}$. Again, we can think
of $Y$ as the space of doubly infinite walks through the graph, now presented
by the edges traversed. The shift map restricted to $Y$ is an edge shift and it
is a shift of finite type: a sufficient list of forbidden words is the set of edge
pairs $ij$ which do not satisfy the head-to-tail rule.

In the sequel, unless otherwise indicated an SFT defined by a matrix $A$
is intended to be the edge shift defined by $A$. We denote this SFT by $S_A$.
Any SFT is isomorphic to an edge shift, because the two-block presentation
of a vertex shift is an edge shift.

The edge shift presentation is very useful. One reason is conciseness: an
edge shift presented by a small matrix (perhaps though with large entries)
may be presentable as a vertex shift only by a large matrix. For example,
the $n$-shift as an edge shift is defined by the matrix $(n)$, as a vertex shift it
is defined by the $n \times n$ matrix with every entry equal to 1.

Another reason is functoriality. Working only with zero-one matrices
rules out some useful matrix operations (such as taking powers) and interpretations. For one of these, first a little preparation.

If $S$ is a subshift, then we let $S^n$ denote the homeomorphism obtained by
iterating $S$ $n$ times. The homeomorphism $S^n$ is isomorphic to a subshift $S^{[n]}$
whose alphabet is the set of $S$-words of length $n$. An isomorphism from $S^n$
to $S^{[n]}$ is given by the map $f$ which sends a point $x$ to the point $y$ such that
for all $k$ in $\mathbb{Z}$,

$$y_k = [x_{kn}...x_{(k+1)n-1}].$$

Claim. Suppose an edge shift $S$ is defined by a matrix $A$. Then the subshift
$S^{[n]}$, after a renaming of symbols, is the edge shift defined by $A^n$.

**Proof.** Let $G$ be the graph with adjacency matrix $A$ and let $G^{[n]}$ be the
graph with adjacency matrix $A^n$. Let these graphs have the same vertex set,
so that $A(i,j)$ is the number of edges from $i$ to $j$ in $G$ and $A^n(i,j)$ is the
number of edges from $i$ to $j$ in $G^{[n]}$.

An element $[e_1 e_2 \ldots e_n]$ in the alphabet of $S^n$ is a path in $G$ of $n$ edges
from some vertex $i$ to some vertex $j$. The number of such paths from $i$ to $j$
is $A^n(i,j)$. So there is a bijection from the alphabet of $S^{[n]}$ to the alphabet
of $S_{A^n}$ (i.e., the edge set of $G^{[n]}$) which respects initial and terminal vertex.
This renaming of symbols defines a one-block isomorphism from $S^{[n]}$ to $S_{A^n}$.

∎

### 3.4.3 Matrices

Because any SFT is isomorphic to an edge SFT $S_A$, all information about the dynamics of $S_A$ is determined by the matrix $A$. So we naturally will try to read dynamical invariants from the matrix $A$. This is the topic of the second lecture.

But it is worth noting right away that infinitely many matrices of arbitrarily large size can define isomorphic SFTs. For example, the $n$-block presentation of $S_A$ is the edge shift defined by the graph $G(n)$ whose vertices are the words of length $n-1$ in $S_A$, and whose edges correspond to the obvious way to the words of length $n$. If the original subshift contains infinitely many points, then as $n$ goes to infinity the size of the adjacency matrix for $G(n)$ must go to infinity.

### 3.4.4 Markov Characterization

An SFT is also called a topological Markov shift, or topological Markov chain. This terminology is appropriate because an SFT can be viewed as the topological support of a finite-state stochastic Markov process, and also as the topological analogue of such a process. This viewpoint was advanced in the seminal 1964 paper of Parry.

Roughly speaking, in a Markov process the past and future are independent if conditioned on the present (or more generally if conditioned on some finite time interval). We can say precisely why an SFT is a topological analogue of this. Suppose the SFT is $n$-step (given by forbidding a certain list of words of length at most $n+1$). Also suppose that $x$ and $y$ are points (doubly infinite sequences) in the SFT such that $x_0...x_{(n-1)} = y_0...y_{n-1}$. Then it follows that the doubly infinite sequence $z = x_{(-\infty, n-1]} y_{[n, +\infty)}$ must also be a point in the SFT. (More precisely, this point $z$ is defined by setting $z_i = x_i$ if $i < n$, and $z_i = y_i$ if $i \geq 0$.)

Put another way: the possibilities for the future (sequences in positive coordinates) and the past (sequences in negative coordinates) are independent conditioned on the near-present (i.e., the word in a certain finite set of coordinates). This by the way is a special case of the "local product structure" of hyperbolic dynamics.

Obviously an edge SFT has the Markov property above. One can check that a subshift isomorphic to a subshift with the Markov property also has the Markov property, and must be SFT.

### 3.4.5 Applications

For completeness I'll mention in the most cursory fashion two ways in which SFT's appear in a natural and useful way.

First, imagining very long tapes of zeros and ones, consider infinite strings of zeros and ones (i.e., points in the full shift on two symbols). It is natural to think of a block code as a machine which takes an input tape and recodes it, and to suppose that somehow the study of block codes may be relevant to efficiently encoding and decoding data. This turns out to be the case.

Second, imagine a homeomorphism (or diffeomorphism) $h$ on some space $X$. One way to study $h$ is by symbolic dynamics. Crudely, cut $X$ into $n$ pieces. Name the pieces $1, 2, ..., n$. To any given point $x$ in $X$ there is associated a sequence $y$ in the full shift on $n$ symbols, where $y$ is defined by setting $y_i$ to be the piece containing $h^i(x)$, for each integer $i$. This gives some set of sequences $y$. The sequence associated to $h(x)$ is the shift of $y$. This establishes some relation between the topological dynamics of the shift space and the dynamics of $h$. Sometimes a relation of this sort is very useful (for example for analyzing $h$-invariant measures or periodic points), when the shift space which arises is a shift of finite type.

A variation on the last theme going back to Hadamard is the study of geodesic flows with symbolic dynamics.

# 3.5   SFT-like Subshifts

## 3.5.1   Sofic Shifts

A subshift is *sofic* (this word is Hebrew for finite) if it is a quotient of an SFT.

Recall that if $f : S \to T$ is a block code, then there is a one-block code $f'$ from a higher block presentation $S'$ of $S$ onto $T$. For $n$ large enough, the $n$-block presentation of an SFT is an edge SFT, and $f'$ is given as a function from the alphabet of $S$ (the edge set) to the alphabet of $T$. We can think of this function as a labelling of the edges of a graph. So, $T$ is a subshift whose domain is the set of sequences which can be read off labels along infinite walks through some labelled graph.

For an example, consider the graph on vertices u and v, with three edges: one edge from u to u labelled 1, one edge from u to v labelled 0, and one edge from v to u labelled 0. The sofic shift corresponding to this labelled graph is the *even system* of Weiss. It is easy to see that the even system does not have the Markov property and therefore gives an example of a sofic shift which is not SFT.

The sofic shifts can be characterized in several natural ways. They are perhaps the most natural "finitely defined" class of subshifts. A natural subclass, with several characterizations, are the AFT ("almost finite type") systems of Marcus. A sofic shift is AFT if it is the quotient of a transitive SFT by a finite-to-one open map. (We will say a subshift is transitive if it has a dense forward orbit.) The even system is AFT.

### 3.5.2  Specification

*Specification* is a concept which comes from smooth hyperbolic dynamics, where it has a complicated formulation. A good introduction to this is [2].

It was A. Bertand who showed that this property has a very simple and natural formulation in the special case of subshifts: a subshift $S$ has the specification property if and only if it has a uniform transition length: i.e., there exists a positive integer $M$ such that for any $S$-words $U$ and $W$ there exists and $S$-word $V$ of length $M$ such that $UVW$ is an $S$-word.

### 3.5.3  Synchronized Systems

A subshift $S$ is *synchronized* if it is transitive with a synchronizing block. A synchronizing block is a word $W$ which gives a weakened version of the Markov property: whenever $x$ and $y$ are points in $S$ such that $x_0...x_{(n-1)} = y_0...y_{n-1} = W$, then $z = x_{(-\infty,n-1]}y_{[n,+\infty)}$ is also a point in S.

### 3.5.4  Coded Systems

A *coded system* is a subshift $S$ for which there is a countable collection of words such that the sequences which are concatenations of these words are a dense subset in $S$. Every synchronized system is coded. The coded systems, introduced with the synchronized systems by Blanchard and Hansel, have several natural characterizations. For example, a subshift is coded iff it contains a dense increasing union of transitive SFT's.

### 3.5.5  The Progression

For mixing subshifts we have a list of implications (which could be lengthened)

$$SFT \implies AFT \implies \text{sofic} \implies \text{specification} \implies \text{synchronized} \implies \text{coded} .$$

(A system with specification must be mixing; apart from specification, the implications hold for merely transitive subshifts.) In an SFT, a point can be built up from words satisfying local constraints. This property grows weaker as we move along the list above.

There are only countably many sofic systems. There are uncountably many shifts with specification.

Specification guarantees a unique measure of maximal entropy with full support, which is a natural limit of certain measures on periodic points. Also specification guarantees the subshift is *almost sofic*: its entropy is the supremum of the entropies of the SFT's it contains as subsystems.

All this fails for synchronized shifts, and even more for general coded shifts. Synchronized systems must still have a large group of symmetries (D. and U. Fiebig), but this breaks down for coded systems.

Coded systems still have dense periodic points, and a transitive system with dense periodic points need not be coded, even if it has positive entropy given by the growth rate of the periodic points (K. Petersen). A transitive subshift with dense periodic points can have little in the way of SFT-like properties.

For the "transsofic" subshifts above, a good start are the papers of Blanchard and Hansel and the Fiebigs, referenced in [7].

# 3.6 Minimal Subshifts

At the other end of the world are the minimal subshifts. A system is minimal if it has no proper subsystem; equivalently, every orbit is dense. An easy argument with Zorn's Lemma shows that every topological dynamical system contains a subsystem which is minimal.

The class of minimal subshifts is absolutely vast. [One indication: Krieger proved that every ergodic finite entropy measure preserving isomorphism of Lebesgue space is isomorphic as a measurable system to some uniquely ergodic minimal system.]

In one approach, these systems are studied with abstract, transcendental structures. This sort of approach is useful in applying topological dynamical ideas to combinatorial number theory.

There are also classes and approaches of a more finite nature. Among the most beautiful and historically rich of these classes are the substitution systems.

# 3.7 Matrix Invariants for SFTS

We recall that a shift of finite type (SFT) can be presented as an edge shift $S_A$. In this lecture we build up a dictionary between various dynamical properties of the SFT and invariants (mostly algebraic) of the matrix $A$. By a matrix invariant of $A$ we will mean a property of $A$ which is the same for matrices $A$ and $B$ which determine isomorphic shifts of finite type. After an initial section on nonnegative matrices, we'll go through the dictionary of corresponding properties by sections as follows.

|   | Dynamics | Matrices |
|---|----------|----------|
| 2. | Transitivity | [irreducible] |
| 3. | Mixing | [primitive] |
| 4. | Entropy | [spectral radius] |
| 5. | Periodic points | [traces of powers] |
| 6. | Zeta function | [nonzero eigenvalues] |
| 7. | Isomorphism | [SSE] |
| 8. | Eventual isomorphism | [SE] |
| 9. | Flow equivalence | [cok($I - A$) + sign] |
| 10. | Relations | |

Throughout this section $A$ will represent a matrix with integral entries. Unless otherwise indicated, we also assume that $A$ is nondegenerate (every row has a nonzero entry and every column has a nonzero entry) and that the entries are nonnegative. This is a matter of avoiding trivialities. If $A$ were nonnegative with $i$th row or column zero, then $A$ would define the same SFT as would the principal submatrix obtained by deleting row $i$ and column $i$ — it is only the "nondegenerate core" of $A$ which carries information about the SFT defined by $A$. For example, the following two matrices define the same SFT:

$$\begin{pmatrix} 1 & 1 \\ 0 & 0 \end{pmatrix} , \begin{pmatrix} 1 & 0 \\ 0 & 0 \end{pmatrix}.$$

Restricting to nondegenerate matrices allows statements about the matrix itself rather than its nondegenerate core.

## 3.7.1  Nonnegative Matrices

The matrix $A$ is *irreducible* if for every $(i,j)$ there exists $n > 0$ such that $A(i,j) > 0$. (This $n$ can depend on $(i,j)$ — consider a cyclic permutation matrix.) The matrix $A$ is *primitive* if there exists $n > 0$ such that $A^n$ has all entries strictly positive.

By the Perron-Frobenius Theorem,

- the spectral radius $\lambda_A$ of $A$ is an eigenvalue of $A$

- if $A$ is irreducible, then $\lambda_A$ has a strictly positive eigenvector

- if $A$ is primitive, then $\lambda_A$ is strictly larger than the modulus of any other eigenvalue.

## 3.7.2 Transitivity

The basic fact is that an SFT $S_A$ is transitive if and only if $A$ is irreducible. (This relies on our standing assumption that $A$ is nondegenerate.) Now we explain.

The word "transitivity" has been used to describe two properties of a dynamical system $(X, T)$:

(a) For all nonempty open sets $U, V$, there exists $n$ such that $(T^n U) \cap V \neq \emptyset$

(b) For all nonempty open sets $U, V$, there exists $n \geq 0$ such that $(T^n U) \cap V \neq \emptyset$

We remark that a standard Baire category argument shows that these properties are equivalent respectively to the following:

($\alpha$) $T$ has a dense orbit (i.e. some set $\{T^n x : n \in \mathbb{Z}\}$ is dense)

($\beta$) $T$ has a dense forward orbit (i.e. some set $\{T^n x : n \in \mathbb{Z}, n \geq 0\}$ is dense)

The properties are in general slightly different. In these lectures, we use transitivity to refer to the property $(b)$.

For a subshift $S$, cylinders give a basis for the topology. It follows that the condition $(b)$ is equivalent to the condition that for any $S$-words $W, W'$ there is an $S$-word of the form $WUW'$. (This tells us that some point in a cylinder set defined by $W$ will visit a cylinder set defined by $W'$.)

For a SFT $S_A$, words correspond to paths in the graph with adjacency matrix $A$. It is routine to check that $A$ is irreducible if and only if there is a path in the graph from any vertex to any other, i.e., if and only if for any two $S$-words $W, W'$ there is an $S$-word of the form $WUW'$.

## 3.7.3 Mixing

The basic fact: $S_A$ is mixing if and only if $A$ is primitive ("only if" relies on our assumption of nondegeneracy of $A$). We explain.

A system $(X, T)$ is *mixing* if for any two nonempty open sets $U, V$ the set $\{n \in \mathbb{N} : (T^{-n} U) \cap V = \emptyset\}$ is finite. For a subshift $T$, this condition will hold if for any two words $W, W'$, for all but finitely many positive integers $n$, there is a $T$-word $WUW'$ such that $U$ is a word of length $n$. Now it is not hard to verify from the graph interpretation that this is equivalent to $A$ being primitive.

The most important class to understand is the class of mixing SFT's. One understands other SFT's by how they are built up from the mixing SFT's. This is analogous to the situation with nonnegative matrices, which one understands by first understanding the primitive case. (Caveat: often the general case of a problem for SFT's follows very easily from the mixing case, but sometimes the generalization is quite difficult.)

### 3.7.4  Entropy

The premier numerical invariant of a dynamical system $S$ is its (topological) entropy, $h(S)$. For a subshift $S$,

$$h(S) = \limsup_{n} (1/n) \log[\#W_n(S)]$$

where $W_n(S)$ is the set of words of length $n$ occuring in sequences of $S$. That is, the entropy is the exponential growth rate of the $S$-words. For a full shift on $n$ symbols, the entropy is $\log(n)$. For an SFT $S_A$,

$$h(S) = \log(\lambda_A)$$

where $\lambda_A$ is the spectral radius of $A$. This follows from the spectral radius theorem because the number of words of length $n$ is the sum of the entries of $A^n$, which is the $L^1$ norm of $A^n$.

### 3.7.5  Periodic Points

The basic fact here is that the periodic point counts for $S_A$ are determined by the sequence $\operatorname{tr}(A^n)$, $n \in \mathbb{N}$.

Let $\pi_n(S)$ be the number of points of least period $n$ in $S$. The sequence $\pi_n(S)$ contains all the information one has from restricting $S$ to its periodic points and forgetting the topology.

Let $t_n(S)$ be the number of fixed points of $S^n$. One can recover the sequence $\pi_n(S)$ from the sequence $t_n(S)$. The sequence $t_n(S)$ is less directly dynamical but it is much better suited to finite summary and algebra.

In a subshift $S$, a fixed point is a sequence with the same symbol in every coordinate. A fixed point of $S^n$ is a sequence which is the infinite concatenation of some word of length $n$. In particular, if the alphabet of $S$ consists of $M$ symbols, then $S$ has at most $M^k$ points of period $n$.

A fixed point $x$ of $S^n$ is determined by the word $x_0 \ldots x_{n-1}$ (which is repeated forever to form $x$). For the edge shift $S_A$, such a word $x_0 \ldots x_{n-1}$ is a path of length $n$ (i.e., a path of $n$ edges) which begins and ends at the same vertex. Since $A^n(i,i)$ is the number of paths of length $n$ which begin and end at vertex $i$, we have

$$t_n(S_A) = \sum_i A^n(i,i) = \operatorname{tr}(A^n).$$

## 3.7.6  Zeta Function

We want a good way to present the periodic point data given by the sequence $t_n(S)$. The favored choice in dynamics for this is the (Artin-Mazur) zeta function of $S$,

$$\zeta_S(z) = \exp \sum_{n=1}^{\infty} \frac{t_n(S)}{n} z^n.$$

For $S = S_A$,

$$\zeta_S(z) = \exp \sum_{n=1}^{\infty} \frac{tr(A^n)}{n} z^n = \left[ \prod_{\lambda} (1 - \lambda z) \right]^{-1} = [\det(I - zA)]^{-1} \quad (3.1)$$

where the product is over the eigenvalues $\lambda$ of $A$, repeated according to multiplicity. (Of course, in this product only the nonzero eigenvalues matter.) So, the inverse zeta function of an SFT is a polynomial with integral coefficients and constant term 1.

For examples, we give some matrices and the corresponding zeta functions:

$$A = (3) \qquad B = \begin{pmatrix} 3 & 1 \\ 0 & 3 \end{pmatrix} \qquad C = \begin{pmatrix} 1 & 4 \\ 4 & 1 \end{pmatrix}$$

$$\zeta_{S_A}(z) = \frac{1}{1 - 3z} \qquad \zeta_{S_B}(z) = \frac{1}{(1 - 3z)^2}$$

$$\zeta_{S_C}(z) = \frac{1}{1 - 2z - 15} = \frac{1}{(1 - 5z)(1 + 3z)}$$

**Proof.** of (3.1) The first equality follows from $t_n(S_A) = \text{tr}(A^n)$.

The last equality follows from dividing the factorization of the characteristic polynomial

$$\det(zI - A) = \prod_{\lambda} (z - \lambda)$$

by $z^k$ (where $A$ is $k$ by $k$) and then replacing $1/z$ with $z$. This equality is saying that the inverse zeta function is the characteristic polynomial "written backwards":

$$\text{if} \quad \det(zI - A) = z^k + a_{k-1}z^{k-1} + a_{k-2}z^{k-2} + \cdots + a_j z^j,$$
$$\text{then} \quad \det(I - zA) = 1 + a_{k-1}z + a_{k-2}z^2 + \cdots + a_j z^{k-j}.$$

The second equality of (3.1) follows from three facts:

(1) $tr(A^n) = \sum_\lambda \lambda^n$,

(2) for any complex number $\lambda$,

$$\exp\left(\sum_{n=1}^{\infty} (1/n)(\lambda z)^n\right) = 1/(1 - \lambda z)$$

(to see this take the derivative of the log of each side)

(3) $\exp(\sum_{n=1}^{\infty} \sum_\lambda (1/n)(\lambda z)^n) = \prod_\lambda \exp(\sum_{n=1}^{\infty} (1/n)(\lambda z)^n)$. ∎

## 3.7.7  Isomorphism

The basic, fundamental result, due to Williams, is that $S_A$ and $S_B$ are isomorphic if and only if the matrices $A$ and $B$ are strong shift equivalent. This is completely general, nondegeneracy of $A$ and $B$ need not be assumed.

Matrices $A$ and $B$ are elementary strong shift equivalent if there are matrices $U, V$ with entries from $\mathbb{Z}_+$ such that $A = UV$ and $B = VU$. The matrices $U, V$ need not be square. $A$ and $B$ are strong shift equivalent (SSE) if they are linked by a finite chain of elementary strong shift equivalences—that is, strong shift equivalence is the transitive closure of elementary strong shift equivalence.

It takes some time to prove that isomorphism of $S_A$ and $S_B$ implies $A$ and $B$ are strong shift equivalent. However the other direction is easy. It suffices to show $S_A$ and $S_B$ are isomorphic under the assumption $A = UV$, $B = VU$.

Let $G_A$, $G_B$ be the directed graphs with adjacency matrices $A, B$. We can name the vertices so that these graphs have no vertices in common. We can also view $U$ as the adjacency matrix for a set of edges with initial vertices in $G_A$ and terminal vertices in $G_B$: now $U(i,k)$ denotes the number of edges drawn from the $A$-vertex $i$ to the $B$-vertex $k$. Similarly let $V(k,j)$ is the number of edges from the $B$-vertex $k$ to the $A$-vertex $j$.

Let lower case letters $a, b, u, v$ represent arcs corresponding to $A, B, U, V$. We have for any $A$-vertices $i$ and $j$,

$$A(i,j) = (UV)(i,j) = \sum_k U(i,k)V(k,j) .$$

This equation has a graphical interpretation. The left hand side $A(i,j)$ is the number of $A$-edges from $i$ to $j$. The right hand side is the number of $UV$ paths from $i$ to $j$.

If follows from this equation (and the other equation $B = VU$) that we may choose bijections of arcs and paths of length 2,

$$\alpha: \{a\} \longleftrightarrow \{uv\}, \qquad \beta: \{b\} \longleftrightarrow \{vu\}$$

which respect initial and terminal vertex.

Now we view a point of $S_A$ as an infinite path $...a_{-1}a_0a_1...$ of edges $a_i$, and we view a point of $S_B$ as an infinite path $...b_{-1}b_0b_1...$ of edges $b_i$. Using the chosen bijections above, we can induce bijections of points as follows.

$$...a_0a_1a_2a_3... \longleftrightarrow ...[u_0v_0][u_1v_1][u_2v_2][u_3v_3]...$$
$$\longleftrightarrow ...[v_0u_1][v_1u_2][v_2u_3][v_3u_4]... \longleftrightarrow ...b_0b_1b_2b_3...$$

Here the left bijection comes from applying $\alpha$ ($a_i \longleftrightarrow u_iv_i$), the middle bijection is the given regrouping, and the right bijection comes from applying $\beta^{-1}$ ($v_iu_{i+1} \longleftrightarrow b_i$). This bijection of points is given by a block code, so it is an isomorphism of $S_A$ and $S_B$.

It is a remarkable fact that ANY isomorphism of edge SFTs can be expressed as a composition of maps of exactly the form above, followed by a power of the shift.

Strong shift equivalence is easy to define but, it turns out, very difficult to understand completely. This led Williams to introduce another matrix relation, shift equivalence, which we consider next.

## 3.7.8 Eventual Isomorphism

Two systems $S$ and $T$ are *eventually isomorphic* if $S^n$ and $T^n$ are isomorphic for all but finitely many $n$.

Nonnegative integral matrices $A$ and $B$ are *shift equivalent* if there are nonnegative integral matrices $U, V$ and a positive integer $\ell$ (called the lag) such that the following equations hold.

$$A^\ell = UV \qquad B^\ell = VU$$
$$AU = UB \qquad BV = VA.$$

Given the equations above, we see that $A^\ell$ and $B^\ell$ are strong shift equivalent. Moreover, so are higher powers, since for any positive integer $n$,

$$U(VA^n) = A^{\ell+n} \quad \text{and} \quad (VA^n)U = (B^nV)U = B^{\ell+n}.$$

Consequently, if $A$ and $B$ are shift equivalent, then the SFT's $S_A$ and $S_B$ are eventually isomorphic. The converse also holds because the shift equivalence of $A^p$ and $B^p$ implies the shift equivalence of $A$ and $B$ if $p$ is a sufficiently large prime. We will skip that proof.

At first glance, shift equivalence may appear to be a more obscure and complicated equivalence relation than strong shift equivalence. In fact, it is just the opposite. Williams introduced the idea of shift equivalence with the intent of reducing strong shift equivalence to a manageable equivalence relation on matrices. He conjectured that shift equivalence implies strong shift equivalence for matrices over $\mathbf{Z}_+$. After many years, this conjecture was finally refuted by Kim and Roush in the reducible case, and very recently in the irreducible case (again by Kim and Roush). We will say a little about how this was done in our fourth lecture.

The crucial irreducible case still remains open for many classes. For example, it is open in the special case where the matrices under consideration have just one nonzero eigenvalue, and it is open in the case that both matrices are two by two, with determinant less than $-1$.

### 3.7.9   Flow Equivalence

Two homeomorphisms are flow equivalent if they are cross sections of a common flow. If $C$ is a square (say, $n \times n$) integral matrix, then by its cokernel $\mathrm{cok}(C)$ we mean the finitely generated abelian group $\mathbf{Z}^n / C\mathbf{Z}^n$.

If $A$ and $B$ are irreducible and neither is a permutation matrix, then $S_A$ and $S_B$ are flow equivalent if and only if

(i) $\det(I - A) = \det(I - B)$   and

(ii) $\mathrm{cok}(I - A) \cong \mathrm{cok}(I - B)$.

Here, the only information in (i) which is not determined by (ii) is the sign of the determinant. This is because for a square integral matrix $C$, the cardinality of $\mathrm{cok}(C)$ is $|\det C|$ if $C$ is nonsingular, and it is infinite if $C$ is singular.

The characterization above is due to Franks, following earlier work of Bowen-Franks and Parry-Sullivan. We do not have time for a proof. For examples, consider the matrices

$$A = \begin{pmatrix} 1 & 1 \\ 4 & 1 \end{pmatrix} \quad B = \begin{pmatrix} 1 & 2 \\ 2 & 1 \end{pmatrix} \quad C = (5) \quad D = \begin{pmatrix} 5 & 2 \\ 2 & 3 \end{pmatrix} \quad E = (25).$$

It is easy to compute
$$\det(I - A) = -4 \quad \text{and} \quad \mathrm{cok}(I - A) \cong \mathbf{Z}/4\mathbf{Z}$$
$$\det(I - B) = -4 \quad \text{and} \quad \mathrm{cok}(I - B) \cong \mathbf{Z}/2\mathbf{Z} \oplus \mathbf{Z}/2\mathbf{Z}$$
$$\det(I - C) = -4 \quad \text{and} \quad \mathrm{cok}(I - C) \cong \mathbf{Z}/4\mathbf{Z}$$
$$\det(I - D) = \phantom{-}4 \quad \text{and} \quad \mathrm{cok}(I - D) \cong \mathbf{Z}/2\mathbf{Z} \oplus \mathbf{Z}/2\mathbf{Z}$$
$$\det(I - E) = -24 \quad \text{and} \quad \mathrm{cok}(I - E) \cong \mathbf{Z}/24\mathbf{Z} \ .$$

Thus $S_A$ and $S_C$ are flow equivalent and no other pair here is. Note, the SFTs $S_A$ and $S_B$ have the same zeta function, and the SFTs $S_A$ and $S_C$ have different entropies. Also, $S_E$ is isomorphic to the square of the $S_C$.

The classification of general (possibly reducible) SFTs up to flow equivalence was completed by D. Huang, but the algebraic invariants are too complicated for us to describe here.

## 3.7.10 Relations

For matrices over $\mathbf{Z}_+$: SSE $\Rightarrow$ SE $\Rightarrow$ same zeta function $\Rightarrow$ same entropy. None of these implications can be reversed.

It also turns out that SE $\Rightarrow$ FE; in the reducible case, this is a difficult result. As we can see from the examples above, the zeta function does not determine the flow equivalence class, and (of course) the flow equivalence class does not determine even the entropy.

# 3.8 Dimension Groups and Shift Equivalence

## 3.8.1 Dimension Groups

### Ordered Groups

An *ordered group* is a pair $(G, G_+)$ such that $G$ is a countable abelian group and $G_+$ is a subset of $G$ (the "positive set") containing 0 such that

- $G_+ + G_+ = G_+$     [$G_+$ is a monoid]
- $G_+ - G_+ = G$     [$G_+$ generates]
- $G_+ \cap -G_+ = \{0\}$     [$((G, G_+)$ is directed]

Example. Let $G = \mathbf{Z}^n$ with the usual positive set $G_+ = \mathbf{Z}_+^n = \{v \in \mathbf{Z}^n : v \geq 0\}$. Such a group is called *simplicial*.

### Direct Limit Groups

Suppose $G_1, G_2, \ldots$ is a sequence of groups and for $n \geq 1$ we are given homomorphisms $\phi_n : G_n \to G_{n+1}$. We call the homomorphisms $\phi_n$ *bonding maps* and picture the information as the diagram

$$G_1 \longrightarrow G_2 \longrightarrow G_3 \longrightarrow G_4 \longrightarrow G_5 \cdots$$

where an arrow $G_n \longrightarrow G_{n+1}$ represents a bonding map $\phi_n$. We will describe the construction of the direct limit group $G$ from the given data.

We let $\phi_{ij}$ denote the composition $\phi_{j-1} \cdots \phi_{i+1} \phi_i$, which takes $G_i$ to $G_j$. Let $X$ be the disjoint union of the $G_n$; concretely,

$$X = \bigcup_n \{(h, n) : h \in G_n,\ n \in \mathbb{N}\}\ .$$

Define a relation $\sim$ by declaring $[h, i] \sim [h', i']$ if for some $j$, $\phi_{ij}(h) = \phi_{i'j}(h')$. (In the picture, this means that $h$ and $h'$ are pushed to a common image at some point to the right.) It is easy to check that $\sim$ is an equivalence relation. As a set, $G$ is the set of equivalence classes for $\sim$.

Note, if an equivalence class contains an element $[h, n]$ of $X$, then it also contains $[\phi_{nj}(h), j]$ for all $j > n$. We can use this to define addition in $G$.

Suppose $g$ and $g'$ are two equivalence classes in $G$. Choose some positive integer $n$ such that $g$ contains some element $[h, n]$ and $g'$ contains some element $[h', n]$. Define $g + g'$ to be the equivalence class containing $[h + h', n]$. It is straightforward to check that $g + g'$ is independent of the particular choice of representatives, and with this operation $G$ is a group.

If each of the groups $G_n$ is actually an ordered group, and each of the bonding maps is order-preserving (maps the positive set of $G_n$ into the positive set of $G_{n+1}$), then we can define the the direct limit order on $G$. Here an equivalence class is in the positive set $G_+$ if it the equivalence class contains an element $[h, n]$ such that $h \in (G_n)_+$. It is easy to verify that $(G, G_+)$ is now an ordered group.

### Dimension Groups

A dimension group is an ordered group which is isomorphic to a direct limit of simplicial groups. So, up to isomorphism, a dimension group is presented by a sequence of homomorphisms of simplicial groups. Such a homomorphism can be presented as matrix multiplication, for example

$$\mathbf{Z}^3 \to \mathbf{Z}^2$$

$$(xyz) \mapsto (xyz) \begin{pmatrix} 1 & 0 \\ 2 & 5 \\ 3 & 1 \end{pmatrix}$$

(We will have our matrices act from the right, on rows, to read composition from left to right.) The condition that the homomorphism of groups be order preserving is exactly the condition that the presenting integral matrix have nonnegative entries.

So: up to isomorphism, a dimension group is presented by a sequence $A_1, A_2, A_3, \ldots$ of nonnegative integral matrices.

The dimension groups also have an abstract, useful characterization due to Effros, Handelman and Shen: an ordered group is a dimension group if and only if it satisfies the following two properties:

Riesz interpolation:  for any $u_1, u_2, v_1, v_2$ from $G$ with all $u_i \leq v_j$, there exists $w$ in $G$ with $u_i \leq w \leq v_j$ for all $i, j$.
Unperforation:if $n \in \mathbb{N}$ and $g \in G$ with $ng \in G_+$, then $g \in G_+$ .

These two properties are what must survive from the lattice ordering of a simplicial group on passage to a direct limit. Notice an ordered group with unperforation must be torsion free. Sometimes unperforation is built into the definition of an ordered group.

The EHS characterization provides a way of recognizing dimension groups independent of matrix constructions. For example, it is trivial to check from the EHS conditions that the group $\mathcal{Q} \oplus \mathcal{Q}$ with positive set $\{(x, y) : x > 0, y > 0\}$ is a dimension group; but constructing a matrix sequence here takes a little thought.

## Uses for Dimension Groups

The dimension groups come to us from $C^*$-algebras, where (among other important uses) they are used to classify the important class of AF (approximately finite) $C^*$-algebras. The dimension groups appear as the $K_0$ groups of these algebras. (The term "dimension group" can be rationalized somewhat by the view that $K_0$ is a generalization of vector space dimension to more complicated modules.) The old exposition [3] of Effros is still a nice introduction.

Dimension groups have played a significant role in the study of SFT's; we will see one example of this in the fourth lecture.

The dimension groups have also played a dramatic role in recent study of topological orbit equivalence of minimal homeomorphisms of the Cantor set, and in the study of substitution systems. Here, if $X$ is the Cantor set and $T$ the minimal homeomorphism, then the dimension group $(G, G_+)$ is given by

$$G = C(X, \mathbb{Z})/(I - T)(X, \mathbb{Z})$$

$$G_+ = C(X, \mathbb{Z}_+)/(I - T)(X, \mathbb{Z})$$

The dramatic results here are beyond the scope of this lecture and I mention their existence for context. (See the lectures of Host for references.) Two remarks are worth making.

First: the order structure of the groups is crucial to their use in the minimal setting. (This is much less true in the SFT setting we are considering now.) One should view the order not as an unwelcome guest to the algebra but as a powerful part of it.

Second: it is not only the definition of the groups which the $C^*$ theory contributes, the groups come to dynamics along with with potent suggestions for their use.

## 3.8.2 Dimension Modules

### Stationary Dimension Groups

A dimension group is *stationary* if it is isomorphic to one which can be defined by a sequence of matrices $A_1, A_2, \ldots$ which are all equal. If this common matrix is $A$, then we can call the dimension group the dimension group of $A$.

There is a very concrete way to visualize this group in terms of linear transformations, which we describe now.

Let $A$ be an $n \times n$ integral matrix. Then $A$ defines a linear transformation on $Q^n$ by matrix multiplication, $v \mapsto vA^n$. The *eventual image* of $A$ is the vector subspace of elements which are in all forward images of $A$,

$$V_A = \bigcap_{k>0} \{vA^k : v \in Q^n\} .$$

Because $Q^n \supset Q^n A \supset Q^n A^2 \supset \cdots \supset Q^n A^n$, we have $\dim(Q^n) \geq \dim(Q^n A) \geq \cdots$ . As soon as $\dim(Q^n A^k) = \dim(Q^n A^{k+1})$ , the restriction of $A$ to $Q^n A^k$ must be an isomorphism of vector spaces, and $V_A = Q^n A^k$. In particular, $V_A = Q^n A^n$.

Now define $G_A = \{v \in V_A : \exists k > 0 \text{ such that } vA^k \in \mathbf{Z}^n\}$ , so $G_A$ is the subgroup of $V_A$ which is eventually mapped by $A$ into the integral lattice. Also define $(G_A)_+ = \{v \in G_A : \exists k > 0 \text{ such that } vA^k \in \mathbf{Z}^n_+\}$ .

It is not hard to check that $(G_A, (G_A)_+)$ is isomorphic to the dimension group of $A$. (Here is an explicit isomorphism from $G \to G_A$: if $g \in G$ and $g$ contains an element $[v, k]$, then send $g$ to $vA^n(A|V_A)^{-(n+k)}$.) It's worth noticing that the pair $(G_{A^n}, (G_{A^n})_+)$ is the same for any positive integer $n$.

Consider some examples.

**Example.** Suppose $A = (2)$. Then $V_A = Q$ and $G_A$ is the group of dyadic rationals, $\{m2^{-k} : m, k \in \mathbf{Z}\}$.

**Example.** Suppose $B = \begin{pmatrix} 1 & 1 \\ 1 & 1 \end{pmatrix}$. Then $V_B = \{q\,(1 \ \ 1) : q \in Q\}$. The action of $B$ on $V_B$ is just multiplication by 2, so an element $q\,(1 \ \ 1)$ of $V_B$ is eventually mapped into $\mathbf{Z}^2$ only if $q$ is an integer divided by a power of 2, i.e., $q$ is a dyadic rational. Thus $G_B$ is isomorphic to $G_A$ of the previous example.

**Example.** Suppose $A$ is $n \times n$ and $\det(A) = +/-1$. Then $G_A \cong \mathbf{Z}^n$.

### The Dimension Module of $A$

Again suppose $A$ is square nonnegative integral. It is an easy exercise to check that multiplication by $A$ defines an automorphism $(\hat{A}$, say) of the ordered group $(G_A, (G_A)_+)$. For an example, think of multiplication by 2 on the dyadic rationals.

There is a more algebraic way of formulating the information given by the "dimension pair" $\{(G_A, (G_A)_+) , \hat{A}\}$.

First we use $\hat{A}$ to make $G_A$ an $L$-module, where $L$ denotes the Laurent polynomials in one variable, $L = \mathbf{Z}[t, t^{-1}]$. Here the action of $L$ on $G_A$ is induced by requiring $t^{-1}$ to act as $\hat{A}$. (We choose $t^{-1}$ over $t$ for concordance with the $\text{cok}(I - tA)$ presentation discussed in Sec. 8.4.1.)

Second we make $L$ an ordered ring by declaring $\mathbb{Z}_+[t, t^{-1}]$ to be its positive set $L_+$. Then $L_+$ sends $(G_A)_+$ to itself and our $L$-module is an ordered module.

We call this ordered $L$-module the *dimension module* of $A$. Fortunately, in the most important SFT case we will be able to forget the order structure here.

We remark that a matrix $A$ and its transpose can define nonisomorphic dimension modules (even disregarding the order). In other words, the modules defined by way of the action of $A$ on row vectors vs. column vectors can be nonisomorphic.

### 3.8.3 Shift Equivalence

**The Relationship**

Recall that square nonnegative integral matrices $A, B$ are shift equivalent if there exist nonnegative integral matrices $U, V$ and a positive integer $\ell$ such that

$$A^\ell = UV \qquad B^\ell = VU$$
$$AU = UB \qquad BV = VA.$$

The basic fact is that $A$ and $B$ are shift equivalent if and only if their dimension modules are isomorphic as ordered modules. That is, there is a group isomorphism from $G_A$ to $G_B$ which sends the positive set $(G_A)_+$ onto $(G_B)_+$ and which intertwines the isomorphisms $\hat{A}$ and $\hat{B}$.

The proof of this fact is routine. Given a shift equivalence as above, the required isomorphism from $G_A$ to $G_B$ is given by matrix multiplication by the matrix $U$. In particular: if $A = B$, then multiplication by $U$ defines an automorphism of the dimension module of $A$. This fact will be important in the next lecture.

For example, with $A = (2), \ell = 5, U = (8)$ and $V = (32)$, the matrix $U$ induces an automorphism on the dyadic rationals $G_A$ which is simply multiplication by 8.

The connection between shift equivalence and dimension groups was made by Krieger. There are deeper aspects to this connection which we will not go into.

**The Primitive Case**

We say that $A$ and $B$ are algebraically shift equivalent over $\mathbb{Z}$ if there are (not necessarily nonnegative) matrices $U, V$ which satisfy the shift equivalence equations for $A$ and $B$. Similarly we can define algebraic strong shift equivalence over $\mathbf{Z}$.

FACT (Parry-Williams): if $A$ and $B$ are primitive, then they are shift equivalent if and only if they are algebraically shift equivalent.

**Proof.** By the Perron-Frobenius Theorem, the spectral radius $\lambda$ of $A$ is an eigenvalue of $A$, for which there are positive left and right eigenvectors, say $x$ and $y$. We choose these so that $xy = 1$. By the Perron-Frobenius Theorem, $\lambda$ is a simple root of the the characteristic polynomial of $A$, and $\lambda$ is strictly greater than the modulus of any other root. It follows that the sequence $[(1/\lambda)A]^n$ converges to the matrix which fixes the dominant eigendirection and annihilates the complementary generalized eigenspaces. This matrix is the positive matrix $yx$.

Now we prove the nontrivial direction. Suppose a pair of integral matrices $U, V$ satisfy the shift equivalence equations for $A$ and $B$. Shift equivalent matrices have the same nonzero eigenvalues, so $A$ and $B$ have the same spectral radius $\lambda$. The vectors $xU$ and $Vy$ are eigenvectors of $B$ with eigenvalue $\lambda$, so each is strictly positive or strictly negative. If one is strictly negative, then so is the other, since $xUVy = xA^\ell y > 0$; so, if necessary after replacing $U$ and $V$ with $-U$ and $-V$, we can assume $xU > 0$ and $Vy > 0$. Since

$$\lim_n [(1/\lambda)A]^n U = yxU > 0 \; ,$$

it follows that $A^n U$ is strictly positive for large $n$. Similarly $VA^n$ is strictly positive for large $n$. The pair $A^n U, VA^n$ give a lag $\ell + 2n$ shift equivalence of $A$ and $B$. ∎

The dimension formulation of the Parry-Williams result is that the dimension modules of $A$ and $B$ are isomorphic as ordered modules if and only if they are isomorphic as unordered modules. So, in the most important case (the case of mixing SFT's), we have the great luxury of ignoring the order.

Although the relationship of SE and SSE over $\mathbf{Z}_+$ is mysterious, the relationship of SE and SSE over $\mathbf{Z}$ is not: over $\mathbf{Z}$, shift equivalence and strong shift equivalence define the same equivalence relation. From a lag $\ell$ SE over $\mathbf{Z}$, it is possible to extract a lag $\ell$ SSE over $\mathbf{Z}$ (Williams).

**Ideal Classes**

We give a useful example of algebraic structure for SE.

Let $p$ be an irreducible monic polynomial with integral coefficients and let $\lambda$ be a root of $p$. Consider the class $\mathcal{M}_\lambda$ of square integral matrices with characteristic polynomial $p$.

Then, there is a bijection between algebraic shift equivalence classes of matrices in $\mathcal{M}_\lambda$, and ideal classes in the ring $\mathbf{Z}[1/\lambda]$. (This is a part of the classification of mixing SFT's up to regular isomorphism.)

This is an analogue of a classical result of Taussky-Todd: there is a bijection between similarity-over-$\mathbb{Z}$ classes of matrices in $\mathcal{M}_\lambda$, and ideal classes of the ring $\mathbb{Z}[\lambda]$. In both cases one checks rather easily that a bijection is given by the map which sends a matrix to the ideal class of the ideal generated by entries of an eigenvector for $\lambda$ which has entries in $\mathbb{Z}[\lambda]$.

The class numbers of the rings $\mathbb{Z}[\lambda]$ and $\mathbb{Z}[1/\lambda]$ are finite and the relations under consideration are computable.

## 3.8.4  Further Developments

### Polynomial Matrices

The presentation of an SFT as a vertex shift allows one to extract algebraic invariants from a defining zero-one matrix. Defining edge shifts with nonnegative integral matrices, one makes a significant advance in conciseness and functoriality of presentation.

There is another advance in conciseness and functoriality gained by presenting SFT's with polynomial matrices. Here is an example illustrating the general situation. Let $B$ be the polynomial matrix

$$B = \left( \begin{array}{cc} 2t & t^3 + t \\ 3t^2 & 0 \end{array} \right).$$

To this $2 \times 2$ matrix, we associate a graph $G$ with two distinguished vertices, say 1 and 2. A term $t^k$ in $B$ corresponds to a path of length $k$ in $G$. From the term $2t$, $G$ acquires two edges from 1 to 1. From the term $t^3$, $G$ acquires a path of three edges from 1 to 2. On this path are two new, intermediate vertices which have no further adjoining edges. Similarly $G$ acquires an edge from 1 to 2 and three paths of length two from 2 to 1.

Clearly we can describe some very complicated graphs with polynomial matrices of small size. Also, the algebra behaves beautifully with this presentation. Let $A$ denote the adjacency matrix of the graph $G$ constructed as above from $B$ (so $tA$ is another polynomial presentation of $G$). Then the SFT $S_A$ has inverse zeta function $\det(I - tA) = \det(I - B)$.

Moreover, if $B$ is $n \times n$, then by matrix multiplication $B$ maps $L^n$ into itself (where $L = Z[t, t^{-1}]$). Let

$$\mathrm{cok}(I - B) = L^n / (L^n (I - B)) .$$

This is an $L$-module and (suitably ordered) it is isomorphic to the dimension module for the SFT $S_A$. (In the case $A = tB$, multiplication by $B$ acts like $t^{-1}$.) If we set $t$ equal to 1, then $L$ becomes $\mathbb{Z}$ and our dimension module becomes the Bowen-Franks invariant of flow equivalence. For a more thorough discussion of all this, see the survey [1].

**SE and SSE for Sofic Systems**

The utility of shift equivalence and strong shift equivalence does not end with SFT's. For at least two other classes of symbolic systems, one can find suitable ordered rings in which the equations of SSE and SE correspond to isomorphism and eventual isomorphism in other cases.

These classes are the irreducible Markov shifts (meaning: irreducible S-FT's with Markov measure) and the sofic systems. In the Markov case, the role played (in the SFT case) by integral matrices is played by matrices whose entries are Laurent polynomials in several variables. In the sofic case, one works in a certain complicated integral semigroup ring and the objects which play the role of matrices are derived from canonical covers of the sofic systems by SFTs.

# 3.9   Automorphisms and Classification of SFTS

## 3.9.1   Automorphisms

**Automorphisms**

An *automorphism* of a dynamical system $(X, S)$ is a homeomorphism $U : X \to X$ such that $US = SU$. So, an automorphism is a self-conjugacy or symmetry of the dynamical system. With the operation of composition, the automorphisms of $S$ are a group, Aut($S$). In this lecture we will always restrict to systems which are nontrivial mixing SFTs (nontrivial means not just a fixed point).

It is difficult to get hold of structure for SFTs, and automorphisms provide one opportunity. How big is the group? How does it act? When are Aut($S$) and Aut($T$) isomorphic groups?

**Aut(S) as an Abstract Group**

It is countable (there are only countably many block codes), but very complicated (containing for example copies of all finite groups and many others).

It is residually finite (this is because the periodic points are dense and for each $n$ the points of period $n$ are a finite invariant set).

The center of Aut($S$) equals the powers of $S$ (Ryan's theorem). Thus, for example, the 2-shift $S_{[2]}$ and the 4-shift $S_{[4]}$ have nonisomorphic automorphism groups. The center of Aut($S_{[4]}$) has a generator which is a square in the group Aut($S_{[4]}$) (because $S_{[4]} = (S_{[2]})^2$). But the center of Aut($S_{[2]}$) does not have a generator which is a square in Aut($S_{[2]}$) (it is not hard to show $S_{[2]}$ has no square root).

One measure of the mystery of Aut($S$) is the following: we know NO way to show Aut($S$) and Aut($T$) are nonisomorphic except by some straightforward application of Ryan's Theorem. For example, we do not know whether Aut($S_{[2]}$) and Aut($S_{[3]}$) are isomorphic as abstract groups.

### Action on Finite Subsystems

Here is a seminal problem posed by Williams in the 1970's (in a 1977 lecture in Boulder and perhaps earlier). Suppose $S$ has fixed points $x$ and $y$. Does there exist an automorphism of $S$ which exchanges them? Williams proposed this as one approach to the classification problem: perhaps different answers to this question would distinguish SFTs defined by shift equivalent matrices.

Well, there's no reason to stop at fixed points. Let $F_n[S]$ denote the (finite) set of points of period at most $n$. Let $S_{<n>}$ be the restriction of $S$ to $F_n[S]$. Now $S_{<n>}$ has its own little group of automorphisms, and we can ask: when does an automorphism of $S_{<n>}$ extend to an automorphism of $S$? Always?

For possible obstructions, we look for useful representations of the automorphism group. By a representation, here we just mean a homomorphism into a group. The only useful representations found so far involve the dimension module or periodic points.

## 3.9.2 Representations

### The Dimension Representation $\rho$

Suppose $\phi$ is an automorphism of $S_A$. Recall that $\phi$ is given by a power of the shift composed with a map constructed from a strong shift equivalence from $A$ to $A$ in the explicit fashion we earlier described. (That is, recall we have already claimed this without proof.) Now, if we have a strong shift equivalence from $A$ to itself, say

$$A = U_1 V_1, \; V_1 U_1 = A_1 \quad A_1 = U_2 V_2, \; V_2 U_2 = A_2 \; \ldots \; A_\ell = U_\ell V_\ell, \; V_\ell U_\ell = A$$

then we get a shift equivalence

$$\begin{aligned} A^\ell &= UV & B^\ell &= VU \\ AU &= UB & BV &= VA \end{aligned}$$

by setting $U = U_1 U_2 \ldots U_\ell$ and $V = V_\ell \ldots V_2 V_1$.

Now, this shift equivalence defines an isomorphism of the dimension module of $S_A$ (from multiplication by $U$) and it turns out that this isomorphism, composed with the isomorphism induced by that power of the shift, is canonical: it depends only on the automorphism $\phi$.

Thus we have a homomorphism $\rho$ from Aut($S_A$) to Aut($M_A$), where Aut($M_A$) is the group of automorphisms of the dimension module. This homomorphism $\rho$ is called the *dimension representation*.

## The Kernel of $\rho$

The group of automorphisms of the dimension module is algebraic, "small" (typically a finitely generated abelian group), and varies from one SFT to another.

   Thus most of the vast complexity of Aut($S$) lies in the kernel of $\rho$. Think of this group as combinatorial, big and somehow homogeneous in its properties across nontrivial mixing SFTs. (This "homogeneity" is not a precise property, but a feeling one gets from experience with the actions of Ker($\rho$).)

   In particular the richness of action of Aut($S$) on finite subsystems comes almost entirely from the action of Ker($\rho$). Constraining the action of Aut($S$) is mostly a matter of finding general constraints on the action of Ker($\rho$); then it is a matter of fine structure to work out the remaining possibilities for a particular SFT.

## Sign

Now we need some algebraic invariants for the action of Aut($S$) on periodic points. Suppose $U$ is an automorphism of $S$. Let $P_n^o(S)$ be the set of points of period exactly $n$ for $S$. This finite set is invariant under $U$.

   $U$ permutes the $S$-orbits of length $n$ (here length means cardinality). If this permutation is even, then define $\text{sign}_n(U) = 0$. If the permutation is odd, then set $\text{sign}_n(U) = 1$. This gives us a homomorphism

$$\text{sign}_n : \text{Aut}(S) \rightarrow \mathbb{Z}/2\mathbb{Z}$$
$$U \mapsto \text{sign}_n(U).$$

(If there are no $S$-orbits of size $n$, then we adopt the convention $\text{sign}_n = 0$.) Since the shift $S$ maps each $S$-orbit to itself, $\text{sign}_n(S) = 0$ for every $n$.

## Gyration

We will now use the action on $P_n^o(S)$ to define a homomorphism from Aut($S$) into $\mathbb{Z}/n\mathbb{Z}$. From each $S$-orbit of length $n$ pick a single point, giving us say $k$ points $x_1, x_2, \ldots x_k$. An automorphism $U$ will send $x_i$ into the orbit of some $x_j$, say $j = \pi(i)$, so for some integer $m(i)$,

$$U : x_i \mapsto S^{m(i)} x_{\pi(i)} .$$

We define

$$\text{gy}_U : \text{Aut}(S) \rightarrow \mathbb{Z}/n\mathbb{Z}$$
$$U \mapsto \sum_i m(i).$$

   It is an easy and pleasant exercise to check that this map is well defined (does not depend on the choice of representatives $x_i$), and is a group homomorphism.

## SGCC

Now we can define the "sign-gyration-compatibility-condition" homomorphism SGCC. This will be a homomorphism

$$\text{SGCC} : \text{Aut}(S) \;\to\; \oplus_n \mathbf{Z}/n\mathbf{Z}$$
$$U \;\mapsto\; \oplus_n \text{SGCC}_n$$

where

$$\text{SGCC}_n : \text{Aut}(S) \;\to\; \mathbf{Z}/n\mathbf{Z}$$
$$U \;\mapsto\; \text{gy}_n(U) + \tfrac{n}{2} \sum_{(n/k)=2^j} \text{sign}_k(U) \;.$$

The last sum is over the positive integers $k$ such that $k < n$ and $n/k$ is a positive integer which is a power of 2. For example, if $n = 12$, then the sum is over $k = 6, 3$. An empty sum is zero.

We will see in a moment that under some conditions SGCC must vanish. This is an astonishing constraint. We do not have time to motivate where in the world it could come from. But, do notice, for $n$ even the condition $\text{SGCC}_n(U)=0$ implies that the action of $U$ on orbits of length $n$ is constrained by its action on orbits of smaller length.

# 3.10 The KRW Factorization Theorem

## 3.10.1 Statement of the Theorem

Kim, Roush and Wagoner ("KRW") proved that there is a homomorphism

$$\text{sgcc}_n : \; \text{Aut}(M_A) \to \oplus_n \mathbf{Z}/n\mathbf{Z},$$

such that $\text{SGCC}_n = \text{sgcc}_n \circ \rho$. Taking the direct sum over $n$, we get $\text{SGCC} = \text{sgcc} \circ \rho$. In other words, SGCC factors through the dimension representation. Especially, SGCC vanishes on the kernel of the dimension representation.

## 3.10.2 Nonsurjectivity of $\rho$

The KRW maps $\text{sgcc}_n$ can be computed (with difficulty: for $n = 2$, KRW computations of examples were computer assisted). With the computability, KRW could produce examples of mixing SFTs $S$ for which $\rho$ is not surjective.

The rough idea for nonsurjective examples is simple: produce (for example) an SFT $S$ in which a certain automorphism $\psi$ of the dimension module has nonzero image under $\text{sggc}_2$, but $S$ has just one fixed point, and no points of period two. An automorphism $\phi$ of $S$ must then satisfy $\mathrm{SGCC}_2(\phi) = 0$, so $\rho(\phi) = \psi$ would give

$$0 = \mathrm{SGCC}_2(\phi) = \mathrm{sgcc}_2(\rho(\phi)) = \mathrm{sgcc}_2(\psi) \neq 0$$

which is a contradiction.

### 3.10.3   A Long Story

The Factorization Theorem did not appear overnight, it was a culmination of a long development also involving Hedlund, Williams, Boyle, Krieger, U. Fiebig, Nasu, Lind and Rudolph. (See the references in [7].) There is also another dramatic recent development on the constructive side in which KRW use innovative constructions with polynomial matrices to show that SGCC = 0 is the *only* constraint on the action of $\mathrm{Ker}(\rho)$ on finite subsystems. (A result of Boyle and Krieger then implies that SGCC = 0 remains the only constraint on the action of $\mathrm{Ker}(\rho)$ on infinite subsystems.)

## 3.11   Classification

### 3.11.1   SE does not imply SSE: the Reducible Case

Kim and Roush were able to use one of the nonsurjectivity examples above to get a counterexample to the shift equivalence conjecture in the reducible case. Here is the rough idea.

Let $A$ be a primitive matrix for which $\rho_A$ is not surjective and consider a pair of reducible matrices $B, C$ with the block forms

$$B = \begin{pmatrix} A & X \\ 0 & A \end{pmatrix} \quad , \quad C = \begin{pmatrix} A & Y \\ 0 & A \end{pmatrix}$$

As we have seen, a strong shift equivalence from $B$ to $C$ induces a shift equivalence from $B$ to $C$. Also, the SSE from $B$ to $C$ restricts to give SSE's of the upper left blocks $A$ and of the lower right blocks $A$. These two SSE's produce isomorphisms from $S_A$ to $S_A$, which we can interpret as automorphisms.

The action of these automorphisms comes from SE's obtained by restriction to diagonal blocks of the SE from $B$ to $C$ (which was induced by the SSE from $B$ to $C$). The matrices involved in the diagonal block SE's are not independent, they satisfy some additional matrix equations arising from the blocks $X$ and $Y$.

Kim and Roush got their counterexample by making very judicious choices of $A$, $X$ and $Y$ to arrange the following:

(i) For $S_A$, the dimension representation is not surjective;

(ii) $B$ and $C$ are shift equivalent; and

(iii) for any shift equivalence of $B$ and $C$, one of the induced diagonal-block shift equivalences from $A$ to $A$ must give an action on the dimension module of $A$ which cannot arise from an isomorphism of $A$.

### 3.11.2 SE does not imply SSE: the Irreducible Case

After the original delivery of these lectures, Kim and Roush obtained a counterexample [4] to the shift equivalence conjecture for irreducible matrices over $\mathbf{Z}_+$. The study of automorphisms of the shift, the dimension representation, and the Factorization Theorem again underly this work; but this work is even more deeply embedded in the developed theory and especially the CW complexes of strong shift equivalences developed over several years by Wagoner. These complexes provide the setting for the work of the Factorization Theorem and the irreducible counterexample of Kim and Roush, and appear to be fundamental to the study and classification of SFTs.

**Acknowledgment.** I thank Pierre-Paul Romagnoli for detailed and helpful comments on the original lectures.

# Bibliography

[1] M. Boyle, *Symbolic Dynamics and Matrices*, in Combinatorial and Graph-Theoretical Problems in Linear Algebra, IMA Volumes in Math. and its Applications **50**, Eds. R.A. Brualdi, S. Friedland and V. Klee, pp. 1–38 (1993)

[2] M. Denker, C. Grillenberger, K. Sigmund, *Ergodic Theory on Compact Spaces*, Springer Lec. Notes in Math **527** (1970).

[3] E.G. Effros, *Dimensions and C*-algebras*, CBMS Reg. Conf. Series in Math **46** (1981).

[4] K.H. Kim, F.W. Roush, *The Williams conjecture is false for irreducible subshifts*, Electron. Res. Announc. Amer. Math. Soc. 3 (1997),105-109 (electronic); Annals of Math., to appear.

[5] K.H. Kim, F.W. Roush, J.B. Wagoner, *Inert actions on periodic points*, Electron. Res. Announc. Amer. Math. Soc. 3 (1997), 55-62 (electronic).

[6] B.P. Kitchens, *Symbolic dynamics. One-sided, two-sided and countable state Markov shifts*, Springer-Verlag (1998).

[7] D. Lind, B. Marcus, *An introduction to Symbolic Dynamics and Coding*, Cambridge University Press (1995).

[8] K. Schmidt, *Dynamical systems of algebraic origin*, Progress in Mathematics **128**, Birkhauser Verlag (1995).

# Chapter 4

# DYNAMICS OF $\mathbb{Z}^d$ ACTIONS ON MARKOV SUBGROUPS

Bruce KITCHENS
IBM T. J. Watson Research Center
Yorktown Heights, NY 10598
U.S.A.

A compact topological group with an expansive $\mathbb{Z}^d$ action of automorphisms can be represented as a Markov subgroup. These notes are an introduction to the study of their dynamical properties. There are two very different cases: one where the group is totally disconnected and the other where the group is connected. Here the concentration is on the disconnected case but some general ideas are mentioned.

## 4.1 Introduction

These lectures contain an introduction to the study of multi-dimensional Markov shifts which have a group structure. They are called Markov subgroups. The alphabet of the Markov subgroup can be any compact group but we will concentrate on the ones with a finite alphabet. Any expansive $\mathbb{Z}^d$ action of a compact group can be represented as a Markov subgroup on a suitable alphabet. The formulation of the problems considered and the results presented here can be found in the papers [6], [7], [8], [9] and [10] (1987-1993). A comprehensive introduction to this subject can be found in the book *Dynamical Systems of Algebraic Origin* by Klaus Schmidt [19]. It contains much more than is presented here and thoroughly treats the case of Markov subgroups with a compact connected alphabet.

The lectures are organized as follows. Section 4.2 contains examples

of one-dimensional Markov subgroups and then proves a structure theorem
(Theorem 4.2.7) for one-dimensional Markov subgroups on a finite alphabet.
Section 4.3 contains a discussion of some decidability questions for general two-dimensional Markov shifts and how they relate to one-dimensional
Markov shifts, tilings and periodic points. Proposition 4.3.5 states that the
general tiling problem is undecidable. This was proved by R. Berger and
the notes do not contain a proof. Then the way these results relate to two-
dimensional Markov subgroups is examined. The section concludes by observing that these decidability problems do not arise for Markov subgroups.
In Section 4.4 an examination of two-dimensional Markov subgroups with ($\mathbb{Z}/$
$2\mathbb{Z}$) as the alphabet is begun. The basic definitions are made and questions
about entropy and directional dynamics are discussed. Section 4.5 contains
the heart of these lectures. To each such Markov subgroup an ideal in the
ring $(\mathbb{Z}/2\mathbb{Z})[x^{\pm 1}, y^{\pm 1}]$ is associated. A systematic study of the relationship
between the dynamical properties of the Markov subgroup and the algebraic
properties of the ideal is begun. Section 4.6 contains a few remarks on the
conjugacy and isomorphism questions about Markov subgroups studied in
Section 4.5. Section 4.7 concludes the lectures by showing how the algebraic
formulation of Section 4.5 can be extended to the study of all expansive $\mathbb{Z}^d$
actions of compact groups.

# 4.2   One-Dimensional Markov Subgroups

In this section we study the structure of expansive automorphisms of compact, totally disconnected groups. Theorem 4.2.7 describes a classification of
their dynamics. These results and some of the ideas developed will be used
when we move to higher dimensional group actions. Some of the results will
be exploited using induction on the dimension of the group actions and some
ideas will be adapted to suit the higher dimensional setting.

**Example 4.2.1** [6] Totally disconnected shift spaces. Take any finite group
$G$ with the discrete topology. We want to consider the space of all doubly
infinite sequences. It is the product space $G^{\mathbb{Z}}$. It is a group when addition
is done coordinate by coordinate. Give the space the product topology. The
topology is induced by the metric where two sequences are close together
if they agree on a large central block. It is given by the formula $d(x, y) =$
$1/2^N$, where $N = \sup\{n : x_i = y_i$ for all $|i| < n\}$. Then $G^{\mathbb{Z}}$ is a compact
topological group which is homeomorphic to the usual Cantor set. The shift
transformation $\sigma : G^{\mathbb{Z}} \to G^{\mathbb{Z}}$ is defined by $\sigma(x)_i = x_{i+1}$. It is a continuous
group automorphism. This is the full shift on the alphabet $G$ and also has a
group structure.

**Example 4.2.2** Finite state Markov subgroups. Begin with the finite group $\mathbf{Z}/4\mathbf{Z}\oplus\mathbf{Z}/2\mathbf{Z}$ and form the full shift on this group as in Example 4.2.1. Define a closed, shift invariant subgroup $X \subseteq (\mathbf{Z}/4\mathbf{Z} \oplus \mathbf{Z}/2\mathbf{Z})^{\mathbf{Z}}$ by a transition rule. The rule is given by the diagram in Figure 4.1 The rule defines for each symbol a *follower set*. They are:

$$f(0,0) = f(2,0) = f(1,1) = f(3,1) = \{(0,0),(2,0),(1,0),(3,0)\}$$
$$f(1,0) = f(3,0) = f(0,1) = f(2,1) = \{(1,1),(3,1),(2,1),(0,1)\}.$$

The new subgroup $X \subseteq (\mathbf{Z}/4\mathbf{Z} \oplus \mathbf{Z}/2\mathbf{Z})^{\mathbf{Z}}$ is $\{x : x_{i+1} \in f(x_i), \text{ all } i \in \mathbf{Z}\}$. One must check the rule is closed under addition (Exercise). With the subset topology $X$ is a compact, totally disconnected topological group and the shift is an automorphism. It is called a *finite state Markov subgroup*.

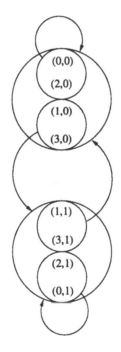

Figure 4.1: Transition graph

**Example 4.2.3** Let $G$ be a finite group. Let $H \subseteq G^k$ be a subgroup of the product of $G$ with itself $k$ times. Define $X_H \subseteq G^{\mathbf{Z}}$ to be the set $X_H = \{x : [x_i, \dots, x_{i+k-1}] \in H \text{ for all } i \in \mathbf{Z}\}$. It is a closed, shift invariant subgroup of

$G^{\mathbb{Z}}$. With the subspace topology $(X_H, \sigma)$ is the $(k-1)$-*step Markov subgroup* defined by $H$. Example 4.2.2 is a particular one-step Markov subgroup. It is defined by the subgroup $H \subseteq (\mathbb{Z}/4\mathbb{Z} \oplus \mathbb{Z}/2\mathbb{Z})^2$

$$((0,0),(0,0)),((0,0),(2,0)),((0,0),(1,0)),((0,0),(3,0))$$
$$((2,0),(0,0)),((2,0),(2,0)),((2,0),(1,0)),((2,0),(3,0))$$
$$((1,0),(1,1)),((1,0),(3,1)),((1,0),(2,1)),((1,0),(0,1))$$
$$((3,0),(1,1)),((3,0),(3,1)),((3,0),(2,1)),((3,0),(0,1))$$
$$((1,1),(0,0)),((1,1),(2,0)),((1,1),(1,0)),((1,1),(3,0))$$
$$((3,1),(0,0)),((3,1),(2,0)),((3,1),(1,0)),((3,1),(3,0))$$
$$((2,1),(1,1)),((2,1),(3,1)),((2,1),(2,1)),((2,1),(0,1))$$
$$((0,1),(1,1)),((0,1),(3,1)),((0,1),(2,1)),((0,1),(0,1)).$$

A one-step Markov subgroup is simply referred to as a Markov subgroup.

We will concentrate on totally disconnected groups but connected ones are also of interest. The next three examples are of automorphism of connected groups.

**Example 4.2.4** Hyperbolic Toral Automorphisms. The space $\mathbb{R}^2$ is a topological group with vector addition the operation. The lattice $\mathbb{Z}^2$ is a discrete subgroup of $\mathbb{R}^2$. The two-dimensional torus $\mathbb{T}^2$ is thought of as the unit square in $\mathbb{R}^2$ with opposite edges identified, $(0, y) \sim (1, y)$ and $(x, 0) \sim (x, 1)$. Vector addition modulo one in each coordinate makes $\mathbb{T}^2$ a group. With the natural topology $\mathbb{T}^2$ is a compact topological group. There is a continuous group homomorphism $\pi$ from $\mathbb{R}^2$ to $\mathbb{T}^2$. It is defined by reduction modulo 1 in each coordinate. More simply, $\mathbb{R}^2/\mathbb{Z}^2 \simeq \mathbb{T}^2$. The $2 \times 2$ matrix $\left(\begin{smallmatrix} 2 & 1 \\ 1 & 1 \end{smallmatrix}\right)$, with integer entries and determinant one defines an automorphism of $\mathbb{R}^2$ which restricts to an automorphism of $\mathbb{Z}^2$. It induces an automorphism of $\mathbb{T}^2$. In $\mathbb{R}^2$ the matrix maps the unit square to a parallelogram with area one. When projected back to the unit square the identifications for the torus are preserved. This is illustrated in Figure 4.2.

**Example 4.2.5** Connected shift spaces. The circle $\mathbb{S}^1$ thought of as $\mathbb{R}/\mathbb{Z}$ with addition modulo one is a compact topological group. We can form the product space $(\mathbb{S}^1)^{\mathbb{Z}}$. It is the space of all doubly infinite sequences with entries in $\mathbb{S}^1$. There is a natural addition of sequences which is defined using coordinate by coordinate addition. Given the product topology $(\mathbb{S}^1)^{\mathbb{Z}}$ is a compact topological group. The shift transformation $\sigma$ from $(\mathbb{S}^1)^{\mathbb{Z}}$ to itself is defined by $(\sigma(x))_i = x_{i+1}$. It is a continuous group automorphism. This is a generalization of the usual shift space with a finite alphabet (as in Example 4.2.1) to a shift space whose alphabet is a compact group.

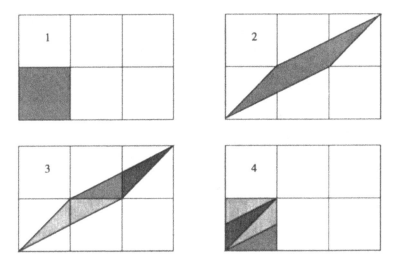

Figure 4.2: Automorphism

**Example 4.2.6** The 2-adic solenoid. Begin with Example 4.2.5. The *2-adic solenoid* is defined to be $\{x \in (\mathbf{S}^1)^{\mathbf{Z}} : 2x_i = x_{i+1}, \text{ all } i \in \mathbf{Z}\}$. It is a closed, shift invariant subgroup of $(\mathbf{S}^1)^{\mathbf{Z}}$. It is a subgroup because the rule is linear. This fits into the framework of Example 4.2.3 by letting

$$H = \{(x, 2x) : x \in \mathbf{S}^1\} \subseteq (\mathbf{S}^1)^2$$

and then the solenoid is $X_H \subseteq (\mathbf{S}^1)^{\mathbf{Z}}$. The solenoid is connected. Locally it is a line segment cross a Cantor set (Exercise). The shift restricted to it is an automorphism. There is also a geometric description of the solenoid. Begin with a solid torus $M$ and define a map $f$ that takes it into itself as shown in Figure 4.3. The image is a two-one torus knot and the cross section is two disks. Let $\tilde{M} = \bigcap f^n(M)$ for $n \in \mathbf{N}$ and $\tilde{f}$ be the induced map on $\tilde{M}$. Then $(\tilde{M}, \tilde{f})$ and $(\mathbf{S}_2, \sigma)$ are topologically conjugate (Exercise).

Let $X$ be a metric space with metric $d$ and let $T$ be a homeomorphism of $X$ to itself. Then $T$ is *expansive* if there is a $c > 0$ such that for any two points $x \neq y$ there exists an $n \in \mathbf{Z}$ with $d(T^n(x), T^n(y)) > c$. Such a $c$ is called an *expansive constant*. If the space is a topological group and the homeomorphism is an automorphism this is equivalent to requiring that there be a neighborhood of the identity $\mathcal{U}$ so that for any two points $x \neq y$ there exists an $n \in \mathbf{Z}$ with $T^n(y) \notin T^n(x) \cdot \mathcal{U}$. Equivalently, there is a neighborhood of the identity $\mathcal{U}$ such that for every $x$ not equal to the identity there exists

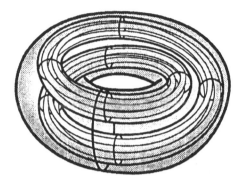

Figure 4.3: Solenoid

an $n$ with $T^n(x) \notin \mathcal{U}$. Such an open set $\mathcal{U}$ is said to *separate points* under $T$. The automorphisms described in Examples 4.2.1, 4.2.2, 4.2.3, 4.2.4 and 4.2.6 are expansive. The automorphism described in Example 4.2.5 is not expansive (Exercises).

The goal of this section is to describe and classify all expansive automorphisms of totally disconnected groups. For this we need the next definitions. If $X$ and $\bar{X}$ are compact topological groups and $T$ and $\bar{T}$ are continuous automorphisms of $X$ and $\bar{X}$ respectively, we say $(X,T)$ and $(\bar{X},\bar{T})$ are *isomorphic* if there is a continuous group isomorphism $\varphi : X \to \bar{X}$ so that $\bar{T} \circ \varphi = \varphi \circ T$. We say $(X,T)$ and $(\bar{X},\bar{T})$ are *conjugate* if there is a homeomorphism $\varphi : X \to \bar{X}$ such that $\bar{T} \circ \varphi = \varphi \circ T$. An automorphism is *transitive* if there is a point $x$ whose orbit is dense in $X$. Let $h(X,T)$ denote the *topological entropy* of $T$ acting on $X$. Will prove the following theorem using a sequence of lemmas.

**Theorem 4.2.7** *If $X$ is a compact, totally disconnected topological group and $T$ is an expansive automorphism then $(X,T)$ is conjugate to one of the following.*

 (i) *An automorphism of a finite group with the discrete topology. This happens if and only if $h(X,T) = 0$.*

 (ii) *A full $m$-shift. This happens if and only if $h(X,T) > 0$ and $T$ is transitive. Then $h(X,T) = \log m$.*

 (iii) *The direct product of an automorphism of a finite group with a full $m$-shift. Then $h(X,T) = \log m$.*

Suppose $G$ is a finite group and $X \subseteq G^{\mathbb{Z}}$ is a closed, shift invariant subgroup of $G^{\mathbb{Z}}$. Let $\mathcal{W}(X,k) = \{[g_0, \ldots, g_{k-1}] \subseteq G^k : g \in X\}$. These are the *words of length $k$* in $X$. For $w \in \mathcal{W}(X,k)$ let $f(w) = \{g \in G : wg \in \mathcal{W}(X, k+1)\}$. This is the *follower set* of $w$. The *predecessor set* of $w$ is the set $\{g \in G : gw \in \mathcal{W}(X, k+1)\}$. Let $e$ be the identity element of $G$ and $[e^k] \in \mathcal{W}(X,k)$ be the word of $k$ $e$'s. The *n-block presentation of $X$* is the closed, shift invariant subgroup of $(\mathcal{W}(X,n))^{\mathbb{Z}}$ defined by $f([g_0, \ldots, g_{n-1}]) = \{[g_1, \ldots, g_{n-1}, g_n] \in \mathcal{W}(X,n) : [g_0, \ldots, g_{n-1}, g_n] \in \mathcal{W}(X, n+1)\}$. We denote the n-block presentation of $X$ by $X^{[n]}$.

The next lemma contains a crucial observation about shift spaces with a group structure.

**Lemma 4.2.8** *Let $X, G, f, \mathcal{W}(X,k)$ and $[e^k]$ be as previously stated; then:*

(i) *$\mathcal{W}(X,k)$ is a subgroup of $G^k$;*

(ii) *for all $k$, $f([e^k])$ is a normal subgroup of $G = \mathcal{W}(X,1)$;*

(iii) *for each $[g_0, \ldots, g_{k-1}] \in \mathcal{W}(X,k)$, $f([g_0, \ldots, g_{k-1}])$ is a coset of $f([e^k])$.*

**Proof.**

(i) $\mathcal{W}(X,k)$ is clearly a subgroup of $G^k$ because $X$ is a subgroup of $G^{\mathbb{Z}}$ with coordinate by coordinate addition from $G$.

(ii) Let $g, g' \in f([e^k])$. In $\mathcal{W}(X, k+1)$

$$[e, \ldots, e, g][e, \ldots, e, g'] = [e, \ldots, e, gg'] \text{ and}$$
$$[e, \ldots, e, g]^{-1} = [e, \ldots, e, g^{-1}],$$

so $f([e^k])$ is a subgroup. If $h \in G$, there is a $[h_0, \ldots, h_{k-1}] \in \mathcal{W}(X,k)$ with $h \in f([h_0, \ldots, h_{k-1}])$ and

$$[h_0, \ldots, h_{k-1}, h] \ [e, \ldots, e, g][h_0, \ldots, h_{k-1}, h]^{-1}$$
$$= [e, \ldots, e, hgh^{-1}].$$

This means $f([e^k])$ is normal in $G$.

(iii) If $[g_0, \ldots, g_{k-1}] \in \mathcal{W}(X,k)$ and $g \in f([g_0, \ldots, g_{k-1}])$ then $g \cdot f([e^k]) \subseteq f([g_0, \ldots, g_{k-1}])$. Conversely, if $h \in f([g_0, \ldots, g_{k-1}])$, then

$$[g_0, \ldots, g_{k-1}, g]^{-1}[g_0, \ldots, g_{k-1}, h] = [e, \ldots, e, g^{-1}h],$$

so $f([g_0, \ldots, g_{k-1}]) \subseteq g \cdot f([e^k])$. ∎

**Lemma 4.2.9** *Let $\bar{G}$ be a finite group and $\bar{X}$ a closed, shift invariant subgroup of $\bar{G}^{\mathbb{Z}}$. Then $\bar{X}$ is isomorphic to a one-step Markov subgroup of $G^{\mathbb{Z}}$ for some finite group $G$.*

**Proof.** Suppose $\bar{G}$ is a finite group and $\bar{X}$ is a closed, shift invariant subgroup of $\bar{G}^{\mathbf{Z}}$. Then

$$G = \mathcal{W}(X, 1) \supseteq f([e]) \supseteq f([e^2]) \supseteq \cdots \supseteq f([e^n]) \supseteq \cdots$$

is a descending chain of normal subgroups. Since $\bar{G}$ is finite there is an $N$ such that $f([e^n]) = f([e^N])$ for all $n \geq N$. Since the other follower sets are cosets of this,

$$f([g_0, \ldots, g_{N-1}]) = f([g_{-j}, \ldots, g_{N-1}]) \quad \text{for all} \quad j \geq 0$$

with $[g_{-j}, \ldots, g_{N-1}] \in \mathcal{W}(X, N + j)$. The $N$-block presentation of $X$ has a one-step transition rule. Setting $X = X^{[N]}$ and $G = \mathcal{W}(X, N)$ completes the proof. ∎

**Lemma 4.2.10** *Let $G$ be a finite group and $X$ a Markov subgroup of $G^{\mathbf{Z}}$. Then:*

(i) *$f : G \to G/f(e)$ is an onto, continuous, group homomorphism with kernel $p(e)$;*

(ii) *$p : G \to G/p(e)$ is an onto, continuous, group homomorphism with kernel $f(e)$.*

**Proof.** The fact that $f : G \to G/f(e)$ is a well-defined, continuous, onto group homomorphism follows immediately from Lemma 4.2.8 To see the kernel of the map $f$ is $p(e)$ observe that $g \in p(e)$ if and only if $f(g)$ contains $e$. But then $f(g) = f(e)$. The same is true when the roles of $p(e)$ and $f(e)$ are reversed. ∎

A Markov subgroup of $G^{\mathbf{Z}}$ is defined by a subgroup $H \subseteq G^2$. In terms of predecessor and follower sets $H = \{(p(g), g) : g \in G\} = \{(g, f(g)) : g \in G\}$.

A standard fact about compact, totally disconnected groups is that every neighborhood of the identity contains an open-closed normal subgroup. See for example [18].

**Lemma 4.2.11** *If $X$ is a compact, totally disconnected group and $T$ is an expansive automorphism then $(X, T)$ is isomorphic to $(\bar{X}, \sigma)$ where $\bar{X}$ is a closed, shift invariant subgroup of $\bar{G}^{\mathbf{Z}}$ and $\bar{G}$ is a finite group.*

**Proof.** Since $T$ is expansive we can apply the previously stated fact about compact, totally disconnected groups to find an open normal subgroup $\mathcal{U}$ which separates points under $T$. Let $\mathcal{P}$ be the finite open-closed partition of $X$ made up of the cosets of $\mathcal{U}$. Let $\bar{G}$ be $X/\mathcal{U}$. Define a map $\varphi$ from $X$ into $\bar{G}^{\mathbf{Z}}$ by saying $(\varphi(x))_i$ is the element of $\bar{G}$ containing $T^i(x)$. Let $\bar{X}$ be the image of $X$. Then $\varphi$ is an isomorphism between $(X, T)$ and $(\bar{X}, \sigma)$. (Exercise) ∎

**Proof of Theorem 4.2.7.** By Lemmas 4.2.9 and 4.2.11. we may assume we have our group and automorphism presented as $(X, \sigma)$, where $X$ is a one-step Markov subgroup of $G^{\mathbf{Z}}$ for some finite group $G$. By Lemma 4.2.10 the map $f : G \to G/f(e)$ is a homomorphism with kernel $p(e)$. When $G$ is finite this implies $|G| = |G/f(e)| \cdot |p(e)|$, where $|\cdot|$ denotes the cardinality of a set. For a finite group $G$ and normal subgroup $f(e)$ we also have $|G| = |G/f(e)| \cdot |f(e)|$. Together these imply $|f(e)| = |p(e)|$ when $G$ is finite. If $|f(e)| = |p(e)| = 1$, define an automorphism $\tau : G \to G$ by $\tau(g) = f(g)$. This is case (*i*). Otherwise we have $|f(e)| = |p(e)| = m$ for some $m$ and all $g \in G$. It means $h(X, T) = \log m$. By Lemma 4.2.8 we know $p(e)$ and $f(e)$ are normal subgroups of $G$. Let $K = p(e) \cap f(e)$. It is a normal subgroup of $G$. Notice what $K$ tells us. The subgroup $K$ is a subgroup of both $f(e)$ and $p(e)$. So for each $g \in K$, $K$ is a subgroup of both $f(g)$ and $p(g)$. This means $K^{\mathbf{Z}}$ is a normal subgroup of $X$. Furthermore, for any $g$, $h$ with $h \in f(g)$, the number of elements $k \in g \cdot f(e)$ with $f(k) = f(g) = h \cdot f(e)$ is $|K|$. There are two possibilities. Either $K \neq \{e\}$ or $K = \{e\}$.

First is the case in which $K \neq \{e\}$. We can write $G \simeq G/K \times K$, where the multiplication in $G/K \times K$ is obtained by viewing $G$ as an extension of $K$ by $G/K$. We are not interested in the algebra here - only the fact that we have a set isomorphism between $G$ and $G/K \times K$. This is the reason we lose the isomorphism and end up with only a conjugacy. Now we have $(X, \sigma)$ is conjugate to $(Y, \sigma) \times (K^{\mathbf{Z}}, \sigma)$, where $Y$ is a Markov subgroup of $(G/K)^{\mathbf{Z}}$. The conjugacy is obtained by the correspondence on the symbol level between $G$ and $G/K \times K$ and by observing that $(G^{\mathbf{Z}}, \sigma)$ is conjugate to $((G/K \times K)^{\mathbf{Z}}, \sigma)$ which is conjugate to $((G/K)^{\mathbf{Z}}, \sigma) \times (K^{\mathbf{Z}}, \sigma)$. The Markov subgroup $Y$ is contained in $(G/K)^{\mathbf{Z}}$. If $G/K = \{e\}$, we are in case (ii). If $G/K \neq \{e\}$ but $f(e) = \{e\}$ in $Y$ we are in case (iii). If $f(e) \neq \{e\}$ in $Y$ we will have in $Y$ that $K = \{e\}$.

Next consider the case in which $K = \{e\}$. We define a Markov subgroup $Y$ of $(G/f(e))^{\mathbf{Z}}$ that is isomorphic to $(X, \sigma)$. To do this define a one-block map from $X$ into $(G/f(e))^{\mathbf{Z}}$ by $g$ goes to $g \cdot f(e)$. The image of $X$ is a Markov subgroup with transition rule which stipulates that $h \cdot f(e)$ can follow $g \cdot f(e)$ if and only if there is a $g\prime$ in $g \cdot f(e)$ with $f(g\prime) = h \cdot f(e)$. Denote the image of $X$ by $Y$. The map from $X$ to $Y$ is invertible by a two-block map because $K = \{e\}$. If $h \cdot f(e)$ can be followed by $g \cdot f(e)$, the $g' \in g \cdot f(e)$ with $f(g') = h \cdot f(e)$ is unique. The map is an isomorphism.

To wind up the proof, we work by induction. Start with $(X, \sigma)$ such that $f(e) \neq \{e\}$ and use the first of the two constructions above to get a conjugate system $(Y, \sigma) \times (K_1^{\mathbf{Z}}, \sigma)$. Either we are done or in $Y$ the subgroup $K = \{e\}$. Then we apply the second of the two constructions above to get $(X_1, \sigma)$. We continue the process alternating the two constructions. Finally we reach the case where $X_k$ is a Markov subgroup of $G_k^{\mathbf{Z}}$ and $f(e) = e$. The result is that $(X, \sigma)$ is conjugate to $(X_k, \sigma) \times (K_k^{\mathbf{Z}}, \sigma) \times \cdots \times (K_1^{\mathbf{Z}}, \sigma)$. This in turn is conjugate to $(G_k, \tau) \times (K^{\mathbb{Z}}, \sigma)$ with $K = K_k \times \cdots \times K_1$. ∎

In the next section we will see why the next corollary is of interest.

**Corollary 4.2.12** *If $X$ is a compact, totally disconnected topological group and $T$ is an expansive automorphism then $(X, T)$ has dense periodic points.*

# 4.3   Decidability in One and Two-Dimensions

In this section we begin an investigation of two-dimensional Markov shifts. First we examine some decidability questions about general two-dimensional Markov shifts. We see that these problems do not arise in one-dimension. We examine some of the relations between the decidability questions and questions about the existence of periodic points. Then we begin our investigation of two-dimensional Markov groups and see that the decidability questions can be answered in that setting.

Let $F$ be a finite set with the discrete topology. Then $F^{\mathbb{Z}}$ is the set of all doubly infinite sequences with entries from $F$ and $F^{\mathbb{Z}^2}$ is the set of all infinite arrays with entries from $F$. With the product topologies both are compact, totally disconnected metric spaces. The shift transformation $\sigma : F^{\mathbb{Z}} \to F^{\mathbb{Z}}$ defined by $(\sigma(x))_i = x_{i+1}$ is a homeomorphism of $F^{\mathbb{Z}}$ to itself. The horizontal shift $\sigma_H : F^{\mathbb{Z}^2} \to F^{\mathbb{Z}^2}$ defined by $(\sigma_H(x))_{(i,j)} = x_{(i+1,j)}$ and the vertical shift $\sigma_V : F^{\mathbb{Z}^2} \to F^{\mathbb{Z}^2}$ defined by $(\sigma_V(x))_{(i,j)} = x_{(i,j+1)}$ are commuting homeomorphisms of $F^{\mathbb{Z}^2}$. Let $L \subseteq F^k$ for some $k$. The *one-dimensional Markov shift* or *Markov shift* $X_L \subseteq F^{\mathbb{Z}}$ defined by the list $L$ is $\{x \in F^{\mathbb{Z}} : [x_i, \ldots, x_{i+k-1}] \in L,$ for all $i \in \mathbb{Z}\}$. It is a closed, shift invariant subset of $F^{\mathbb{Z}}$ and $\sigma$ restricted to it is a homeomorphism. If $L \subseteq F^{k \times k}$ for some $k$ then the *two-dimensional Markov shift* $X_L \subseteq F^{\mathbb{Z}^2}$ defined by the list $L$ is

$$\left\{ x \in F^{\mathbb{Z}^2} : \begin{bmatrix} x_{(i,j+k-1)} & \cdots & x_{(i+k-1,j+k-1)} \\ x_{(i,j)} & \cdots & x_{(i+k-1,j)} \end{bmatrix} \in L, \text{ for } (i,j) \in \mathbb{Z}^2 \right\}.$$

It is a closed shift invariant subset of $F^{\mathbb{Z}^2}$ and $\sigma_H$ and $\sigma_V$ restricted to it are commuting homeomorphisms.

Given a list $L \subseteq F^k$ or $L \subseteq F^{k \times k}$ there are many natural questions. Some of the most basic are the following.

(2.1)  Is $X_L$ empty or nonempty?

(2.2)  Does $X_L$ contain a periodic point? (A point is periodic if it has a finite orbit.)

(2.3)  In the $\mathbb{Z}^2$ case does $X_L$ contain a directionally periodic point? (A point $x$ is periodic in the $(r, s)$ direction if $(\sigma_H^r \sigma_V^s)^n(x) = x$ for some $n > 0$.)

(2.4) Given an allowable block, is there a point in $X_L$ containing it as a subblock? (A block is allowable if every subblock of the right size is in $L$.)

(2.5) Are the periodic points dense in $X_L$ ?

(2.6) In the $\mathbf{Z}^2$ case, is there a direction where the directional periodic points are dense?

Question 2.1 is often called the *tiling problem*. We will see it is closely related to questions 2.2 and 2.3. Question 2.4 is often called the *extension problem*. If it can be answered then 2.1 can be answered. We will see it is closely related to questions 2.5 and 2.6.

Suppose $L \subseteq F^k$ and we want to answer the above questions about $X_L$. Form the *directed graph* $G_L$ corresponding to $L$. It has the elements of $L$ as vertices and an edge from $[i_1, \ldots, i_k]$ to $[j_1, \ldots, j_k]$ when $j_t = i_{t+1}$ for $1 \leq t \leq k - 1$. Points in $X_L$ correspond to the doubly infinite paths in $G_L$. It is easy to see $X_L$ is nonempty if and only if $G_L$ contains a cycle. That's the same as $X_L$ containing a periodic point. So the tiling problem is easily solved in one dimension. It is equally easy to solve the extension problem in one-dimension by inspection of $G_L$. It is also easily seen if the periodic points are dense in $X_L$ by examining $G_L$ (Exercises).

**Example 4.3.1** Let $F = \{1, 2, 3\}$ and $L = \{[1, 1], [1, 2], [2, 1], [2, 3]\}$. Draw $G_L$. Answer questions 2.1, 2,2, 2,4 and 2,5 for $X_L$.

This leads to the next statement.

**Observation 4.3.2** For one-dimensional Markov shifts:

(i) $X_L$ is nonempty if and only if contains a periodic point;

(ii) there are algorithms to decide the tiling and extension problems for one-dimensional Markov shifts;

(iii) there is an algorithm to determine whether or not the the periodic points are dense in $X_L$.

Now we turn to two-dimensional Markov shifts and examine these questions in this setting. The questions were formulated Hao Wang in 1961 [20]. The tiling problem was formulated in terms of unit squares with colored edges. In [20] Wang also proved Proposition 4.3.3 and conjectured that any set of tiles which tiles the plane admits a periodic tiling.

**Proposition 4.3.3** *Let* $X_L$ *be a two-dimensional Markov shift defined by* $L \subseteq F^{k \times k}$.

(i) $X_L$ *contains a periodic point if and only if it contains a directional periodic point.*

(ii) *If every nonempty Markov shift contains a periodic point then there is an algorithm to decide the tiling problem.*

**Proof.** (i) Suppose $X_L$ contains a point $x$ with $\sigma_H^p(x) = x$ with $p > k$. Consider the infinite vertical strip of $x$ between $(0, n)$ and $(p + k - 2, n)$ for $n \in \mathbf{Z}$. This is illustrated in Figure 4.4.

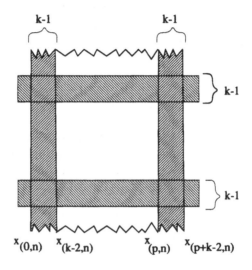

Figure 4.4:

Observe $[x_{(0,n)}, \dots, x_{(k-2,n)}] = [x_{(p,n)}, \dots, x_{(p+k-2,n)}]$ for all $n$. Consider the horizontal strips of width $k - 1$ between $x_{(0,n)}$ and $x_{(p+k-2,n)}$. An infinite number of the strips occur in $x$. There are only a finite number distinct strips of this size which occur in $X_L$. Two of the strips occurring in $x$ must agree. If we consider the rectangle cut out by two horizontal $k-1$ strips which agree and the vertical strip of $x$ we have a rectangle whose top and bottom $k - 1$ strips agree and whose left and right $k - 1$ strips agree. We can tile the plan with this rectangle overlapping by $k - 1$ and obtain a periodic point in $X_L$. This argument works regardless of the direction where the point is periodic.

(ii) Suppose every nonempty $X_L$ contains a periodic point. Examine the $k \times k$ squares in $L$ to see if there is one whose top and bottom $k - 1$ strips agree and whose left and right hand $k - 1$ strips agree. If there is then it will tile the plane and produce a periodic point in $X_L$ as in (i). If there isn't such a $k \times k$ square, form all admissible $(k+1) \times (k+1)$ squares and inspect each to see if it is periodic as described. If not move on to the allowable $(k+2) \times (k+2)$ squares. Continue this process. Either we find a periodic square or at some point we can no longer make larger squares. If we find a periodic square $X_L$ is nonempty. We cannot go on finding arbitrarily large squares without finding a periodic square because then compactness would imply that $X_L$ is nonempty but doesn't contain a periodic point. ∎

**Proposition 4.3.4** *Let $X_L$ be a two-dimensional Markov shift defined by $L \subseteq F^{k \times k}$.*

*(i) It is possible to have a direction $(r, s) \in \mathbf{Z}^2$ where the $\sigma_H^r \sigma_V^s$ directional periodic points are dense in $X_L$ but still the periodic points are not dense in $X_L$.*

*(ii) If every nonempty $X_L$ has dense periodic points then there is an algorithm to decide the extension problem.*

**Proof.**

(i) Construct an example by taking a one-dimensional Markov shift without dense periodic points (Exercise). The points in $X_L$ have the points from the one-dimensional Markov shift horizontally and are constant vertically. This mean every point is fixed by $\sigma_V$ but $X_L$ does not have dense periodic points.

(ii) We mimic the proof of Proposition 4.3.3 (ii). Start with an admissible block and begin constructing all squares containing it. At each stage examine for a periodic square. If the periodic points are dense we must either find a periodic square containing the block in question or at some stage we are not be able to build larger squares. ∎

Next we see that for two-dimensional Markov shifts the answers to questions 2.1 to 2.2 are dramatically different than for one-dimensional Markov shifts.

**Proposition 4.3.5**

*(i) There exist nonempty two-dimensional Markov shifts containing no periodic points.*

*(ii) The problem of determining whether or not $X_L$ is nonempty from $L \subseteq F^{k \times k}$ is undecidable.*

Robert Berger proved Proposition 4.3.5 in 1966 [2]. To prove 4.3.5(i) he produced a set of tiles which could tile the plane but did not admit any periodic tilings. The set of tiles he constructed contains more than twenty thousand tiles. By 1969 Raphael Robinson [17] had found a set of six tiles that would tile the plane only aperiodically. His tiles had geometric shapes rather than colored edges. These tiles have the additional property that any tiling contains every admissible finite patch infinitely often. In dynamical terms he had constructed a two-dimensional, minimal,uncountable, Markov shift (minimal means every orbit is dense). In one-dimension, minimal subshifts have almost the opposite dynamics from subshifts of finite type. In the 1970's Roger Penrose came up with his much publicized pair of geometric tile which tile only aperiodically.

Now we return to spaces which are groups. Let the finite set we begin with be a finite group $G$. As in Section 4.2 the sequence space $G^{\mathbf{Z}}$ is a compact, topological group. By the same reasoning $G^{\mathbf{Z}^2}$ is a compact, topological group. As in Section 4.2 the transformation $\sigma$ is an automorphism and now $\sigma_H$ and $\sigma_V$ are commuting automorphisms of $G^{\mathbf{Z}^2}$. If the list we choose $H \subseteq G^k$ or $H \subseteq G^{k \times k}$ is a subgroup then the resulting space $X_H$ will be a *one or two-dimensional Markov subgroup*. We saw in Section 4.2 that one-dimensional Markov subgroups have a very special structure. We will see that two-dimensional Markov subgroups have much more structure than arbitrary two-dimensional Markov shifts. In particular, we will see that Proposition 4.3.5 does not apply to two-dimensional Markov subgroups.

If $X$ is a compact, topological group and $T$ is an automorphism of $X$ we say $(X, T)$ satisfies the *descending chain condition* if every sequence

$$X \supseteq X_1 \supseteq X_2 \supseteq \ldots \supseteq X_n \supseteq \ldots$$

of closed, $T$ invariant subgroups stops. That is, there is an $N$ so that $X_N = X_n$ for all $n \geq N$. Compare this to the condition on follower sets in the proof of Lemma 4.2.9. In the rest of this section we take the results of Section 4.2 and use induction and the descending chain condition to generalize the results to higher dimensions.

**Lemma 4.3.6** *If $G$ is a finite group then $(G^{\mathbf{Z}}, \sigma)$ satisfies the descending chain condition.*

**Proof.** We know and used in the proof of Lemma 4.2.9 that a finite group $G$ with the identity map obviously satisfies the descending chain condition. Suppose

$$G^{\mathbf{Z}} \supseteq X_1 \supseteq X_2 \supseteq \ldots \supseteq X_n \supseteq \ldots$$

is a descending chain of closed shift invariant subgroups. We saw in the proof of Lemma 4.2.9 that each $X_n$ has a final follower set for the words $[e^m]$.

For each $X_n$ there is an $N_n$ so that $f_{X_n}([e^{N_n}]) = f_{X_n}([e^m])$ for all $m \geq N_n$. Denote this $f_{X_n}([e^{N_n}])$ by $f(X_n)$. It is a normal subgroup of $G$. Then

$$G \supseteq f(X_1) \supseteq f(X_2) \supseteq \ldots \supseteq f(X_n) \supseteq \ldots$$

Since $G$ satisfies the descending chain condition there is a final $f(X_N)$. This means $X_N = X_n$ for all $n \geq N$ (Exercise) ∎

Next we generalize Lemmas 4.3.6 and 4.2.9 to two-dimensional Markov subgroups. Let $X$ be a compact, topological group with $S$ and $T$ a pair of commuting automorphisms. We say $(X, S, T)$ satisfies the *descending chain condition* if every sequence

$$X \supseteq X_1 \supseteq X_2 \supseteq \ldots \supseteq X_n \supseteq \ldots$$

of closed, $S$ and $T$ invariant subgroups stops. We basically reprove Lemma 4.3.6 using the descending chain condition on $(X, \sigma)$ rather than the finiteness of the group $G$.

**Proposition 4.3.7** *If $G$ is a finite group then:*

(i) $(G^{\mathbf{Z}^2}, \sigma_H, \sigma_V)$ *satisfies the descending chain condition;*

(ii) *if $X \subseteq G^{\mathbf{Z}^2}$ is a closed, shift invariant subgroup then it is a Markov subgroup.*

**Proof.** Suppose

$$G^{\mathbf{Z}^2} \supseteq X_1 \supseteq X_2 \supseteq \ldots \supseteq X_n \supseteq \ldots$$

is a descending chain of closed, invariant subgroups. For each $n$ let $A_n$ denote the set of all width one, infinite vertical strips which occur in $X_n$. So $A_n \subseteq G^{\mathbf{Z}}$, thought of vertically, is a closed, $\sigma_V$ invariant subgroup. By Lemma 4.2.9 we know it is Markov subgroup. Let $e_V$ denote the vertical strip of all $e$'s in $G^{\mathbf{Z}}$. It is the identity element in $A_n$. For a block $[a_1, \ldots, a_k]$ in $X_n$ with $a_i \in A_n$ let $f_{X_n}([a_1, \ldots, a_k])$ be the set $\{a \in A_n : [a_1, \ldots, a_k, a]$ occurs in $X_n\}$. It is a nonfinite version of the previous follower sets. Let $[e_V^m]$ denote the block of $m$ $e_V$'s. Note that $f_{X_n}([e_V^m])$ is a closed, $\sigma_V$ invariant subset of $A_n$. As in Lemma 4.2.8 we see it is a closed, normal subgroup of $A_n$ and every $f_{X_n}([a_1, \ldots, a_k])$ is a coset. Then

$$G^{\mathbf{Z}} \supseteq A_n \supseteq f_{X_n}([e_V]) \supseteq \ldots \supseteq f_{X_n}([e_V^m]) \supseteq \ldots$$

is a chain of closed, $\sigma_V$ invariant subgroups. By Lemma 4.3.6 there is a final $f([e_V^{N_n}])$ Let $f(X_n) = f([e_V^{N_n}])$ denote this final follower set in $X_n$. Since the $X_n$ are nested we get a new chain

$$G^{\mathbf{Z}} \supseteq f(X_1) \supseteq f(X_2) \supseteq \ldots \supseteq f(X_n) \supseteq \ldots$$

of closed, $\sigma_V$ invariant subgroups. Again we apply Lemma 4.3.6 to get a final subgroup $f(X_N)$. Since every $f([a_1, \ldots, a_N])$ is a coset of $f(X_N)$ we conclude that $X_N = X_n$ for all $n \geq N$ (Exercise).

(ii) This is a simple application of (i). Suppose $X \subseteq G^{\mathbf{Z}^2}$. For each $n$ let $H_n \subseteq G^{n \times n}$ be the subgroup of $n \times n$ blocks that occur in $X$. Then

$$G^{\mathbf{Z}^2} \supseteq X_{H_1} \supseteq X_{H_2} \supseteq \cdots \supseteq X_{H_n} \supseteq \cdots$$

is a descending chain of closed, $(\sigma_H, \sigma_V)$ invariant subgroups with $X = \cap X_n$. By (i) we conclude $X = X_N$ for some $N$.

We state the next result without proof. The proof once again uses induction to go from the one-dimensional result, Corollary 4.2.12, to two dimensions.

**Proposition 4.3.8** *If $G$ is a finite group and $X \subseteq G^{\mathbf{Z}^2}$ is a closed, invariant subgroup the periodic points are dense in $X$.*

**Corollary 4.3.9** *If $G$ is a finite group and $H \subseteq G^{k \times k}$ is a subgroup the tiling and extension problems are decidable for $X_H$.*

**Proof.** Use Proposition 4.3.8 and apply Proposition 4.3.4.

# 4.4  Markov Subgroups of $(\mathbf{Z}/2\mathbf{Z})^{\mathbf{Z}^2}$

Here we will study a special class of two-dimensional Markov subgroups. The alphabet is the group $\mathbf{Z}/2\mathbf{Z}$. The compact group is $(\mathbf{Z}/2\mathbf{Z})^{\mathbf{Z}^2}$ with coordinate by coordinate addition. The transformations $\sigma_H$ and $\sigma_V$ are automorphisms. Let $F \subseteq \mathbf{Z}^2$ be a finite set and define a group homomorphism (group character) $\chi_F : (\mathbf{Z}/2\mathbf{Z})^{\mathbf{Z}^2} \to \mathbf{Z}/2\mathbf{Z}$ by $\chi_F(x) = \sum_{(i,j) \in F} x_{(i,j)}$. Let $ker\chi_F$ denote the kernel of $\chi_F$ and $X_F = \cap_{(n,m) \in \mathbf{Z}^2} \{x \in (\mathbf{Z}/2\mathbf{Z})^{\mathbf{Z}^2} : \sigma_H^n \sigma_V^m(x) \in ker\chi_F\}$. If we think of $F$ as a shape then the Markov subgroup $X_F$ consists of the points where the sum over the entries of $F$ is zero (modulo 2) no matter where you place $F$ on $x$.

**Example 4.4.1** $F$ is:

If $X$ is a compact topological group there is a unique measure which is invariant under translation. See for example [5]. By that we mean that if $U \subseteq X$ and $x \in X$ then the measure of $U$ is the same as the measure of $xU$, the translate of $U$. This measure is *Haar* measure and it is preserved by any automorphism. When $X_F$ is a Markov subgroup Harr measure is easily described. We denote the Haar measure by $\mu$. Suppose $E \subseteq \mathbf{Z}^2$ is a finite set and $\pi_E : X \to (\mathbf{Z}/2\mathbf{Z})^E$ is the projection map. Then $\pi_E(X) \subseteq (\mathbf{Z}/2\mathbf{Z})^E$ is a subgroup. Let $k_E = |\pi_E(X)|$, it will be a power of two. If $A \in \pi_E(X)$, then $A_E = \{x \in X : \pi_E(x) = A\}$, and $\mu(A_E) = 1/k_E$. Let $F \subseteq \mathbf{Z}^2$ be another finite set, denote by $0 \in \pi_E(X)$ the element of all zeros, and note that $\pi_F(0_E) \subseteq \pi_F(X)$ is a subgroup. For $A \in \pi_E(X)$, $\pi_F(A_E)$ is a coset of $\pi_F(0_E)$ in $\pi_F(X)$, and $|\pi_{E\cup F}(X)| = k_E|\pi_F(0_E)|$. If $A \in \pi_E(X), B \in \pi_F(X)$, and $A_E \cap B_F \neq \emptyset$, then $\mu(A_E \cap B_F) = \mu(A_E)|\pi_F(0_E)|^{-1} = \left(k_E \, |\pi_F(0_E)|\right)^{-1}$.

Finite sets $E, F \subseteq \mathbf{Z}^2$ are *independent* for $X$ if $\pi_F(0_E) = \pi_F(X)$ or equivalently, if $\pi_E(0_F) = \pi_E(X)$. A finite collection of finite sets $E_1, \cdots, E_k \subseteq \mathbf{Z}^2$ is *independent* if for each partition $\{\{i_1, \cdots, i_m\}, \{j_1, \cdots, j_n\}\}$ of $\{1, \cdots, k\}$, the sets $E_{i_1} \cup \ldots \cup E_{i_m}$ and $E_{j_1} \cup \ldots \cup E_{j_n}$ are independent.

The next observation is a consequence of the previous discussion.

**Observation 4.4.2** For finite $E, F \subseteq \mathbf{Z}^2$, $E$ and $F$ are independent if and only if, for any $A \in \pi_E(x), B \in \pi_F(x), \mu(A_E \cap B_F) = \mu(A_E) \cdot \mu(B_F)$. If $E$ and $F$ are not independent, then either $\mu(A_E \cap B_F) = 0$ or $\mu(A_E \cap B_F) > \mu(A_E) \cdot \mu(B_F)$.

Let $X_F$ be a Markov subgroup defined by a finite set $F \subseteq \mathbf{Z}^2$ and let $(r, s) \in \mathbf{Z}^2$ with $(r, s) \neq (0,0)$, $\gcd(r, s) = 1$. The map $\sigma_H^r \sigma_V^s : X_F \to X_F$ is an automorphism of a compact totally disconnected group. We recall two results on ergodic properties of such automorphisms.

The first result concerns group characters. Let $X$ be a compact, abelian, topological group. A *group character* of $X$ is a group homomorphism from $X$ into the circle $\mathbf{S}^1$. Two group characters can be multiplied. If $\chi$ and $\psi$ are group characters then $\chi\psi$ is the group character defined by $\chi\psi(x) = \chi(x)\psi(x)$ in the circle. The collection of all group characters forms a group under this operation (Exercise). This group is called the *character group* or *dual group* of $X$ and is denoted by $X^\wedge$. If $T$ is an automorphism of $X$ it induces an automorphism $T^\wedge$ of $X^\wedge$ defined by $T^\wedge(\chi)(x) = \chi(T(x))$ (Exercise). For a thorough discussion of group characters see [5].

**Theorem 4.4.3** [3] *$(X, T, \mu)$ is ergodic if and only if $(X^\wedge, T^\wedge)$, the character group and induced automorphism, has no nontrivial periodic orbits.*

**Theorem 4.4.4** [13], [15] *$(X, T, \mu)$ is ergodic if and only if it is Bernoulli.*

Also keep in mind Theorem 4.2.7

Given a shape (*i.e.* a finite set) $F \subseteq \mathbf{Z}^2$ we define the convex hull, $C(F)$, as a subset of $\mathbf{R}^2$ in the usual way. It is a solid polygon. To each face of the polygon we associate a vector in $\mathbf{Z}^2$ by going clockwise around the polygon and making each face a vector that we are traversing from tail to head. For any shape, the sum of these vectors is zero. We assign to each face $f$ its *size* $\rho = \rho(f)$ by taking the vector $v_f = (r, s)$ corresponding to the face $f$ and letting $\rho = gcd\{r, s\} + 1$. Geometrically, the size of a face is the number of lattice points that lie on the face in the convex hull. For $(r, s) \neq 0$ let $\ell_{(r,s)}$ be the line through $(0, 0)$ and $(r, s)$ We say that $F$ has *width* $w$ in the $(r, s)$ *direction* if

$$ w = |\{\ell' \in \ell_{(r,s)} + \mathbf{Z}^2 : \ell' \cap C(F) \neq \emptyset\}|. $$

**Example 4.4.5** For 4.4.1a there are three faces each of size 2 in the $(-1, 0), (0, 1)$, and $(1, -1)$ directions. It has width 2 in the $(0, 1)$ direction and width 3 in the $(1, 1)$ direction.

**Proposition 4.4.6** *If $F$ has width $w$ in direction $(r, s)$, where $(r, s) \neq (0, 0)$ and $gcd(r, s) = 1$, then either:*

i) *$F$ doesn't have a face in the $(r, s)$ or $(-r, -s)$ directions in which case $\sigma_H^r \sigma_V^s$ is expansive on $X_F$ and $(X_F, \sigma_H^r \sigma_V^s)$ is topologically conjugate to a full $2^{w-1}$ shift; or*

(ii) *$F$ has a face in one of the directions $(r, s)$ or $(-r, -s)$, in which case $\sigma_H^r \sigma_V^s$ is not expansive on $X_F$, and $(X_F, \sigma_H^r \sigma_V^s)$ has entropy $\log 2^{w-1}$.*

**Proof.** By applying an element of $GL(2, \mathbf{Z})$ we may assume that $(r, s) = (1, 0)$. For each $k, \ell \geq 0$, let $\pi_{-k}^\ell : X_F \to (\mathbf{Z}/2\mathbf{Z})^{\mathbf{Z} \times \{-k, \dots, \ell\}}$ be the projection map $\left(\pi_{-k}^\ell(x)\right)_{(i,j)} = x_{(i,j)}$ for $(i, j) \in \mathbf{Z} \times \{-k, \dots, \ell\}$. For all $k, \ell$, $\sigma_H$ acts expansively on $\pi_{-k}^\ell(X_F)$, so by Theorem 4.2.7 $\left(\pi_{-k}^\ell(X_F), \sigma_H\right)$ is topologically conjugate to an automorphism of a finite group cross a full shift. This means $(X_F, \sigma_H)$ is topologically conjugate to $\varprojlim (G_k \times \Lambda_k, \tau_k \times \sigma_k)$, an inverse limit of such systems. We also see that $\pi_0^{w-2}(X_F)$ consists of all possible "strips" of 0's and 1's of width $w - 1$, i.e. that $\pi_0^{w-2}(X_F)$ is the full $2^{w-1}$ shift.

For case (i), notice that if we place $F$ in a strip of width $w - 1$ as well as possible, one element, a "corner" of $F$, will stick out of either the bottom or the top. In $\mathbf{Z}^2$ put down an arbitrary strip of width $w - 1$ in the subset $\mathbf{Z} \times \{0, \dots, w - 2\}$. Place $F$ so that a corner sticks out above the strip. Because all but one of the coordinates of $F$ are specified there is a unique solution for the entry above the strip. Sliding $F$ along shows that a specified strip on $\mathbf{Z} \times \{0, \dots, w-2\}$ determines a strip on $\mathbf{Z} \times \{0, \dots, w-1\}$. We may do the same working downwards. We see that each strip on $\mathbf{Z} \times \{0, \dots, w-2\}$

determines a point in $X_F$. This means for all $\ell \geq w-2, k \geq 0$, $\left(\pi^\ell_{-k}(X_F), \sigma_H\right)$ is topologically conjugate to a full $2^{w-1}$ shift. An examination of Example 4.4.1a in the $(1,1)$ direction will make this argument clear: it is a $2^2$ shift.

For case (ii), suppose $F$ has a face on the bottom but not on the top, as in example 4.4.1a. When we place $F$ in a strip of width $w-1$ as well as possible either a corner will stick out of the top or a face will stick out the bottom. Since a corner sticks out the top we see that the projection map $\pi^\ell_0(X_F) \to \pi^{w-2}_0(X_F)$ for $\ell \geq w-2$ is a topological conjugacy. In fact for any $k, \ell$ sufficiently large $\pi^\ell_{-k}(x_F)$ is a full $2^{w-1}$ shift. But, notice that the map $\pi^{w-2}_{-1}(X_F) \to \pi^{w-2}_0(X_F)$ is not a conjugacy, it is a $2^{\rho-1}$ to 1 factor map, where $\rho$ is the size of the bottom face. This follows because when we fix the coordinates $\mathbb{Z} \times \{0, \dots, w-2\}$ and go to extend this to the $\mathbb{Z} \times \{-1, \dots, w-2\}$ coordinates, we have some choice. We are free to choose $\rho - 1$ consecutive entries and then the entire row is determined. In this case we see that $\underleftarrow{\lim}$ $\left(\pi^k_{-k}(X_F), \sigma_H\right)$ is an inverse limit of $2^{w-1}$ shifts where each bonding map is of degree $2^{\rho-1}$. This means $(X_F, \sigma_H)$ is not expansive but has entropy $\log 2^{w-1}$. It is clearly not expansive because this choice we have extending downward means we can find two different points in $x_F$ that agree on arbitrarily large horizontal strips.

The other case of (ii) is where $F$ has horizontal faces on the top and bottom, as in example 4.4.1e. Now when extending up or down we have a choice and $\pi^\ell_{-k}(X_F)$ need not be conjugate to a full shift. By Theorem 4.2.7 it must be a finite group automorphism cross a full shift. The map $\pi^{w-1}_{-1}(X_F) \to \pi^{w-2}_0(X_F)$ is $2^{\rho-1} \cdot 2^{\rho'-1}$ to one, where $\rho$ is the size of the bottom face and $\rho'$ the size of the top face. This means $\pi^\ell_{-k}(X_F)$ is always a finite group automorphism cross a $2^{w-1}$ shift. Because of the inverse limit, $(X_F, \sigma_H)$ will have entropy $\log 2^{w-1}$ and is not expansive. ∎

**Corollary 4.4.7** *If $F$ is not linear, every directional entropy is positive.*

In the previous construction it is useful to examine the image of Haar measure under the projection maps, $\pi^\ell_{-k}$. If $F$ has one or no faces in the $(\pm 1, 0)$ direction then $\pi^\ell_{-k}(X_F)$ is a conjugate to a full shift and the image of Haar measure is the equidistributed Bernoulli measure. In this case, $(X_F, \sigma_H)$ is an inverse limit of Bernoulli shifts and by [16] we know that an inverse limit of Bernoulli's is Bernoulli. If $F$ has two faces in the $(\pm 1, 0)$ direction then $\pi^\ell_k(X_F)$ is conjugate to $(G \times \Lambda, \tau \times \sigma)$. The image of Haar measure on the finite group $G$ is the equidistributed measure, the image of Haar measure on the full shift $\Lambda$ is the equidistributed Bernoulli measure, and the image on $G \times \Lambda$ is the product of these. If the $G$ is trivial for all, $\pi^\ell_k$, then again $(X_F, \sigma_H)$ is Bernoulli. If $G$ is not trivial for some $\pi^\ell_{-k}$ then $(X_F, \sigma_H)$ is not ergodic. Later we will identify exactly the directions where $\sigma^r_H \sigma^s_V$ is not ergodic. Notice there can be at most a finite number of these directions.

We will see that Example 4.4.1b is Bernoulli in every direction but Example 4.4.1e is not ergodic in the $(1, 0)$ or $(0, 1)$ directions.

**Proposition 4.4.8** *If $F \neq \emptyset$, then the two dimensional topological entropy of $X_F$ is equal to zero.*

**Proof.** We want to count the number of elements in $\pi_{\overline{N}}(X_F)$ where $\overline{N}$ is an $N \times N$ square. For large $N$ the projection of the point $x \in X_F$ to the boundary of width $w$ of the square $\overline{N}$ determines $x$ completely in $\overline{N}$.

Therefore, $^\# \pi_{\overline{N}}(X_F)$ is bounded by $2^{4wN}$, and the entropy, $h(X_F)$, is equal to

$$\lim_N \frac{1}{N^2} \log {}^\#\pi_{\overline{N}}(X_F) \leq \lim_N \frac{1}{N^2} \log 2^{4wN} = 0. \quad \blacksquare$$

Next we examine the periodic points of a Markov subgroup. As in Section 4.3 we say a point $x \in X_F$ is *periodic in the $(r, s)$ direction* if $(\sigma_H^r \sigma_V^s)^p(x) = x$ for some $p \neq 0$. We say that $x \in X$ is *periodic* if the orbit of $x$ under $\sigma_H$ and $\sigma_V$ is finite. We prove a special case of Proposition 4.3.8.

**Proposition 4.4.9** *For any finite $F \subseteq \mathbf{Z}^2$ the periodic points of $X_F$ are dense.*

**Proof.** If $F = \emptyset$ this is obvious. If $F \neq \emptyset$ then by Proposition 4.4.6 we may choose $(r, s) \in \mathbf{Z}^2$ so that $\sigma_H^r \sigma_V^s$ is expansive on $X_F$, and $(X_F, \sigma_H^r \sigma_V^s)$ is topologically conjugate to a full $2^{w-1}$ shift. This means that the periodic points for $\sigma_H^r \sigma_V^s$ are dense in $X_F$ and that there are a finite number of points of each period. The automorphism $\sigma_H^r \sigma_V^{s+1}$ permutes the points of period $p$ for $\sigma_H^r \sigma_V^s$. Each will have a finite orbit for $\sigma_H^r \sigma_V^{s+1}$. Every periodic point for $\sigma_H^r \sigma_V^s$ will be periodic for $\sigma_H^r \sigma_V^{s+1}$, which means it will have a finite orbit under $\sigma_H$ and $\sigma_V$. $\blacksquare$

## 4.5 Markov Subgroups Polynomial Rings

Next we study the connection between Markov subgroups and polynomial rings. We start with the group $(\mathbf{Z}/2\mathbf{Z})^{\mathbf{Z}^2}$. A finite subset $F \subseteq \mathbf{Z}^2$ defines a group character $\chi_F$ of $(\mathbf{Z}/2\mathbf{Z})^{\mathbf{Z}^2}$ by $\chi_F(x) = \Sigma x_{(i,j)}$, where the sum is over $(i, j) \in F$. Conversely, every character is defined by a finite subset of $\mathbf{Z}^2$ in this way. The character group of $(\mathbf{Z}/2\mathbf{Z})^{\mathbf{Z}^2}$ can be identified with the finite subsets of $\mathbf{Z}^2$. The finite subsets of $\mathbf{Z}^2$ can in turn be identified with the Laurent polynomial ring $\mathbf{Z}/2\mathbf{Z}[x^{\pm 1}, y^{\pm 1}]$. Each finite subset $F \subseteq \mathbf{Z}^2$ corresponds to the polynomial with support $F$. From this point we will think of the character group of $(\mathbf{Z}/2\mathbf{Z})^{\mathbf{Z}^2}$ as the $\mathbf{Z}/2\mathbf{Z}[x^{\pm 1}, y^{\pm 1}]$ and the group operation is polynomial addition (Exercise). For a polynomial $q \in \mathbf{Z}/2\mathbf{Z}[x^{\pm 1}, y^{\pm 1}]$ and $z \in (\mathbf{Z}/2\mathbf{Z})^{\mathbf{Z}^2}$ we will denote by $q(z)$ the character $q$ applied to $z$. The automorphisms $\sigma_H$ and $\sigma_V$ on $(\mathbf{Z}/2\mathbf{Z})^{\mathbf{Z}^2}$ induce automorphisms of $\mathbf{Z}/2\mathbf{Z}[x^{\pm 1}, y^{\pm 1}]$. The automorphism induced by $\sigma_H$ is multiplication by $x$ and $\sigma_V$ induces multiplication by $y$ (Exercises). In the terminology we have been using, a shape $F \subseteq \mathbf{Z}^2$ corresponds to a polynomial $p_F$, $\langle p_F \rangle$ denotes the ideal generated by $p_F$ and

$$
\begin{aligned}
X_F &= \{z \in (\mathbf{Z}/2\mathbf{Z})^{\mathbf{Z}^2} : p_F(\sigma_H^i \sigma_V^j z) = 0 \quad \text{for all } (i, j) \in \mathbf{Z}^2\} \\
&= \{z \in (\mathbf{Z}/2\mathbf{Z})^{\mathbf{Z}^2} : (x^i y^j p_F)(z) = 0 \quad \text{for all } (i, j) \in \mathbf{Z}^2\}.
\end{aligned}
$$

Let $I$ be an ideal in $\mathbf{Z}/2\mathbf{Z}[x^{\pm 1}, y^{\pm 1}]$. The *annihilator of $I$* is the subset of $(\mathbf{Z}/2\mathbf{Z})^{\mathbf{Z}^2}$ which is $\{z : q(z) = 0$ for all $q \in I\}$. It is denoted by $I^\perp$. It is closed, shift invariant subgroup of $(\mathbf{Z}/2\mathbf{Z})^{\mathbf{Z}^2}$ (Exercise). Conversely, if $X \subseteq (\mathbf{Z}/2\mathbf{Z})^{\mathbf{Z}^2}$ is a closed, shift invariant, subgroup then its annihilator $X^\perp \subseteq \mathbf{Z}/2\mathbf{Z}[x^{\pm 1}, y^{\pm 1}]$ is $\{q \in \mathbf{Z}/2\mathbf{Z}[x^{\pm 1}, y^{\pm 1}] : q(z) = 0$ for all $z \in X\}$. It is an ideal. Not every ideal in the ring $\mathbf{Z}/2\mathbf{Z}[x^{\pm 1}, y^{\pm 1}]$ is principal but every ideal is finitely generated. So every Markov subgroup $X$ is of the form $X = X_{F_1} \cap \ldots \cap X_{F_n}$ for a finite collection of shapes $\{F_1, \ldots, F_n\}$. We have proved the following basic theorem.

**Theorem 4.5.1** *The Markov subgroups of $(\mathbf{Z}/2\mathbf{Z})^{\mathbf{Z}^2}$ are in one-to-one correspondence with the ideals in $\mathbf{Z}/2\mathbf{Z}[x^{\pm 1}, y^{\pm 1}]$. Any Markov subgroup is the intersection of finitely many subgroups defined by shapes.*

From now on we will be dealing with ideals $I \subseteq \mathbf{Z}/2\mathbf{Z}[x^{\pm 1}, y^{\pm 1}]$. There are two operations on ideals we will use. One is addition where $I + J$ is the ideal generated by the sums of elements in $I$ and in $J$. The other operation is intersection $I \cap J$. They are related by the fact that $(I + J)^\perp = I^\perp \cap J^\perp$ and $(I \cap J)^\perp = I^\perp + J^\perp$ (Exercises). For basic properties of ideals we refer to [1]. Our aim now is to make a correspondence between algebraic properties of ideals $I \subseteq \mathbf{Z}/2\mathbf{Z}[x^{\pm 1}, y^{\pm 1}]$ and dynamical properties of the associated Markov subgroups.

Let $I$ be an ideal then $I^\perp$ is its associated Markov subgroup. The character group of $I^\perp$ can be identified with quotient group $\mathbf{Z}/2\mathbf{Z}[x^{\pm1}, y^{\pm1}]/I$ (Exercise).

**Proposition 4.5.2** *A continuous, shift commuting homomorphism $\varphi$ from $I^\perp$ into $(\mathbf{Z}/2\mathbf{Z})^{\mathbf{Z}^2}$ is defined by a polynomial $q$ with $\varphi(z)_{(i,j)} = q\big(\sigma_H^i \sigma_V^j(z)\big)$.*

**Proof.** The map $\varphi : I^\perp \to \mathbf{Z}/2\mathbf{Z}$ defined by $z \mapsto \varphi(z)_{(0,0)}$ is a group character, so is defined by a polynomial $q$. The rest follows because $\varphi$ is shift commuting. ∎

Our notation will be to use $q$ interchangeably as a polynomial, a group character, and a map on Markov subgroups.

**Proposition 4.5.3** *Let $I$ be an ideal and $q \in \mathbf{Z}/2\mathbf{Z}[x^{\pm1}, y^{\pm1}]$. We have an exact sequence*

$$0 \to (I + \langle q \rangle)^\perp \to I^\perp \xrightarrow{q} J^\perp \to 0,$$

*where $J = \{r : rq \in I\}$.*

**Proof.** The sequence is exact if the map

$$(I + \langle q \rangle)^\perp \to I^\perp$$

is one-to-one and the map

$$I^\perp \xrightarrow{q} J^\perp$$

is onto.

$(I + \langle q \rangle)^\perp = I^\perp \cap \langle q \rangle^\perp$ which is clearly the kernel of $q$. Next observe that

$$q(I^\perp)^\perp = \{r : rq(x) = 0 \ \forall x \in I^\perp\} = \{r : rq \in I\}. \blacksquare$$

**Lemma 4.5.4** $0 \to \langle q \rangle^\perp \cap \langle p \rangle^\perp \to \langle p \rangle^\perp \xrightarrow{q} \langle p \rangle^\perp \to 0$ *is exact if and only if $\langle q \rangle^\perp \cap \langle p \rangle^\perp = \langle p, q \rangle^\perp$ is finite.*

**Proof.** By Proposition 4.4.6 we may pick $(r, s) \in \mathbf{Z}^2$ with $gcd(r, s) = 1$ and so that $(\langle p \rangle^\perp, \sigma_H^r \sigma_V^s)$ is topologically conjugate to a full shift. The polynomial $q$ defines a shift commuting map from $(\langle p \rangle^\perp, \sigma_H^r \sigma_V^s)$ to itself. A well known theorem of Hedlund [4] asserts that this is onto if and only if it is finite-to-one. ∎

**Lemma 4.5.5** $\langle p \rangle^\perp \cap \langle q \rangle^\perp = \langle p, q \rangle^\perp$ *is finite if and only if $gcd(p, q) = 1$.*

**Proof.** Consider the exact sequence $0 \to \langle q \rangle^\perp \cap \langle p \rangle^\perp \to \langle p \rangle^\perp \to J^\perp \to 0$. The ideal $J = \{r : rq \in \langle p \rangle\}$ is equal to $\langle p \rangle$ if and only if $gcd(p,q) = 1$. Now apply Lemma 4.5.4. ∎

For an ideal $I \subseteq \mathbf{Z}/2\mathbf{Z}[x^{\pm 1}, y^{\pm 1}]$ we define $gcd(I)$, as expected, to be the greatest common factor of all elements of $I$.

**Lemma 4.5.6** *If $q = gcd(I)$ then $\langle q \rangle / I$ and $I^\perp / \langle q \rangle^\perp$ are finite.*

**Proof.** Since $(I^\perp / \langle q \rangle^\perp)^\wedge \simeq \langle q \rangle / I$, either both are finite or both are infinite. We use induction on the number of generators for $I$. If $I = \langle f_1, f_2 \rangle$, $gcd(I) = q, f_1 = qp_1, f_2 = qp_2$, then $\langle f_1, f_2 \rangle = \langle q \rangle \langle p_1, p_2 \rangle \subseteq \langle q \rangle$ and

$$0 \to \langle q \rangle^\perp \to \langle f_1, f_2 \rangle^\perp \overset{q}{\to} \langle p_1, p_2 \rangle^\perp \to 0.$$

By Lemma 4.5.5, $\langle p_1, p_2 \rangle^\perp \simeq \langle f_1, f_2 \rangle^\perp / \langle q \rangle^\perp$ is finite.

Next suppose the lemma is true when the number of generators is $n$, and $I = \langle f_1, \ldots, f_{n+1} \rangle$. Let $q' = gcd(\langle f_1, \ldots, f_n \rangle)$ and $q = gcd(q', f_{n+1}) = gcd(I)$. By the induction hypothesis $\langle q' \rangle / \langle f_1, \ldots, f_n \rangle$ and $\langle q \rangle / \langle q', f_{n+1} \rangle$ are finite. Note that

$$\langle q', f_{n+1} \rangle / \langle f_1, \ldots, f_{n+1} \rangle \simeq \langle q' \rangle + \langle f_{n+1} \rangle / \langle f_1, \ldots, f_n \rangle + \langle f_{n+1} \rangle$$

has cardinality less than or equal to the cardinality of $\langle q' \rangle / \langle f_1, \ldots, f_n \rangle$. As a set, $\langle q \rangle / I \simeq \langle q \rangle / \langle q', f_{n+1} \rangle \times \langle q', f_{n+1} \rangle / I$ and so is finite. ∎

Every Markov subgroup $X \subseteq (\mathbf{Z}/2\mathbf{Z})^{\mathbf{Z}^2}$ has a unique minimal, open, shift invariant subgroup, which we call the *irreducible component of the identity*. It is equal to the maximal subgroup of $X$ with a dense orbit. If $X$ is the irreducible component of the identity then $X$ is *transitive* or *irreducible*.

**Proposition 4.5.7** *If $q = gcd(I)$ then $\langle q \rangle^\perp$ is the irreducible component of the identity.*

**Proof.** $\langle q \rangle^\perp$ is an invariant subgroup of $I^\perp$. It is irreducible because there is an $(r,s) \in \mathbf{Z}^2$ so that $(\langle q \rangle^\perp, \sigma_H^r \sigma_V^s)$ is conjugate to a full shift (Proposition 4.4.6). The group $I^\perp / \langle q \rangle^\perp$ is a finite homomorphic image of $I^\perp$ with an induced $\mathbf{Z}^2$ action. ∎

**Proposition 4.5.8** *$I^\perp$ is transitive if and only if $I$ is principal.*

**Proof.** This follows immediately from Proposition 4.5.7. ∎

**Proposition 4.5.9** *$I^\perp$ is infinite if and only if $gcd(I) \neq 1$.*

**Proof.** If $q = gcd(I)$ then $I^\perp / \langle q \rangle^\perp$ is finite by Lemma 4.5.6. If $q = 1$, $I^\perp \simeq I^\perp / \langle q \rangle^\perp$. If $q \neq 1$ then $\langle q \rangle^\perp \subseteq I^\perp$ is infinite. ∎

**Remark 4.5.10** With reference to Lemma 3.11 we note that it is not true in general that

$$0 \to \langle q \rangle^\perp \cap I^\perp \to I^\perp \xrightarrow{q} I^\perp \to 0$$

if and only if $\langle q \rangle^\perp \cap I^\perp$ is finite. To see this let $p_1, p_2, q$ be non-trivial polynomials with $gcd(p_1, p_2 q) = 1$. Then

$$0 \to \langle p_1 \rangle^\perp \cap \langle p_1 q, p_2 q \rangle^\perp \to \langle p_1 q, p_2 q \rangle^\perp \xrightarrow{p_1} \langle q \rangle^\perp \to 0$$

and $\langle p_1 \rangle^\perp \cap \langle p_1 q, p_2 q \rangle^\perp \subseteq \langle p_1, p_2 q \rangle^\perp$ is finite by Proposition 4.5.9.

**Proposition 4.5.11** *Let* $p = p_1 \cdots p_\ell$ *with* $p_i$ *irreducible over* $\mathbf{Z}/2\mathbf{Z}$. *Then* $\langle p \rangle^\perp$ *is not ergodic in the* $(r, s)$ *direction if and only if some* $p_i$ *is a polynomial in* $x^r y^s$.

**Proof.** By Theorem 4.4.3, $\langle p \rangle^\perp$ is nonergodic in the $(r, s)$ direction if and only if there is a character with a finite orbit under $\sigma_H^r \sigma_V^s$. This means there exists a $q \in \mathbf{Z}/2\mathbf{Z}[x^{\pm 1}, y^{\pm 1}]$, $q \notin \langle p \rangle$ and $t > 0$ such that $q(1 + (x^r y^s)^t) \in \langle p \rangle$, so that $\sigma_H^r \sigma_V^s$ is nonergodic if and only if $1 + (x^r y^s)^t$ is either in $\langle p \rangle$ or a zero divisor for $\langle p \rangle$, i.e. if and only if one of the $p_i$'s divides $1 + (x^n y^m)^t$. If a polynomial $p_i$ divides $1 + (x^r y^s)^t$ then it is a polynomial in $\mathbf{Z}/2\mathbf{Z}[x^r y^s]$. Conversely, if $p_i$ is in $\mathbf{Z}/2\mathbf{Z}[x^r y^s]$ then it divides $1 + (x^r y^s)^t$ for some $t$. ∎

**Proposition 4.5.12** $I^\perp$ *is ergodic if and only if* $I$ *is principal.*

**Proof.** If $I$ is principal then by Proposition 4.5.11 there are ergodic directions. If $I$ is not principal we set $q = gcd(I)$ and note that $I^\perp / \langle q \rangle^\perp$ is a non-trivial, non-ergodic, finite factor. Hence $I^\perp$ is not ergodic. ∎

**Proposition 4.5.13** $I^\perp$ *is infinite and contains no proper infinite Markov subgroups if and only if* $I = \langle p \rangle$ *is principal for some irreducible polynomial* $p \neq 1$.

**Proof.** If $I$ is not principal let $q = gcd(I)$. If $q = 1$, $I^\perp$ is finite by Lemma 4.5.5. If $q \neq 1$ then $\langle q \rangle^\perp$ is a proper, infinite, Markov subgroup of $I^\perp$.

If $I = \langle p \rangle$ and $p$ factors as $p = p_1 \cdots p_n$ then $\langle p \rangle \subseteq \langle p_i \rangle$ and $\langle p_i \rangle^\perp$ is a proper infinite Markov subgroup of $I^\perp$. Conversely, suppose $I^\perp$ contains a proper infinite Markov subgroup $J^\perp$. If $q' = gcd(J)$, then $\langle q' \rangle^\perp \subseteq J^\perp \subseteq I^\perp$ is infinite. This means $I \subseteq \langle q' \rangle$ and $q'$ divides $gcd(I)$. ∎

**Proposition 4.5.14** $\langle p \rangle^\perp$ *is an ergodic subgroup of* $I^\perp$ *if and only if* $p$ *divides* $gcd(I)$.

**Proof.** Let $q = gcd(I)$. If $p$ divides $q$ then $\langle p \rangle^\perp \subseteq \langle q \rangle^\perp \subseteq I^\perp$. If $\langle p \rangle^\perp$ is an ergodic subgroup of $I^\perp$ then $I \subseteq \langle p \rangle$ and $p$ divides $gcd(I)$. ∎

**Remark 4.5.15** There may be non-ergodic infinite subgroups. For example, let $I = \langle q_1^2 q_2 \rangle$ and $J = \langle q_1^2, q_1 q_2 \rangle$. Then $I$ is a proper subideal of $J$, $J^\perp \subseteq I^\perp$, and $gcd(J)$ divides $gcd(I)$.

**Proposition 4.5.16** $I^\perp \cap J^\perp = \{0\}$ *if and only if $I$ and $J$ are relatively prime.*

**Proof.** $I^\perp \cap J^\perp = (I + J)^\perp = \{0\}$ if and only if $I + J = \mathbf{Z}/2\mathbf{Z}[x^{\pm 1}, y^{\pm 1}]$ which is the definition of relatively prime. ∎

**Proposition 4.5.17** $I^\perp \cap J^\perp$ *is infinite if and only if $gcd(I, J) \neq 1$.*

**Proof.** Let $q = gcd(I, J) = gcd(I + J)$. Then $I^\perp \cap J^\perp = (I + J)^\perp$ is infinite if and only if $q \neq 1$ by Proposition 4.5.9. ∎

The rest of this section is concerned with the mixing properties of ergodic subgroups. The question of higher order mixing was motivativation for F. Ledrappier's original example [11]. The subgroup $\langle p \rangle^\perp$ is *mixing of order $n$* (or *$n$-mixing*) if, for any $n$ measurable sets $U_1, \dots, U_n$ the measure $\mu\big(\sigma_H^{r_1}\sigma_V^{s_1}(U_1) \cap \dots \cap \sigma_H^{r_n}\sigma_V^{s_n}(U_n)\big)$ goes to $\Pi_i\mu(U_i)$ as $d\big((r_i, s_i), (r_j, s_j)\big)$ goes to infinity.

The subgroup $\langle p \rangle^\perp$ is *mixing on a shape* $E = \{(r_1, s_1), \dots, (r_n, s_n)\} \subseteq \mathbf{Z}^2$ if for any $n$ measurable sets $U_1, \dots, U_n$ in $\langle p \rangle^\perp$ the measure $\mu\big((\sigma_H^{r_1}\sigma_V^{s_1})^t(U_1) \cap \dots \cap (\sigma_H^{r_n}\sigma_V^{s_n})^t(U_n)\big)$ goes to $\Pi\mu(U_i)$ as $t$ goes to infinity.

The subgroup $\langle p \rangle^\perp$ is *mixing on* $E = \{(r_1, s_1), \dots, (r_n, s_n)\}$ *at times $a2^{lt}$* if, for any $n$ measurable sets $U_1, \dots, U_n$ the measure $\mu\big((\sigma_H^{r_1}\sigma_V^{s_1})^{a2^{lt}}(U_1) \cap \dots \cap (\sigma_H^{r_n}\sigma_V^{s_n})^{a2^{lt}}(U_n)\big)$ goes to $\Pi\mu(U_i)$ as $t$ goes to infinity. We sometimes say *mixing on a polynomial $q$* instead of on a set, when $S(q) = E$.

Mixing properties with respect to Haar measure are tied to the independence properties of subsets of $\mathbf{Z}^2$ as discussed in before and in Observation 4.4.2. Let $\overline{N} \subseteq \mathbf{Z}^2$ be the $N \times N$ square and $E = \{(r_1, s_1), \dots, (r_n, s_n)\} \subseteq \mathbf{Z}^2$, $A_1, \dots, A_n \in \pi_{\overline{N}}(\langle p \rangle^\perp)$. Consider for each $t \in \mathbf{N}$ the measure $\mu\big((\sigma_H^{r_1}\sigma_V^{s_1})^t A_1 \cap \dots \cap (\sigma_H^{r_n}\sigma_V^{s_n})^t A_n\big)$. Observation 4.4.2 and the properties of Haar measure imply that $\mu\big((\sigma_H^{r_1}\sigma_V^{s_1})^t A_1 \cap \dots \cap (\sigma_H^{r_n}\sigma_V^{s_n})^t A_n\big)$ is either zero or larger than a constant $c$ times $\Pi\mu(A_i)$. The constant $c > 1$ is independent of $t$ when $(\sigma_H^{r_1}\sigma_V^{s_1})^{-t}\overline{N}, \dots, (\sigma_H^{r_n}\sigma_V^{s_n})^{-t}\overline{N}$ are not independent. This means $\langle p \rangle^\perp$ is mixing on $E$ if and only if, for every $N \in \mathbf{N}$ there exists a $t_0$ so that $(\sigma_H^{r_1}\sigma_V^{s_1})^{-t}\overline{N}, \dots, (\sigma_H^{r_k}\sigma_V^{s_k})^{-t}\overline{N}$ are independent for all $t \geq t_0$. Next we examine independence of subsets in terms of the algebra of the ideal $\langle p \rangle$. The shape $S(q) \subseteq \mathbf{Z}^2$ of a polynomial $q \in \mathbf{Z}/2\mathbf{Z}[x^{\pm 1}, y^{\pm 1}]$ was defined to be the elements of $\mathbf{Z}^2$ where $q$ has nonzero coefficients.

**Observation 4.5.18** A finite collection of disjoint finite sets $E_1, \dots, E_n \subseteq \mathbf{Z}^2$ is independent for $\langle p \rangle^\perp$ if and only if there do not exist polynomials $q_1, \dots, q_n \in \mathbf{Z}/2\mathbf{Z}[x^{\pm 1}, y^{\pm 1}]$ with $S(q_i) \subseteq E_i$ for all $i = 1, \dots, n$, $q_i \notin \langle p \rangle$ for some $i$, and $q_i + \dots + q_n \in \langle p \rangle$.

This observation, which is clear from the definition of $\langle p \rangle^{\perp}$ and independence, illustrates the relationship between the mixing properties of $\langle p \rangle^{\perp}$ and the algebraic properties of $\langle p \rangle$.

We will examine the problem of mixing on shapes. The proof of Theorem 4.5.23 can be found in [10]. We prove some partial results and give examples which indicate why these conditions arise.

We restrict our attention to ergodic subgroups. By Proposition 4.5.12 these are the ones defined by a principal ideal. Proposition 4.5.11 answers the question of directional mixing. A subgroup $\langle p \rangle^{\perp}$ is two-mixing if and only if no irreducible factor of $p$, over $\mathbf{Z}/2\mathbf{Z}$, is a polynomial in $x^r y^s$ any $(r, s) \in \mathbf{Z}^2$. That theorem is proved in [7]. To discuss the higher order mixing problem we need some facts about finite fields. For more information see [12]. Let $\mathbf{F}_2 = \mathbf{Z}/2\mathbf{Z}$, the finite field with two elements. For each $l \geq 1$, let $\mathbf{F}_{2^l}$ be the field with $2^l$ elements, and let $\overline{\mathbf{F}_2}$ be the algebraic closure of $\mathbf{F}_2$. We are interested in factoring polynomials over these fields. For all $l \geq 1$, the rings $\mathbf{F}_{2^l}[x, y]$ and $\overline{\mathbf{F}_2}[x, y]$ are unique factorization domains and $\mathbf{F}_2[x, y] \subseteq \mathbf{F}_{2^l}[x, y] \subseteq \overline{\mathbf{F}_2}[x, y]$. For a polynomial $p = a_0 x^{r_0} y^{s_0} + \ldots + a_n x^{r_n} y^{s_n} \in \mathbf{F}_2[x, y]$, $a_i \neq 0$, we say the shape of $p$, $S(p) \subseteq \mathbf{Z}^2$, is $\{(r_0, s_0), \ldots, (r_n, s_n)\}$. We may assume our shapes are contained in the positive quadrant of $\mathbf{Z}^2$, i.e. that their corresponding polynomials have no negative exponents.

We begin with the following observation due to F. Ledrappier.

**Observation 4.5.19** If $q = q_1 q_2 \in \langle p \rangle, q_1 \notin \langle p \rangle$, then $\langle p \rangle^{\perp}$ is not mixing on $q_2$ at times $2^t$.

**Proof.** Let $S(q_2) = \{(r_1, s_1), \ldots, (r_k, s_k)\}$, observe that $q_1 q_2^{2^t} = \Sigma q_1 (x^{r_i} y^{s_i})^{2^t} \in \langle p \rangle$, and apply Observation 4.5.18. ∎

**Example 4.5.20** (Ledrappier): Example 4.4.1a is ergodic in every direction and so by Theorem 2.4 in [7] it is mixing of order two. It is not mixing on $1 + x + y$ because it is in $\langle 1 + x + y \rangle^{\perp}$. This means it is not mixing of order three.

The next lemmas generalize this observation. Their proofs are sketched.

**Lemma 4.5.21** Let $p \in \mathbf{F}_2[x, y]$ be an irreducible polynomial. If $f = a_0 + a_1(x^{r_1} y^{s_1}) + \ldots + a_n(x^{r_n} y^{s_n}) \in \mathbf{F}_{2^l}[x, y]$ divides $p$ in $\mathbf{F}_{2^l}[x, y]$ then there are $c_0, \ldots, c_n \in \mathbf{F}_2[x, y]$, not all in $\langle p \rangle$, so that $c_0 + c_1(x^{r_1} y^{s_1})^{2^{lt}} + \ldots + c_n(x^{r_n} y^{s_n})^{2^{lt}} \in \langle p \rangle$ for all $t$. This means $\langle p \rangle^{\perp}$ is not mixing on $S(f)$ at times $2^{lt}$.

**Proof.** Let $f$ be defined as in the statement. Consider the two rings $\mathbb{F}_2[x,y]/\langle p \rangle \subseteq \mathbb{F}_{2^l}[x,y]/\langle f \rangle$. Define the matrix

$$F = \begin{bmatrix} 1 & (x^{r_1}y^{s_1}) & \ldots & (x^{r_n}y^{s_n}) \\ 1 & (x^{r_1}y^{s_1})^{2^l} & \ldots & (x^{r_n}y^{s_n})^{2^l} \\ \vdots & \vdots & & \vdots \\ 1 & (x^{r_1}y^{s_1})^{2^{ln}} & \ldots & (x^{r_n}y^{s_n})^{2^{ln}} \end{bmatrix}$$

$$\bar{a} = \begin{bmatrix} a_0 \\ a_1 \\ \vdots \\ a_n \end{bmatrix}.$$

Observe that $F\bar{a} = 0$ in $(\mathbb{F}_{2^l}[x,y]/\langle f \rangle)^{n+1}$ and hence that $det\ F \in \langle p \rangle$. This produces the desired $c_0, \ldots, c_n \in \mathbb{F}_2[x,y]$. ∎

For a polynomial $g = b_0 + b_1(x^{r_1}y^{s_1}) + \ldots + b_m(x^{r_m}y^{s_m}) \in \mathbb{F}_2[x,y]$ and $a \in \mathbb{N}$ we set $g^{(a)} = b_0 + b_1(x^{r_1}y^{s_1})^a + \ldots + b_m(x^{r_m}y^{s_m})^a$.

**Lemma 4.5.22** *Suppose that $f = a_0 + a_1(x^{r_1}y^{s_1}) + \ldots + a_n(x^{r_n}y^{s_n}) \in \mathbb{F}_{2^l}[x,y]$ and that $f^{(a)}$ divides $p^{(b)}$ in $\overline{\mathbb{F}_2}[x,y]$. Then there are $c_0, \ldots, c_n \in \mathbb{F}_2[x,y]$ and a sequence of $t$ going to infinity so that $c_0 + c_1(x^{r_1}y^{s_1})^t + \ldots + c_n(x^{r_n}y^{s_n})^t \in \langle p \rangle$ for all $t$ in in this sequence. This means that $\langle p \rangle^{\perp}$ is not mixing on $S(f)$.*

**Proof.** Let $f^{(a)}, p^{(b)}$ be as stated and in $\mathbb{F}_{2^l}[x,y]$. Consider the isomorphism

$$\mathbb{F}_2[x,y]/\langle p \rangle \xrightarrow{\varphi_b} \mathbb{F}_2[x^b,y^b]/\langle p^{(b)} \rangle \subseteq \mathbb{F}_2[x,y]/\langle p^{(b)} \rangle \subseteq \mathbb{F}_{2^l}[x,y]/\langle f^{(a)} \rangle,$$

where $\varphi_b(g) = g^{(b)}$ for every $g \in \mathbb{F}_2[x,y]$. From Lemma 4.5.21 we know there are $d_0, \ldots, d_n \in \mathbb{F}_2[x,y]$ so that $d_0 + d_1(x^{r_1}y^{s_1})^{a2^{lt}} + \ldots + d_n(x^{r_n}y^{s_n})^{a2^{lt}} \in \langle p^{(b)} \rangle$ for all $t$. For an infinite sequence of $k \in \mathbb{N}$ we have $a2^{lk} = c\ (mod\ b)$ for some $c$. Define the matrix $F$ and the vector $\bar{d}$ by

$$F = \begin{bmatrix} 1 & (x^{r_1}y^{s_1})^{a2^{lk_0}-c} & \ldots & (x^{r_n}y^{s_n})^{a2^{lk_0}-c} \\ 1 & (x^{r_1}y^{s_1})^{a2^{lk_1}-c} & \ldots & (x^{r_n}y^{s_n})^{a2^{lk_1}-c} \\ \vdots & \vdots & & \vdots \\ 1 & (x^{r_1}y^{s_1})^{a2^{lk_n}-c} & \ldots & (x^{r_n}y^{s_n})^{a2^{lk_n}-c} \end{bmatrix}$$

$$\bar{d} = \begin{bmatrix} d_0 \\ d_1(x^{r_1}y^{s_1})^c \\ \vdots \\ d_n(x^{r_n}y^{s_n})^c \end{bmatrix}.$$

For any $k_0, \ldots, k_n$ in our sequence, $F\overline{d} = 0$ in $\mathbb{F}_2[x, y]/\langle p^{(b)} \rangle$, and $\det F \in \mathbb{F}_2[x^b, y^b]$. This means $\det F$ is a zero divisor in $\mathbb{F}_2[x^b, y^b]/\langle p^{(b)} \rangle$ and implies that we can find $c_0^{(b)}, \ldots, c_n^{(b)} \in \mathbb{F}_2[x^b, y^b]$ with $c_0^{(b)} + c_1^{(b)}(x^{r_1} y^{s_1})^{2^{lk}-c} + \cdots + c_n^{(b)}(x^{r_n} y^{s_n})^{2^{lk}-c} \in \langle p^{(b)} \rangle$. By taking $t = (2^{lk} - c)/b$ for each $k$ in our sequence we obtain a sequence of $t$ going to infinity for which $c_0^{(b)} + c_1^{(b)}(x^{r_1} y^{s_1})^t + \cdots + c_n^{(b)}(x^{r_n} y^{s_n})^t \in \langle p \rangle$. ∎

   These lemmas give an indication of why the following conditions are sufficient. The proof of the necessity of the conditions relies on an unpublished theorem due to D.W. Masser [14]. The full proof can be found in [9].

**Theorem 4.5.23** *The Markov subgroup $\langle p \rangle^{\perp}$ fails to be mixing on a shape $S \subseteq \mathbb{Z}^2$ if and only if there exist $a$, $b \in \mathbb{N}$ and $f \in \mathbb{F}_{2^l}[x, y]$ with $S(f) \subseteq S$ so that $f^{(a)}$ and $p^{(b)}$ have a common factor. Moreover, $a, b, l$ can be explicitly bounded as a function of $p$.*

**Example 4.5.24** Let $p = 1 + x + y + x^2 + y^2 + xy$. This is irreducible in $\mathbb{F}_2[x, y]$ but factors in $\mathbb{F}_{2^2}[x, y]$ into $(1 + \alpha x + \alpha^2 y)(1 + \alpha^2 x + \alpha y)$, where $\mathbb{F}_{2^2} = \{0, 1, \alpha, \alpha^2\}$ with $\alpha^3 = 1$ and $\alpha + \alpha^2 = 1$. We can solve for $c_0, c_1, c_2$ and obtain that $(x + y) + (1 + y)x^{2^{2t}} + (1 + x)y^{2^{2t}} \in \langle p \rangle$ for all $t$. We see that $\langle p \rangle^{\perp}$ fails to be mixing on $S = \{(0,0), (0,1), (1,0)\}$.

**Example 4.5.25** The polynomial $p = 1 + x + y + x^2 + y^2 + xy + x^3 + x^2 y + xy^2 + y^3$ is absolutely irreducible, but $p^{(3)}$ is divisible by $f = 1 + x + y$. So $\langle p \rangle^{\perp} +$ fails to be mixing mixing on $S = \{(0,0), (0,1), (1,0)\}$.

**Example 4.5.26** Let $p = 1 + x^3 + y^3$. It is absolutely irreducible but for $f = 1 + x + y$, $f^{(3)} = p$. So $\langle p \rangle^{\perp}$ fails to be mixing on $S = \{(0,0), (1,0), (0,1)\}$.

**Example 4.5.27** Let $p$ be as in Example 4.5.26. Let $f = (1 + \alpha x + \alpha^2 y)(1 + \alpha x) = 1 + \alpha^2 + \alpha^2 x^2 + xy$, so $S(f) = \{(0,0), (0,1), (2,0), (1,1)\}$. Then $\langle p \rangle^{\perp}$ fails to be mixing on $S(f)$ even though $f^{(a)}$ never divides $p^{(b)}$ and $\langle p \rangle^{\perp}$ is mixing on every subset of $S(f)$.

   Although we can relate the factors of $p^{(b)}$ in $\overline{\mathbb{F}_2}[x, y]$ to mixing on a shape, we are not able to answer the most basic question about mixing: given $p$, what is $\langle p \rangle^{\perp}$'s maximal order of mixing? We end this section with the following conjecture.

**Conjecture 4.5.28** Let $p \in \mathbb{F}_2[x, y]$ and let $k$ be the least number of nonzero terms that appear in a polynomial in $\langle p^{(b)} \rangle$ for $b \in \mathbb{N}$. Then $\langle p \rangle^{\perp}$ is mixing of order $k - 1$ but not of order $k$.

# 4.6 Conjugacy and Isomorphism in $(\mathbb{Z}/2\mathbb{Z})^{\mathbb{Z}^2}$

We would like to know when two subgroups are topologically conjugate or
measurably isomorphic. We would also like to know when one subgroup is a
factor, either measurable or topological, of another. We cannot answer these
questions in any generality. In this section are two simple observations and
two intriguing conjectures.

First we observe that no nontrivial subgroup can be a measurable factor
of $(\mathbb{Z}/2\mathbb{Z})^{\mathbb{Z}^2}$. If $I \neq \{0\}$ then there is a $p \in I, p \neq 0$ and $I^\perp$ fails to be mixing
on $p$ at times $2^t$ by Observation 4.5.19. But $(\mathbb{Z}/2\mathbb{Z})^{\mathbb{Z}^2}$ is mixing of all orders
so $I^\perp$ cannot be a factor of $(\mathbb{Z}/2\mathbb{Z})^{\mathbb{Z}^2}$.

Similar reasoning leads to the following observation.

**Observation 4.6.1** Suppose that $q = 1 + x + y$, $p \in \mathbb{Z}/2\mathbb{Z}[x^{\pm 1}, y^{\pm 1}]$, $\langle p \rangle^\perp$ is
ergodic in the $x$ and $y$ directions, and $\varphi : \langle p \rangle^\perp \to \langle q \rangle^\perp$ a continuous factor
map. Then $\varphi$ is a group homomorphism.

**Proof.** Let $\varphi$ be an $(2\ell + 1)^2$ block map and let $L \subseteq \mathbf{Z}^2$ be the $(2\ell + 1)^2$
square centered at the origin. Let $A, B, O \in \pi_L(\langle p \rangle^\perp)$, where $O$ is the square
of all $0$'s. Let $A_{(0,0)} = \{x \in \langle p \rangle^\perp : \pi_L(x) = A\}$ and define $B_{(0,0)}$ and $\mathcal{O}_{(0,0)}$
similarly. Since $\langle p \rangle^\perp$ is ergodic in the $x$ and $y$ directions, the sets $L$ and
$L + (2^t, 0)$, as well as the sets $L$ and $L + (0, 2^t)$, are independent in $\langle p \rangle^\perp$ for
all sufficiently large $t$. Fix such a $t$ and choose $x \in A_{(0,0)} \cap \sigma_H^{-2^t}(O_{(0,0)})$ and
$y \in B_{(0,0)} \cap \sigma_V^{-2^t}(O_{(0,0)})$. Then $\varphi(x)_{(2^t,0)} = \varphi(y)_{(0,2^t)} = 0$ because $\varphi$ must map
the fixed point of all zeros in $\langle p \rangle^\perp$ to the only fixed point in $\langle q \rangle^\perp$.

Note that $\varphi(x)_{(0,0)} = \varphi(A), \varphi(y)_{(0,0)} = \varphi(B)$, where we are using $\varphi$ as a
map on both blocks and points. Since $q^{2^t} = 1 + x^{2^t} + y^{2^t} \in \langle q \rangle$, $\varphi(x)_{(0,0)} +$
$\varphi(x)_{(2^t,0)} + \varphi(x)_{(0,2^t)} = \varphi(A) + \varphi(x)_{(0,2^t)} = 0$ and $\varphi(y)_{(0,0)} + \varphi(y)_{(0,2^t)} +$
$\varphi(y)_{(0,2^t)} = \varphi(B) + \varphi(y)_{(2^t,0)} = 0$. Next consider the point $x + y \in \langle p \rangle^\perp$,
whose image satisfies that $\varphi(x + y)_{(0,0)} + \varphi(x + y)_{(2^t,0)} + \varphi(x + y)_{(0,2^t)} =$
$\varphi(A + B) + \varphi(y)_{(2^t,0)} + \varphi(x)_{(0,2^t)} = 0$. But comparing the three equations we
see that $\varphi(A + B) = \varphi(A) + \varphi(B)$. This means $\varphi$ is a homomorphism. $\blacksquare$

We would like to apply this type of argument to arbitrary $p$ and $q$ but
because we do not understand the mixing properties of these subgroups well
enough, we are unable to do it.

**Conjecture 4.6.2** *If $p, q \in \mathbb{Z}/2\mathbb{Z}[x^{\pm 1}, y^{\pm 1}], p, q \neq 0$ and $\varphi : \langle p \rangle^\perp \to \langle q \rangle^\perp$
is a measurable factor map then either $\varphi$ is a homomorphism or $\tau \circ \varphi$ is a
homomorphism, where $\tau : \langle q \rangle^\perp \to \langle q \rangle^\perp$ is the map that interchanges $0$ and $1$.
This can happen only when $q$ has an even number of non-zero coefficients.*

Because of the results in Section 4.5 the structure of shift commuting
group homomorphisms is easy to understand. In particular we would obtain
the following.

**Corollary 4.6.3** *to Conjecture 4.6.2. If $\langle p \rangle^{\perp}$ and $\langle q \rangle^{\perp}$ are measurably isomorphic then $p = q$.*

# 4.7  General $\mathbf{Z}^d$ Actions

In this section we will describe the background needed to study the dynamics of general expansive actions on compact abelian groups. A complete introduction to this is in [19].

First we need to discuss some more about character theory for general compact abelian groups. This enlarges on the discussion in Sections 4.4 and 4.5. A complete discussion can be found in [5]. Let $G$ be a compact abelian group. A group character is a continuous group homomorphism $\chi : G \to \mathbf{S}^1$. The set of all group characters for $G$ is denoted by $G^{\wedge}$. It is a group where the operation is defined to be pointwise multiplication of maps. The group $G^{\wedge}$ is given the compact-open mapping topology. That is, an open set of maps is defined for each pair of sets $(K, U)$ where $K$ is a compact subset of $G$ and $U$ is an open subset of $\mathbf{S}^1$. The open set of $G^{\wedge}$ defined by $(K, U)$ is the set of maps which take $K$ into $U$. With this operation and topology $G^{\wedge}$ is a countable discrete group (Exercise). In this setting the character group is also referred to as the *dual group* of $G$. A crucial property of the character group of a compact group is that its character group *separates the points of $G$*. This means that if $x \neq y \in G$ then there is a $\chi \in G^{\wedge}$ with $\chi(x) \neq \chi(y)$.

If $G$ is a countable discrete group we can also consider the group characters of $G$ with pointwise multiplication and the compact-open topology. Then the character group $G^{\wedge}$ is a compact group.

**Exercise.** Show the character group of $\mathbf{Z}$ is $\mathbf{S}^1$.

The character group of a countable discrete group also separates points. The Pontriagin Duality Theorem asserts that if $G$ is a compact or countable discrete group then $(G^{\wedge})^{\wedge}$ is canonically isomorphic to $G$.

In our discussion we will always think of the the circle $\mathbf{S}^1$ as an additive group. Addition is well defined modulo one. This allows us to think of the character group as an additive group also. That is consistent with Section 4.5 where we took $(\mathbf{Z}/2\mathbf{Z})^{\mathbf{Z}^2}$ as our compact group and $\mathbf{Z}/2\mathbf{Z}[x^{\pm 1}, y^{\pm 1}]$ to be the character group.

**Example 4.7.1** The $n$-dimensional torus is a compact abelian group and its character group thought of additively is $\mathbf{Z}^n$.

**Example 4.7.2** The space $(\mathbf{S}^1)^{\mathbf{Z}}$ of Example 4.2.5 is a compact abelian group and its character group can be thought of as $\mathbf{Z}[x^{\pm 1}, y^{\pm 1}]$ (Exercise).

**Example 4.7.3** The 2-adic solenoid of Example 4.2.6 is a compact abelian group. What is its character group?

**Example.** The character group of $\mathbf{Z}$ is $\mathbf{S}^1$.

Let $G$ be a compact abelian group with character group $G^\wedge$. If $H$ is a closed subgroup of $G$ then it is also a compact abelian group. The annihilator of $H$ is the set of $\{\chi \in G^\wedge : \chi(x) = 0 \text{ for all } x \in H\}$. It is a subgroup of $G^\wedge$ and is denoted by $H^\perp$. This was discussed for our special case in Section 4.5. The dual group of $H$ is isomorphic to the quotient group $H/H^\perp$ (Exercise). Apply this to Example 4.7.3.

Now we generalize the notion of group action, expansiveness and the descending chain condition to $\mathbf{Z}^d$ actions. Let $X$ be a compact abelian group and $Aut(X)$ its automorphism group. Let $T : \mathbf{Z}^d \to Aut(X)$ be a one-to-one group homomorphism of $\mathbf{Z}^d$ into $Aut(X)$. Then we say $\mathbf{Z}^d$ *acts of $X$*. For $\gamma \in \mathbf{Z}^d$ we denote by $T_\gamma$ the corresponding automorphism of $X$. We say a $\mathbf{Z}^d$ action is *expansive* if there a neighborhood $\mathcal{U}$ of the identity and for any $x \in X$ not equal to the identity there is a $\gamma \in \mathbf{Z}^d$ with $T_\gamma(x) \notin \mathcal{U}$. We say a $\mathbf{Z}^d$ action satisfies the *descending chain condition* if every chain

$$X \supseteq X_1 \supseteq X_2 \supseteq \ldots \supseteq X_n \supseteq \ldots$$

of closed, $\mathbf{Z}^d$ invariant subgroup stops. These are straight forward generalizations of previous definitions.

**Exercise.** Show any expansive $\mathbf{Z}^d$ action satisfies the descending chain condition. Show $(\mathbf{S}^1)^\mathbf{Z}$ of Example 4.2.5 is not expansive but satisfies the descending chain condition.

Now we explain the general framework for studying $\mathbf{Z}^d$ actions of compact abelian groups satisfying the descending chain condition. Let $X$ be a compact group with a $\mathbf{Z}^d$ action which satisfies the descending chain condition. The character group of $X$ is a countable group. Enumerate the elements $\chi_1, \chi_2, \ldots$. The kernel of each $\chi_n$ is a closed, normal subgroup of $X$. Since the character group separates the points of $X$ we have

$$\bigcap_n \ker(\chi_n) = \{0\}$$

where $\ker(\chi_n)$ is the kernel of $\chi_n$ and 0 is the identity element of $X$ (Exercise). Let

$$X_1 = \bigcap_{\gamma \in \mathbf{Z}^d} \ker(\chi_1 \circ T_\gamma).$$

Then $X_1$ is a closed, invariant, normal subgroup of $X$. Define by induction

$$X_n = \left( \bigcap_{\gamma \in \mathbf{Z}^d} \ker(\chi_n \circ T_\gamma) \right) \cap X_{n-1}.$$

Then each $X_n$ is a closed, invariant, normal subgroup and we have the descending chain

$$X \supseteq X_1 \supseteq X_2 \supseteq \ldots \supseteq X_n \supseteq \ldots$$

The descending chain condition nows shows there is a $N$ with $X_N = \{0\}$.

Use the first $N$ characters to define a homomorphism $\bar{\chi} : X \to \mathbb{T}^N$ the N-dimensional torus by $\bar{\chi}(z) = (\chi_1(z), \ldots, \chi_N(z))$. This in turn defines a one-to-one group homomorphism $\varphi$ of $X$ into $(\mathbb{T}^n)^{\mathbf{Z}^d}$ by $\varphi(z)_\gamma = \bar{\chi}(T_\gamma(z))$. It is a group homomorphism by construction and it is one-to-one since $X_N = \{0\}$. The space $(\mathbb{T}^n)^{\mathbf{Z}^d}$ is a $d$-dimensional shift space with alphabet $\mathbb{T}^n$. There are $d$ shifts acting and we denote the $\mathbf{Z}^d$ action by $\sigma_\gamma$ for $\gamma \in \mathbf{Z}^d$. By construction we have $T_\gamma = \sigma_\gamma$ for all $\gamma \in \mathbf{Z}^d$. Let $\bar{X}$ denote the image of $X$ in $(\mathbb{T}^n)^{\mathbf{Z}^d}$. Then the $\mathbf{Z}^d$ actions $(X, T)$ and $\bar{X}, \sigma)$ are isomorphic. We can proceed to use the descending chain condition on $(\mathbb{T}^n)^{\mathbf{Z}^d}$ to see that $(\bar{X}, \sigma)$ is a Markov subgroup and is defined by a finite rule.

Next we relate these Markov subgroups to the algebra of certain modules. This is a generalization of relationship between the Markov subgroups and ideals in Sections 4.5 and 4.6.

We need to review a little more algebra before proceeding. Let $\mathbf{Z}^N[x_1^{\pm 1}, \ldots, x_d^{\pm 1}]$ denote the Laurant polynomials in the indeterminates $x_1^{\pm 1}, \ldots, x_d^{\pm 1}$ with coefficients in $\mathbf{Z}^N$. It is a group with addition of polynomials as usual. It is also a $\mathbf{Z}[x_1^{\pm 1}, \ldots, x_d^{\pm 1}]$ module. That is, we can multiply any element of $\mathbf{Z}^N[x_1^{\pm 1}, \ldots, x_d^{\pm 1}]$ by an element of $\mathbf{Z}[x_1^{\pm 1}, \ldots, x_d^{\pm 1}]$ to get another element of $\mathbf{Z}^N[x_1^{\pm 1}, \ldots, x_d^{\pm 1}]$. The character group of $\mathbb{T}^N$ is isomorphic to $\mathbf{Z}^N$ (Exercise). In turn the character group of $(\mathbb{T}^N)^{\mathbf{Z}^d}$ is isomorphic to $\mathbf{Z}^N[x_1^{\pm 1}, \ldots, x_d^{\pm 1}]$ (Exercise). Moreover, multiplication by $x_1, \ldots, x_d$ is the dual action to the $\mathbf{Z}^d$ action on $(\mathbb{T}^N)^{\mathbf{Z}^d}$ given by $\sigma$.

Now suppose we have a $\mathbf{Z}^d$ Markov subgroup $(X, \sigma)$ contained in $((\mathbb{T}^n)^{\mathbf{Z}^d}, \sigma)$ Just as in Section 4.5 we see that the annihilator of $X$ is a $\mathbf{Z}[x_1^{\pm 1}, \ldots, x_d^{\pm 1}]$ submodule of $\mathbf{Z}^N[x_1^{\pm 1}, \ldots, x_d^{\pm 1}]$. Conversely, if $\mathcal{M}$ is a $\mathbf{Z}[x_1^{\pm 1}, \ldots, x_d^{\pm 1}]$ submodule of $\mathbf{Z}^N[x_1^{\pm 1}, \ldots, x_d^{\pm 1}]$ then its annihilator $\mathcal{M}^\perp$ is a $\sigma$ invariant Markov subgroup of $((\mathbb{T}^n)^{\mathbf{Z}^d}, \sigma)$. We conclude the with a summary of this discussion.

**Theorem 4.7.4** *If $X$ is a compact abelian group with a $\mathbf{Z}^d$ action which satisfies the descending chain condition. Then there is an $N$ and a $\mathbf{Z}[x_1^{\pm 1}, \ldots, x_d^{\pm 1}$ submodule $\mathcal{M}$ of $\mathbf{Z}^N[x_1^{\pm 1}, \ldots, x_d^{\pm 1}]$ so that $X$ is the character group of $\mathbf{Z}^N[x_1^{\pm 1}, \ldots, x_d^{\pm 1}]/\mathcal{M}$ and the induced $\mathbf{Z}^d$ action is the action induced by multiplication by $x_1, \ldots, x_d$.*

The goal as in Section 4.5 is to exploit the interplay between the dynamics of the Markov subgroup and the algebra of the module.

Exercise. Find the modules corresponding to the Markov subgroups in Examples 4.2.1 through 4.2.6.

# Bibliography

[1] M. F. Atiyah and I. G. MacDonald, *Introduction to Commutative Algebra*, Addison Wesley, Reading, Mass. (1969).

[2] R. Berger, *The Undecidability of the Domino Problem*, Mem. AMS **66** (1966).

[3] P. Halmos, *On Automorphisms of Compact Groups*, Bull. Am. Math. Soc. **49**, pp. 619–624 (1943).

[4] G. A. Hedlund, *Endomorphisms and Automorphisms of the Shiftdynamical System*, Math. Sys. Th. **3**, pp. 320–375 (1969).

[5] E. Hewitt, K. Ross, *Abstract Harmonic Analysis*, Academic Press, Inc. and Springer Verlag (1963).

[6] B. Kitchens, *Expansive Dynamics on Zero-Dimensional Groups*, Ergodic Th. and Dynam. Sys. **7**, pp. 249–261 (1987).

[7] B. Kitchens, K. Schmidt, *Automorphisms of Compact Groups*, Ergodic Th. and Dynam. Sys. **9**, pp. 691–735 (1989).

[8] B. Kitchens, K. Schmidt, *Periodic Points, Decidability and Markov Subgroups*, Dynamical Systems, Proceeding of the special Year, J. C. Alexander, Springer-Verlag, **1342**, pp. 440–454 (1988).

[9] B. Kitchens, K. Schmidt, *Markov Subgroups of* $(\mathbb{Z}/2\mathbb{Z})^{\mathbb{Z}^2}$, Contemporary Mathematics, **135**, pp. 265-283 (1992).

[10] B. Kitchens, K. Schmidt, *Mixing sets and Relative Entropies for Higer - Dimensional Markov Shifts*, Ergodic Th. and Dynam. Sys. **13**, pp. 705-735 (1993).

[11] F. Ledrappier, *Un Champ Markovian peut être d'Entropie Nulle and Mélangeant*, C. R. Acad. Sc. Paris, Ser. **A 287**, pp. 561–562 (1978).

[12] R. Lidl, H. Niederreiter, *Finite Fields*, Addison-Wesley (1983).

[13] D. Lind, *The Structure of Skew Products with Ergodic Group Automorphisms*, Israel J. Math. **28**, pp. 205–248 (1977).

[14] W. Masser, letters to D. Berend, 12 September 1985 and 19 September 1985.

[15] G. Miles, R. K. Thomas, *The Breakdown of Automorphisms of Compact Topological Groups*, Studies in Probability and Ergodic Theory, Advances in Mathematics Supplementary Studies **2**, pp. 207–218, Academic Press, New York-London (1978).

[16] D. Ornstein, *Ergodic Theory, Randomness, and Dynamical Systems,* Yale University Press (1974).

[17] R. Robinson, *Undecidability and Nonperiodicity for Tilings of the Plane,* Inventiones Math. **12** (1971).

[18] J. Rotman, *An Introduction to the Theory of Groups,* 3rd ed. Allyn and Bacon (1984).

[19] K. Schmidt, *Dynamical Systems of Algebraic Origin,* Birkhäuser (1995).

[20] H. Wang, *Proving Theorems by Pattern Recognition-II, Bell System,* Bell Systems Tech. J. **40** (1961).

# Chapter 5

# ASYMPTOTIC LAWS FOR SYMBOLIC DYNAMICAL SYSTEMS

*Zaqueu COELHO*
*Dept. of Mathematics, Statistics and Operational Research*
*Nottingham Trent University*
*Burton Street, Nottingham NG1 4BU*
*United Kingdom*

These are notes of a mini-course given in the Summer School on Symbolic Dynamics, Temuco (Chile), January 1997. The aim is to present some results concerning asymptotic limit laws for the occurrence of asymptotically rare events associated with some symbolic dynamical systems. The results will be mainly presented for shifts of finite type, but there will be left questions of whether these results remain valid for more general or possibly intrinsically different subshifts.

## 5.1  Introduction

It is natural in Probability Theory to study the asymptotic behaviour of random events, when these events are subject to conditions which make them vary in probability. Special concern has been given to studying how long one has to wait in order to see the occurrence of these events. The typical example of this approach, is a simple random walk on the states $\{0, 1, \cdots, N\}$ with a reflecting barrier at $N$ and an absorbing barrier at 0. In this case, one is interested to know how the law of the random time for absorption behaves asymptotically when $N$ diverges. Rescaling these times by $N \log(N)$ gives rise to a limit law of an exponential random variable (cf. [2]). Hence, in general, the usual question is to decide which scale is best suited to describe the asymptotic behaviour, and then to show whether there is convergence in law to some well known distribution.

These problems have inspired the general study of asymptotic limit laws for the occurrence times of certain events which have asymptotically zero probability. These events are commonly known as asymptotically rare events. In these notes, we give a summary of recent developments in this subject within the dynamical systems framework. Later we specialise to symbolic dynamical systems, where there is a sketch of the technique involved in obtaining certain limit laws of occurrence times of the rare event corresponding to visiting a subsystem of a symbolic dynamical system. The latter is work in progress of P. Collet and the author [9].

These notes are organised in the following way. In the Preliminaries we show how these problems arise in the study of Markov Chains. Then we introduce a general setup in which all of these problems can be phrased in Ergodic Theory, and we give some motivation coming from the consideration of different types of ergodic dynamical systems. In Section 5.4 we summarise some important properties of shifts of finite type and Hölder equilibrium states, which will be used in Section 5.5 for the study of asymptotic law of visiting times to a subsystem of finite type. In the Final Remarks we leave some questions of whether these results can be extended to different types of symbolic dynamical systems.

The present notes are summaries of the four lectures given by the author in the Summer School on Symbolic Dynamics, Temuco (Chile), January 1997. Due to other heavy commitments, the author was unable to prepare this final version prior to the course, for which he apologises to the organisers of the School for any inconvenience caused.

# 5.2   Preliminaries

We show a series of examples in which the concept of asymptotically rare event appears naturally. The reader should be familiar with the classical result that the Binomial distribution (subject to a combined asymptotic limit) gives rise to the Poisson distribution. We make a brief exposition of this fact in what follows and later on, we relate this initial step with more general asymptotic limits.

## 5.2.1   Bernoulli Trials

Let $X_i \in \{0, 1\}$, for $i = 1, 2, \cdots$, denote a sequence of independent random variables defined on a standard probability space $(\Omega, \mathbb{P})$. We suppose $X_i$

corresponds to the independent repetition of a particular random experiment where the possible observed outcomes are failure in the $i$-th attempt $X_i = 0$, or success in the $i$-th attempt $X_i = 1$. By repetition of the experiment we mean that the distributions of the $X_i$'s are identical, therefore there is a probability $0 < p < 1$ of success and consequently a probability $1 - p$ of failure. For fixed $n > 1$, let $Y_n$ be the number of successes in $n$ successive repetitions of the experiment, i.e.

$$Y_n = X_1 + \cdots + X_n .$$

Using a simple counting method of the number of possibilities we see that $Y_n$ has the Binomial distribution, i.e.

$$\mathbb{P}(Y_n = k) = \binom{n}{k} p^k (1 - p)^{n-k} ,$$

for each $k = 0, 1, \cdots, n$. Now it is clear that if we let $n$ diverge to infinity and maintain $k$ fixed, the probability of having exactly $k$ successes in $n$ attempts of the experiment tends to zero. However, if we allow the probability of successes $p$ to converge to zero with $n$, then it becomes "asymptotically rare" to have successes when $n$ diverges and therefore we may have a chance of getting positive asymptotic probability of having exactly $k$ successes in $n$ attempts. This is in fact the classical result that if we take $p = p_n$ such that $p_n \sim \lambda/n$ for some $\lambda > 0$, then indeed

$$\mathbb{P}(Y_n = k) = \binom{n}{k} p_n^k (1 - p_n)^{n-k}$$

$$\sim \binom{n}{k} \left(\frac{\lambda}{n}\right)^k \left(1 - \frac{\lambda}{n}\right)^{n-k} \longrightarrow \frac{\lambda^k}{k!} e^{-\lambda} ,$$

which is the Poisson distribution of parameter $\lambda$.

We mentioned this classical result to illustrate the concept of asymptotically rare event. However, from the point of view of our study in these notes, this example has a more intriguing structure since in order to get convergence of the distribution of $Y_n$, the random variables $Y_n$ needed to be defined in different probability spaces. To be more precise, we introduce the language of shift spaces which will be needed later and translate this result in the framework of symbolic dynamics.

Let $\mathcal{A} = \{0, 1, \cdots, \ell - 1\}$ be an alphabet of $\ell$ numbers (corresponding maybe to different states of a random discrete dynamical system, or the finite possible outcomes of a random experiment). The forward sequence of outcomes is the space

$$\mathcal{A}^{\mathbb{N}} = \{(x_0, x_1, x_2, \cdots): x_i \in \mathcal{A}, \forall i \in \mathbb{N}\} .$$

There is a natural map (the shift) $\sigma\colon \mathcal{A}^{\mathbb{N}} \to \mathcal{A}^{\mathbb{N}}$ acting on the sequences by

$$\sigma(x_0, x_1, x_2, \cdots) \ = \ (x_1, x_2, \cdots) \,.$$

We put a topology on this space by defining the cylinder sets

$$C[i_0, i_1, \cdots, i_m]_s \ = \ \{(x_0, x_1, x_2, \cdots) \in \mathcal{A}^{\mathbb{N}}\colon \ x_j = i_j \,, \text{ for } j = 0, 1, \cdots, m\}$$

to be a base of open sets. This is the Tychonov product topology on $\mathcal{A}^{\mathbb{N}}$ induced by the discrete topology on $\mathcal{A}$. The space $\mathcal{A}^{\mathbb{N}}$ is a compact metrisable space in this topology and $\sigma$ is a continuous surjective map. A finite Borel measure $\mu$ on $\mathcal{A}^{\mathbb{N}}$ is uniquely determined by its values on the cylinders sets $C[i_0, i_1, \cdots, i_m]_s$.

Now we interpret the classical result mentioned earlier. In this case $\mathcal{A} = \{0, 1\} = \{\text{"failure", "success"}\}$, and $\mu$ is the product measure on $\mathcal{A}^{\mathbb{N}}$ generated by the measure $(1 - p, p)$ on $\{0, 1\}$, which will be denoted by $(1 - p, p)^{\mathbb{N}}$. The measure of the cylinder sets are

$$\mu\big(C[i_0, i_1, \cdots, i_m]_s\big) \ = \ p^{\sum_{j=0}^{m} i_j} (1 - p)^{m+1 - \sum_{j=0}^{m} i_j} \,.$$

Defining $f\colon \mathcal{A}^{\mathbb{N}} \to \{0, 1\}$ by $f(x_0, x_1, x_2, \cdots) = x_0$, we have

$$X_i(x) \ = \ f(\sigma^i x) \,,$$

for $i = 0, 1, \cdots$, is a sequence of i.i.d. random variables defined on the probability space $(\mathcal{A}^{\mathbb{N}}, \mu)$ and the random variables

$$Y_n \ = \ X_0 + X_1 + \cdots + X_{n-1} \ = \ \sum_{i=0}^{n-1} f(\sigma^i x) \ = \ x_0 + x_1 + \cdots + x_{n-1}$$

have the Binomial distribution $B(n, p)$, i.e.

$$\mu\{x \in \mathcal{A}^{\mathbb{N}}\colon \ Y_n(x) = k\} \ = \ \binom{n}{k} p^k (1 - p)^{n-k} \,.$$

Hence the classical result corresponds to considering a sequence of probability measures $\mu_n$ on $\mathcal{A}^{\mathbb{N}}$ as $(1 - p_n, p_n)^{\mathbb{N}}$ and it states that, for all fixed $k = 0, 1, \cdots$,

$$\lim_{n \to \infty} \mu_n\{x \in \mathcal{A}^{\mathbb{N}}\colon \ Y_n(x) = k\} \ = \ \frac{\lambda^k}{k!} e^{-\lambda} \,,$$

whenever $p_n \sim \lambda/n$ for some $\lambda > 0$.

Here both measure and random variable are varying with $n$. We will consider in the sequel asymptotic limit laws, where the probability space $(\mathcal{A}^{\mathbb{N}}, \mu)$ remains fixed and the problem will be to find rescalings of a sequence of random variables of type $Y_n$ in order to guarantee convergence in distribution.

## 5.2.2   Occurrence and Waiting Times

We begin with a simple example. Let us record the times of occurrence of success in the independent repetition of an experiment. So given a random sequence of outcomes

$$x = (x_0, x_1, x_2, \cdots) \in \mathcal{A}^{\mathbb{N}} \quad (\mathcal{A} = \{0, 1\})$$

we are interested in $n$ such that $x_n = 1$. Let

$$\tau(x) \ = \ \tau^{(1)}(x) \ = \ \inf\left\{n \geq 0 \colon x_n = 1\right\}$$

and define, for $j \geq 2$,

$$\tau^{(j)}(x) \ = \ \inf\left\{n > \tau^{(j-1)}(x) \colon x_n = 1\right\}.$$

If $\mu$ denotes the measure $(1-p, p)^{\mathbb{N}}$ on $\mathcal{A}^{\mathbb{N}}$ then

$$\mathbb{P}(\tau = k) \ = \ \mu\{x \in \mathcal{A}^{\mathbb{N}} \colon \tau(x) = k\} \ = \ p\,(1-p)^k\,,$$

for all $k \geq 0$. Also it is an exercise to show that, for fixed $k > 0$,

$$\mathbb{P}\left(\tau^{(j)} - \tau^{(j-1)} = k \,\Big|\, \tau^{(j-1)} = k_{j-1}, \cdots, \tau^{(1)} = k_1\right)$$
$$= \ \mathbb{P}(\tau^{(j)} - \tau^{(j-1)} = k) \ = \ p\,(1-p)^{k-1}\,,$$

for all $k_1 < \cdots < k_{j-1}$. Therefore the *inter-occurrence* times $1 + \tau^{(1)}, \tau^{(2)} - \tau^{(1)}, \tau^{(3)} - \tau^{(2)}, \cdots$ are independent and identically distributed with geometric distribution of parameter $p$. Consequently $\tau^{(j)}$ has the Negative Binomial distribution $BN(j, p)$.

Suppose now we have two random sequences of outcomes corresponding to the repetition of an experiment with $\ell$ possible different results. Let $x = (x_0, x_1, x_2, \cdots)$ and $y = (y_0, y_1, y_2, \cdots)$ denote the sequence of outcomes ($x$ independent of $y$). We say there is a *match* at time $n$ if $x_n = y_n$. We may think of matching at time $n$ as a success and no matching is failure. We then note that

$$\mathbb{P}(x_n = y_n) \ = \ \mathbb{P}(x_0 = y_0) \ = \ p\,,$$

for all $n > 0$. Suppose $0 < p < 1$ and write $q = 1 - p$. Now consider the problem of having a successive sequence of matchings. Define $\Phi \colon \mathcal{A}^{\mathbb{N}} \times \mathcal{A}^{\mathbb{N}} \to \mathbb{N}$ by

$$\Phi(x, y) \ = \ k \quad \text{if} \quad x_0 = y_0, x_1 = y_1, \cdots, x_{k-1} = y_{k-1}, x_k \neq y_k\,,$$

and consider the random variables $Z_n = \Phi(\sigma^n x, \sigma^n y)$.

**Exercise 5.2.1** *Show that $Z = \{Z_n : n \geq 0\}$ is a Markov chain on the countable states $\{0, 1, \cdots\}$ with transition probabilities*

$$\mathbb{P}(Z_n = k - 1 \mid Z_{n-1} = k) = 1 \quad \text{if} \quad k \geq 1,$$
$$\mathbb{P}(Z_n = k \mid Z_n = 0) = p^k q \quad \text{if} \quad k \geq 0.$$

The Markov chain $Z$ has a graph representation given by Figure 5.1. This Markov chain is positive recurrent with stationary distribution given by $\pi_j = p^j q$.

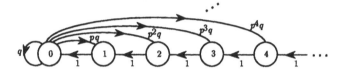

Figure 5.1: The Markov chain $Z$

Consider now the waiting time $W_k$ to have $k$ matchings in a row. Then $W_k = (k - 1) + \tau_k$, where

$$\tau_k = \inf\{n \geq 0 : Z_n \geq k\}.$$

The conditional distribution of $\tau_k$ given that we start with no matching is

$$\mathbb{P}(\tau_k = m \mid Z_0 = 0) = \mathbb{P}(Z_m \geq k, Z_{m-1} < k, \cdots, Z_1 < k \mid Z_0 = 0)$$
$$= \sum_{j \geq k} \mathbb{P}(Z_m = j, Z_{m-1} < k, \cdots, Z_1 < k \mid Z_0 = 0)$$
$$= \sum_{j \geq k} \mathbb{P}(Z_m = j, Z_{m-1} = 0, Z_{m-2} < k, \cdots, Z_1 < k \mid Z_0 = 0).$$

For $m = 1$ we have

$$\mathbb{P}(\tau_k = 1 \mid Z_0 = 0) = \sum_{j \geq k} \mathbb{P}(Z_1 = j \mid Z_0 = 0) = \sum_{j \geq k} p^j q = p^k,$$

for $m = 2$,

$$\mathbb{P}(\tau_k = 2 \mid Z_0 = 0) = p^k q,$$

and for $m > 2$,

$$\mathbb{P}(\tau_k = m \mid Z_0 = 0)$$
$$= \sum_{j \geq k} (p^j q) \, \mathbb{P}(Z_{m-1} = 0, Z_{m-2} < k, \cdots, Z_1 < k \mid Z_0 = 0)$$
$$= p^k \, \mathbb{P}(Z_{m-1} = 0, Z_{m-2} < k, \cdots, Z_1 < k \mid Z_0 = 0).$$

Hence if we define the constants $\nu_0(k, 1) = 1$, $\nu_0(k, 2) = q$ and for $m > 2$,

$$\nu_0(k, m) = \mathbb{P}(Z_{m-1} = 0, Z_{m-2} < k, \cdots, Z_1 < k \mid Z_0 = 0),$$

we obtain in short

$$\mathbb{P}(\tau_k = m \mid Z_0 = 0) = p^k \nu_0(k, m).$$

The expected time to have $k$ successes in a row is then

$$
\begin{aligned}
\mathbb{E}(W_k \mid Z_0 = 0) &= (k - 1) + \mathbb{E}(\tau_k \mid Z_0 = 0) \\
&= (k - 1) + \sum_{m \geq 1} m \, p^k \nu_0(k, m).
\end{aligned}
$$

Here we note that $\mathbb{E}(\tau_k \mid Z_0 = 0)$ diverges with $k$, since the event corresponding to the random path $Z_m$ visiting one of the states $\{k, k+1, \cdots\}$ is "asymptotically rare" with $k$. Therefore in order to have a chance to get convergence in distribution of the sequence of random variables $\tau_k$ we rescale it by its expectation, so we consider

$$
\begin{aligned}
\mathbb{P}\left(\frac{\tau_k}{\mathbb{E}(\tau_k)} \leq t \,\Big|\, Z_0 = 0\right) &= \sum_{m \leq t\,\mathbb{E}(\tau_k)} \mathbb{P}(\tau_k = m \mid Z_0 = 0) \\
&= \sum_{m \leq t\,\mathbb{E}(\tau_k)} p^k \nu_0(k, m).
\end{aligned}
$$

Now we see that for $m > 2$ we have

$$\nu_0(k, m) = \sum_{i=0}^{m-2} \sum_{\substack{0 = j_0 < j_1 < \cdots < j_i \leq m-2 \\ j_s - j_{s-1} < k}} p^{m-2-i} q^i.$$

The latter corresponds to fixing the number of visits to the state $0$ and then summing over the possible different number of loops. The following result can be proved using renewal theory.

**Theorem 5.2.2** *Asymptotic Exponential Law for $\tau_k$. For every $t > 0$ we have*

$$\lim_{k \to \infty} \mathbb{P}\big(\tau_k > t\,\mathbb{E}(\tau_k) \mid Z_0 = 0\big) = e^{-t}.$$

*In fact,*

$$\lim_{k \to \infty} \mathbb{P}\big(\tau_k > t\,\mathbb{E}(\tau_k)\big) = e^{-t}.$$

*Moreover, the scaling $\mathbb{E}(\tau_k)$ satisfies $\mathbb{E}(\tau_k) \sim \lambda\, p^{-k}$ with $0 < \lambda < 1$ depending only on $p$.*

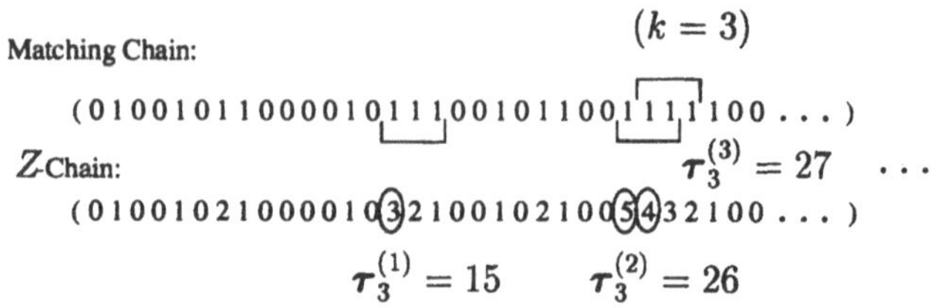

Matching Chain:                                          $(k = 3)$

$(0\,1\,0\,0\,1\,0\,1\,1\,0\,0\,0\,0\,1\,0\,\underbrace{1\,1\,1}\,0\,0\,1\,0\,1\,1\,0\,0\,\overbrace{1\,1\,1}\,1\,0\,0\,\dots\,)$

Z-Chain:                                                         $\tau_3^{(3)} = 27$  $\cdots$

$(0\,1\,0\,0\,1\,0\,2\,1\,0\,0\,0\,0\,1\,0\,③2\,1\,0\,0\,1\,0\,2\,1\,0\,0⑤④3\,2\,1\,0\,0\,\dots\,)$

$\tau_3^{(1)} = 15$          $\tau_3^{(2)} = 26$

Figure 5.2: Example of a matching chain and translation into the $Z$-chain

We will give a sketch proof of a much stronger result in the more general setting of equilibrium states on shifts of finite type, and the above result will be a consequence of that.

To finish this section we mention the problem of subsequent $k$ matchings (for an example see Figure 5.2).

How about the asymptotic distribution of the inter-occurrence times $\tau_k^{(2)} - \tau_k^{(1)}$, $\tau_k^{(3)} - \tau_k^{(2)}$, $\cdots$, in this context? Intuitively, they should not be asymptotically exponential (after rescaling), since when we "hit" $k$ matchings for the first time we may have hit more than $k$ matchings in a row.

The situation would be different if we were interested in the occurrence of say no match, $k$ matchings and no match in succession (i.e. studying the times when the string

$$0\,\underbrace{1\,1\cdots1}_{k\ \text{times}}\,0 \tag{5.1}$$

occurs). In the latter case, for fixed $k$, every hit does not mean hitting more than $k$ successes in a row, and then the inter-occurrence times rescaled by expectation should converge to an exponential law of rate 1. (This can be shown to be true also using renewal theory.) If we denote by $T_k^{(j)}$ the times of occurrence of the string in (5.1) then the following result holds. (Define $T_k^{(0)} \equiv 0$.)

**Theorem 5.2.3** *Asymptotic Exponential Law for Inter-occurrence Times. For every $t > 0$ and $j = 1, 2, \cdots$, we have*

$$\lim_{k \to \infty} \mathbb{P}\left(T_k^{(j)} - T_k^{(j-1)} > t\,\mathbb{E}(T_k^{(j)} - T_k^{(j-1)}) \mid Z_0 = 0\right) = e^{-t}.$$

*In fact,*

$$\lim_{k \to \infty} \mathbb{P}\left(T_k^{(j)} - T_k^{(j-1)} > t\,\mathbb{E}(T_k^{(j)} - T_k^{(j-1)})\right) = e^{-t}.$$

*Moreover, the scaling $\mathbb{E}(T_k^{(j)} - T_k^{(j-1)})$ satisfies $\mathbb{E}(T_k^{(j)} - T_k^{(j-1)}) \sim q^{-2}\,p^{-k}$ for all $j$, and the random variables $D_k^{(j)} = T_k^{(j)} - T_k^{(j-1)}$ are asymptotically independent when $k$ diverges.*

The above result is a particular case of a more general result which we will discuss in later sections of these notes.

In order to relate this problem with the classical result mentioned in Section 5.2.1 we will compute the asymptotic law associated with the number of hits in the event no match, $n$ matchings in a row and no match. Let, for fixed $n > 0$, $\Psi_n : \mathcal{A}^{\mathbf{N}} \times \mathcal{A}^{\mathbf{N}} \to \{0, 1\}^{\mathbf{N}}$ be defined by

$$\Psi_n(x, y) = \omega = (\omega_0, \omega_1, \cdots),$$

where $\omega_s = 1$ iff $x_s \neq y_s$, $x_{s+1} = y_{s+1}, \cdots, x_{s+n} = y_{s+n}$, and $x_{s+n+1} \neq y_{s+n+1}$. Then the functions $X_i : \mathcal{A}^{\mathbf{N}} \times \mathcal{A}^{\mathbf{N}} \to \{0, 1\}$ defined by $X_i(x, y) = \omega_i$ form a sequence of identically distributed random variables on the probability space $(\mathcal{A}^{\mathbf{N}} \times \mathcal{A}^{\mathbf{N}}, \mu \times \mu)$ with distribution $(1 - p_n, p_n)$, where $p_n$ is the probability of the event no match, $n$ matchings in a row and no match, i.e. $p_n = p^n q^2$. (We note that $X_i$ is *not* an independent sequence!) Let

$$Y_s = X_1 + \cdots + X_s.$$

Defining $N_n = [t\, q^{-2} p^{-n}]$, where $[t]$ denotes the integer part of $t$, we obtain as a direct consequence of Theorem 5.2.3.

**Corollary 5.2.4** *Given $t > 0$, for every fixed $k \geq 0$, we have*

$$\lim_{n \to \infty} (\mu \times \mu)\big\{(x, y) \in \mathcal{A}^{\mathbf{N}} \times \mathcal{A}^{\mathbf{N}} : Y_{N_n}(x, y) = k\big\} = \frac{t^k}{k!} e^{-t}.$$

Therefore, an analogue of the classical result also exists in this context, since the above gives an asymptotic Poisson limit law for the number of hits to the event no match, $n$ matchings in a row and no match, in time $N_n \sim t\, q^{-2} p^{-n}$. (This is *not true* if we considered the event $n$ matchings in a row!)

The study of sequence matching is far reaching and has been an active research field in recent years in the case of weakly dependent random sequences. In the independent case, there are also many interesting questions one can ask, for instance, what is the asymptotic distribution of the maximum number of matchings between two samples of the same length when the length diverges. An example of this approach is given in Arratia et al. [3], which contains a refinement of earlier results by Erdös & Rényi [18], and uses the famous Chen-Stein method to prove Poisson approximation results.

In the following sections we will rephrase the initial problem in a general setup. Then we will mention some results from different dynamical systems which will serve as motivation for the consideration of these problems. Next we will specialise the problem to shifts of finite type endowed with a Hölder equilibrium state, where we will spend some time explaining the ingredients behind the technique which allows us to prove these asymptotic distribution results for "well"-behaved symbolic dynamical systems. In the Final Remarks we discuss how the general results of Section 5.5 prove the comments in this Section.

## 5.3 General Setup and Motivation

Let $(\Omega, \mathfrak{B}, \mu)$ be a standard probability space and let $f \colon \Omega \to \Omega$ be a *measure-preserving* map (i.e. $\mu(f^{-1}A) = \mu(A)$, $\forall A \in \mathfrak{B}$). This means that given a point $\omega \in \Omega$ randomly chosen according to $\mu$, the probability to visit a set $A \in \mathfrak{B}$ in $n$-steps is $\mu(A)$. Recall the Poincaré Recurrence

$$\mu\{x \in A \colon \ f^n x \in A, \text{ for some } n > 0\} \ = \ \mu(A) \, ,$$

i.e. $\mu$-almost every point in $A$ returns to $A$. In these notes we will assume that $\mu$ is *ergodic* for $f$ (i.e. $\nexists A \in \mathfrak{B}$, $0 < \mu(A) < 1$ such that $\mu(A \triangle f^{-1}A) = 0$). This is equivalent to every fixed $A \in \mathfrak{B}$ with $\mu(A) > 0$ satisfies

$$\mu\{\omega \in \Omega \colon \ f^n \omega \in A, \text{ for some } n > 0\} \ = \ 1 \, ,$$

i.e. $\mu$-almost every point in the whole space visits $A$. So it makes sense to talk about visiting times to a fixed set $A$ of positive measure for $\mu$-almost every point in $\Omega$. Hence fix $A \in \mathfrak{B}$ with $\mu(A) > 0$, and define $\tau \colon \Omega \to \mathbf{N}$ by

$$\tau(x) \ = \ \tau^{(1)}(x) \ = \ \inf\{n > 0 \colon \ f^n x \in A\} \, .$$

The function $\tau = \tau^{(1)}$ will be called the *first visiting time* of $A$. (Note that $\tau(x) < \infty$, $\mu$-a.e. $x$.) Since by Poincaré Recurrence $\mu$-a.e. $x$ in $A$ returns to $A$ we can also define the *$j$-th visiting time* of $A$, which is a function $\tau^{(j)} \colon \Omega \to \mathbf{N}$ given by

$$\tau^{(j)}(x) \ = \ \inf\{n > \tau^{(j-1)}(x) \colon \ f^n x \in A\} \, ,$$

for $j \geq 2$.

### 5.3.1 Return Maps and Expected Return Times

Let $T_A \colon A \to A$ be the induced map of $f$ on $A$, i.e.

$$T_A(x) \ = \ f^{\tau(x)}(x)$$

is the *first return map* to the set $A$. We note that $T_A$ preserves the restriction of the probability measure $\mu$ on $A$. Therefore, $T_A$ preserves the conditional probability measure $\mu_A$ defined by

$$\mu_A(B) \ = \ \frac{\mu(B \cap A)}{\mu(A)} \, , \quad \forall B \in \mathcal{B} \, .$$

In this context we recall the following result which shows that the expected return time to $A$ *given that we start in* $A$ is exactly $1/\mu(A)$.

**Kac's Theorem.** *Let* $(\Omega, \mathcal{B}, \mu, f)$ *be an ergodic dynamical system and let* $A \in \mathcal{B}$ *be a set of positive measure, then*

$$\mathbb{E}_A(\tau) \;=\; \int_A \tau(x)\, d\mu_A(x) \;=\; \frac{1}{\mu(A)}\,.$$

We will give an idea of the proof in the case $f$ has an inverse $\mu$-a.e.. (In the general case one should use the natural extension of $f$.)

**Proof.** Define, for $i \geq 1$, the sets $A_i = \{x \in A : \tau(x) = i\}$. Then $\mu(A) = \mu(\cup_{i\geq1} A_i)$ and we may consider the Rokhlin Tower depicted in Figure 5.3. The map $f$ takes a point in $A$ and moves one step up in the ladder, until it gets to the top and then it returns back to $A$. In every step, $f$ preserves the measure $\mu$ of any of the sets $A_i$ on the base. Defining the sets $B_0 = A$, and

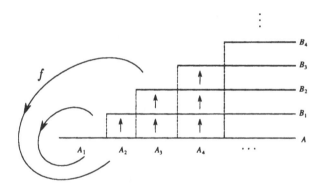

Figure 5.3: The Rokhlin tower of $f$ induced on $A$

for $j > 0$,

$$B_j \;=\; f(B_{j-1}) \setminus \left( \bigcup_{i=0}^{j-1} B_i \right),$$

and using the fact that $\mu$ is ergodic for $f^{-1}$ we have

$$\mu\!\left( \cup_{j\geq0} B_j \right) \;=\; \mu\!\left( \cup_{n\geq1} f^n A \right) \;=\; \mu(\Omega)\,.$$

Note also that the restriction of $T_A$ to each $A_i$ satisfies $T_A|_{A_i} = f^i|_{A_i}$. Finally, we have

$$\int_A \tau(x)\,d\mu(x) \;=\; \sum_{m\geq 1} m\,\mu\{x\in A:\ \tau(x)=m\}$$

$$=\; \sum_{m\geq 1} m\,\mu(A_m) \;=\; \mu\big(\cup_{j\geq 0} B_j\big) \;=\; 1\,.$$

Therefore,

$$\int_A \tau(x)\,d\mu_A(x) \;=\; \frac{1}{\mu(A)}\ \blacksquare$$

## 5.3.2   Asymptotically Rare Events

Suppose we are given a sequence of subsets $A_k \in \mathfrak{B}$ with $\mu(A_k) > 0$, but such that

$$\lim_{k\to\infty} \mu(A_k) \;=\; 0\,.$$

We will say in this case that $\{A_k\}$ is *asymptotically rare*, since the expected time to wait for the occurrence of $A_k$ is in general divergent with $k$. However, *it is not always* the case that

$$\lim_{k\to\infty} \mathbb{E}(\tau_k)\,\mu(A_k) \;=\; 1\,,$$

where $\tau_k$ denotes the first visiting time of $A_k$. We are interested in the asymptotic distribution of these visiting times, i.e. in the non-negative functions

$$\widetilde{F}_k(t) \;=\; \mu\Big\{x\in\Omega:\ \frac{\tau_k(x)}{\mathbb{E}(\tau_k)}\leq t\Big\}\,.$$

Since the measure of the events $A_k$ are usually easier to compute then $\mathbb{E}(\tau_k)$, it is usually taken as a scale $c_k$ for these times. Therefore the general problem is then to find a sequence of positive numbers $c_k \to 0$ such that there exists the limit

$$F(t) \;=\; \lim_{k\to\infty} F_k(t) \;=\; \lim_{k\to\infty} \mu\{x\in\Omega:\ c_k\,\tau_k(x)\leq t\}\,,$$

for all $t > 0$. We note that necessarily we must have

$$\int_0^\infty t\,dF(t) \;=\; \lim_{k\to\infty} c_k\,\mathbb{E}(\tau_k)\,.$$

Therefore if we wish to have finite expectation of the limit distribution, we should choose $c_k$ such that the latter limit exists.

### 5.3.3   Known Results and Motivation

In this section we will discuss special sequences of asymptotically rare events in certain ergodic dynamical systems, where the limit distribution $F(t)$ is known to exist. The idea here is only to motivate the study of this problem for symbolic dynamical systems, which will be done in the later sections. Therefore, we will not attempt to be exhaustive. (For instance, there are many results related to this subject in the framework of continuous time Markov chains and certain Particle Systems, which we will not mention cf. [2], [21]. Also missing is probably the first result on Poisson limit laws associated to dynamical systems, due to Doeblin ([15], 40), who considers certain random variables associated to the continued fraction expansion.)

**Toral Automorphisms**

What we discuss here is essentially contained in Pitskel ([23], 91). Let $\mathcal{T}^n = \mathbb{R}^n/\mathbb{Z}^n = S^1 \times \cdots \times S^1$ denote the $n$-torus with its natural structure as cartesian $n$-product of circles $S^1 = \{z \in \mathbb{C}\colon |z| = 1\}$ viewed as a Lie group. Every continuous group endomorphism $f\colon \mathcal{T}^n \to \mathcal{T}^n$ has the form

$$f(z_1, \cdots, z_n) = (z_1^{\alpha_{11}} \cdots z_n^{\alpha_{1n}}, \cdots, z_1^{\alpha_{n1}} \cdots z_n^{\alpha_{nn}}),$$

where the constants $\alpha_{ij} \in \mathbb{Z}$. Let $A$ denote the matrix with entries $[\alpha_{ij}]$. The map $f$ is an automorphism if and only if $\det(A) = 1$ or $-1$. Additionally, if $A$ has no eigenvalues of modulus one, then $f$ is called a *hyperbolic automorphism*. The standard example is given by the matrix

$$A = \begin{pmatrix} 2 & 1 \\ 1 & 1 \end{pmatrix},$$

in this case $f(z, w) = (z^2 w, z w)$. Every surjective continuous endomorphism of $\mathcal{T}^n$ preserves $\mu$ the product measure induced by the Lebesgue measure on each of the circles $S^1$, i.e. $\mu$ is the normalised Haar measure on $\mathcal{T}^n$.

Let now $f$ be a hyperbolic automorphism. Fixing a base point $x \in \mathcal{T}^n$ and given $\varepsilon > 0$ we consider the first visiting time to the ball $B_\varepsilon(x)$ of radius $\varepsilon$ centred at $x$, i.e. we define $\tau\colon \mathcal{T}^n \to \mathbb{N}$ by

$$\tau_\varepsilon(\omega) = \inf\{m > 0\colon f^m \omega \in B_\varepsilon(x)\}.$$

Define $A_k = B_{\varepsilon_k}(x)$ for some sequence $\varepsilon_k \to 0$. (Also write $\tau_k$ for $\tau_{\varepsilon_k}$.) Pitskel shows the following result.

**Theorem 5.3.1** *There exists $\varepsilon_k \to 0$ such that*

$$F(t) = \lim_{k\to\infty} F_k(t) = \lim_{k\to\infty} \mu\{\omega \in \mathcal{T}^n\colon \mu(A_k)\,\tau_k(\omega) \le t\} = 1 - e^{-t},$$

*for $\mu$-a.e. $x \in \mathcal{T}^n$ and all $t > 0$.*

Using Markov partitions (cf. [4]), this result has a consequence for finite s-
tate Markov Chains. We recall that given a general ergodic dynamical system
$(\Omega, \mathfrak{B}, \mu, f)$, a finite partition of $\Omega$ into the disjoint subsets $B_1, \cdots, B_\ell \in \mathfrak{B}$,
defines a map $\varphi \colon \Omega \to \mathcal{A}^{\mathbf{N}}$ by

$$\varphi(\omega) = (\omega_0, \omega_1, \cdots) ,$$

where $\mathcal{A} = \{1, \cdots, \ell\}$ and $\omega_n = j$ if $f^n \omega \in B_j$. We then have the following
commutative diagram

$$
\begin{array}{ccc}
\Omega & \xrightarrow{\ f\ } & \Omega \\
\varphi \downarrow & & \downarrow \varphi \\
\mathcal{A}^{\mathbf{N}} & \xrightarrow{\ \sigma\ } & \mathcal{A}^{\mathbf{N}}
\end{array}
$$

where $\sigma$ denotes the shift map. Projecting the measure $\mu$ on $\mathcal{A}^{\mathbf{N}}$ by $\mu \mapsto \nu = \mu \circ \varphi^{-1}$ we obtain an ergodic shift-invariant measure on $\mathcal{A}^{\mathbf{N}}$.

**Remark 5.3.2** In the precise construction of Markov partitions there is a
need to consider $\mathcal{A}^{\mathbf{Z}}$, i.e. the space of "double-sided" sequences in $\ell$-symbols.
This is because in general one would like $\varphi$ to be invertible $\nu$-a.e. Here
we will only consider the one-sided case, since we are only interested in the
motivation of the result of Pitskel for Markov Chains. In fact, Pitskel's
approach is to consider the above result for finite state Markov Chains and
transfer the result for automorphisms of $\mathcal{T}^n$. For the special case considered
here, when $\Omega = \mathcal{T}^n$, $f$ being a hyperbolic automorphism, and $\mathcal{A}^{\mathbf{Z}}$ replacing
$\mathcal{A}^{\mathbf{N}}$, there exists a finite partition of $\mathcal{T}^n$ such that $\varphi$ is invertible $\nu$-a.e. and
$\nu$ is a Markov measure on $\mathcal{A}^{\mathbf{Z}}$ (cf. [1]).

Rephrasing the result of Pitskel we have the following consequence for
Markov Chains. Let $\mu$ be a Markov measure on $\mathcal{A}^{\mathbf{N}}$, i.e. there exists a $\ell \times \ell$ stochastic matrix $P = [P(i,j)]$ with stationary probability vector $p = (p_1, \cdots, p_\ell)$ such that

$$\mu\big(C[i_0, \cdots, i_m]_s\big) = p_{i_0} P(i_0, i_1) \cdots P(i_{m-1}, i_m) ,$$

for every cylinder $C[i_0, \cdots, i_m]_s$ in $\mathcal{A}^{\mathbf{N}}$. Suppose the matrix $P$ is irreducible
and aperiodic. Fixing $x = (x_0, x_1, \cdots) \in \mathcal{A}^{\mathbf{N}}$, consider the sequence of cylinders

$$A_k = C[x_0, \cdots, x_k]_0 ,$$

then this can be interpreted as a sequence of asymptotically rare neighbour-
hoods of $x \in \mathcal{A}^{\mathbf{N}}$. So if $\tau_k$ denotes the first time we see the string $[x_0 \cdots x_k]$
then Pitskel's result implies

**Theorem 5.3.3** *For $\mu$-a.e. $x \in \mathcal{A}^{\mathbb{N}}$, there exists a subsequence $k_j \to \infty$ such that*

$$F_{k_j}(t) = \mu\{\omega \in \mathcal{A}^{\mathbb{N}}: \ \mu(A_{k_j}) \tau_{k_j}(\omega) \le t\}$$

*has the limit $F(t) = 1 - e^{-t}$, for every $t > 0$, when $k_j$ goes to infinity.*

In fact the above result holds for the sequence $k$ as we will see later. Note that for $x = (1, 1, 1, \cdots)$ and

$$A_k = C[\underbrace{1, \cdots, 1}_{(k+1) \text{ times}}]_0 \ ,$$

the above result would partially solve one of the matching problems mentioned in Section 5.2.2. However, since the above result holds only for $\mu$-a.e. $x$ we cannot apply it just yet.

**Hyperbolic Diffeomorphisms**

Pitskel's result was generalised by Hirata ([20], 93) for Hyperbolic Dynamical Systems. (The latter has also a consequence for shifts of finite type, which we will talk about later.) Let $M$ be a smooth compact Riemannian manifold and $f: M \to M$ a diffeomorphism. Let $\Omega \in M$ denote the *non-wandering set* of $f$, i.e. the set of points $x \in M$ such that $\mathcal{U} \cap \left( \cup_{n>0} f^n \mathcal{U} \right) \neq \emptyset$ for every neighbourhood $\mathcal{U}$ of $x$. We say that $\Omega$ is (uniformly) *hyperbolic* if there exist $C > 0$, $0 < \rho < 1$ and a continuous splitting of the tangent bundle $T_x M = E_x^s \oplus E_x^u$ into the so-called stable $E^s$ and unstable $E^u$ bundles, respectively, satisfying

(i) $Df_x(E_x^s) = E_{fx}^s$, $Df_x(E_x^u) = E_{fx}^u$ ;

(ii) $\|Df_x^n(v)\| \le C \rho^n \|v\|$, $\|Df_x^{-n}(w)\| \le C \rho^n \|w\|$ ,

for all $n > 0$, and all $v \in E_x^s$, $w \in E_x^u$, at all points $x \in \Omega$. We say that $f$ is an *Axiom A diffeomorphism* if $\Omega$ is hyperbolic and the set of periodic points for $f$ are dense in $\Omega$. The map $f$ is said to be an *Anosov* diffeomorphism if $M$ is hyperbolic. We note here that the conditions defining Axiom A diffeomorphisms do not depend on the Riemannian metric chosen on $M$. However, the constants $C$ and $\rho$ do depend on the metric.

Under the hypothesis of topological mixing of $f$ restricted to $\Omega$ (i.e. for every pair of non-empty open sets $U, V$ in $\Omega$, with respect to the induced topology of $M$ in $\Omega$, there exists $N > 0$ such that $U \cap f^{-n}V \neq \emptyset$, for all $n > N$), and assuming $f$ is Axiom A, there exists a unique probability measure $\mu$ supported on $\Omega$ such that

$$\lim_{n\to\infty} \frac{1}{n} \sum_{i=0}^{n-1} \text{vol}(f^{-i}B) = \mu(B) \ ,$$

for every Borel set $B$ in $M$ with $\mathrm{vol}(\partial B) = 0$. Here $\mathrm{vol}(B)$ denotes the normalised volume measure on $M$ and $\partial B$ is the boundary of $B$. The measure $\mu$ is $f$-invariant and it is called the *Bowen-Sinai-Ruelle* measure of $f$. (See Bowen [4] for the ergodic properties of these maps and for the construction of Markov Partitions.)

With the assumptions of topological mixing and Axiom A on $f$, Hirata proves a Poisson limit law for visiting times which we now describe. With respect to the Riemannian structure on $M$, consider $B_\varepsilon(x)$, the $\varepsilon$-ball centred at $x$, and define the visiting times to this ball

$$\tau_\varepsilon(\omega) \;=\; \tau_\varepsilon^{(1)}(\omega) \;=\; \inf\{n > 0\colon\, f^n\omega \in B_\varepsilon(x)\}\,,$$

and for $j \geq 2$,

$$\tau_\varepsilon^{(j)}(\omega) \;=\; \inf\{n > \tau_\varepsilon^{(j-1)}(\omega)\colon\, f^n\omega \in B_\varepsilon(x)\}\,.$$

Hirata's result is the following.

**Theorem 5.3.4** *For $\mu$-a.e. $x \in \Omega$, as $\varepsilon \to 0$ we have*

$$F_\varepsilon(t) \;=\; \mu\{\omega \in \Omega\colon\, \mu(B_\varepsilon(x))\,\tau_\varepsilon(\omega) \leq t\} \to 1 - e^{-t}\,,$$

*and also for $j \geq 2$,*

$$F_\varepsilon^{(j)}(t) \;=\; \mu\{\omega \in \Omega\colon\, \mu(B_\varepsilon(x))\left(\tau_\varepsilon^{(j)}(\omega) - \tau_\varepsilon^{(j-1)}(\omega)\right) \leq t\} \to 1 - e^{-t}\,.$$

*Moreover, the random variables $\tau_\varepsilon^{(1)}$, $\tau_\varepsilon^{(2)} - \tau_\varepsilon^{(1)}$, $\cdots$, $\tau_\varepsilon^{(j)} - \tau_\varepsilon^{(j-1)}$, $\cdots$ are asymptotically independent.*

As we will see in Section 5.5.1, the above result means that the visiting times to an $\varepsilon$-ball $B_\varepsilon(x)$ centred at a fixed $\mu$-generic point $x$, converges in law, as $\varepsilon \to 0$, to a Poisson point process of rate 1, when the times are rescaled by $\mu(B_\varepsilon(x))^{-1}$.

### Piecewise-Expanding Maps of an Interval

Here we describe the result of Collet and Galves ([13], 93), which is also a Poisson limit law in the context of piecewise-expanding maps. Let $f\colon [0,1) \to [0,1)$ be a measurable map. We say that $f$ is *piecewise-expanding* if there exists a partition $0 = a_0 < a_1 < \cdots < a_{\ell-1} < a_\ell = 1$ such that $f$ is $C^\infty$ restricted to each $(a_{j-1}, a_j)$, $f$ has a $C^\infty$-extension to $[a_{j-1}, a_j]$, and there exists $\rho > 1$ such that $|f'(x)| > \rho$ everywhere $f$ is differentiable.

Assuming $f$ to be eventually piecewise-expanding (i.e. there exists $m > 0$ such that $f^m$ is piecewise-expanding) and topologically mixing, there exists a unique probability measure $\mu$ on $[0,1)$ which is $f$-invariant and absolutely continuous with respect to Lebesgue measure. Let $h(x)$ be the Radon-Nikodym derivative $d\mu(x)/dx$. We know that $h(x) > 0$ for Lebesgue a.e. $x$. Here we will also assume that ess $\inf\{h(x)\} > 0$.

Let $I_k$ be a sequence of intervals satisfying $I_k \cap f^s I_k = \emptyset$, for $s = 1, \cdots, m_k$, where $m_k$ diverges as $k \to \infty$. This condition guarantees that $I_k$ does not contain periodic points with uniform bounded period on $k$. Under these hypotheses Collet and Galves prove

**Theorem 5.3.5** *The visiting times to $I_k$, rescaled by $\mu(I_k)^{-1}$, converge in law to a Poisson point process of rate 1.*

We say $f$ is a *Markov* map if the image of each interval $(a_{j-1}, a_j)$ is a union of these intervals. Such maps have a natural coding into the symbolic dynamical systems using the map $\varphi$ described in Section 5.3.3, and using $B_j = [a_{j-1}, a_j)$, $j = 1, \cdots, \ell$ as the finite partition of $\Omega = [0,1)$. It can be shown that the measure $\mu$ projects onto $\mathcal{A}^{\mathbb{N}}$ as a Markov measure. Also any finite state Markov Chain can be obtained in this manner by considering some Markov map $f$. Therefore the above result has a direct consequence for finite state Markov Chains as follows.

Let $\mu$ be a Markov measure on $\mathcal{A}^{\mathbb{N}}$ associated to an irreducible and aperiodic matrix $P$. Let $\sigma$ be the shift on $\mathcal{A}^{\mathbb{N}}$. Consider a sequence of cylinders

$$A_k = C[x_0^k, \cdots, x_k^k]_0 ,$$

where $x_i^j \in \{1, \cdots, \ell\}$ are chosen such that

$$A_k \cap \sigma^s A_k = \emptyset ,$$

for $s = 1, \cdots, m_k$, where $m_k$ diverges with $k$. (For instance, the sequence

$$A_1 = C[0,1,0]_0 , \quad A_2 = C[0,1,1,0]_0 , \quad \cdots , A_k = C[0,\underbrace{1,\cdots,1}_{k},0]_0 , \quad \cdots ,$$

which appeared in Section 5.2.2 related to the occurrence of the event "no match, $k$ matchings in a row, and no match", satisfies this condition.) Therefore the following result proves Theorem 5.2.3.

**Corollary 5.3.6** *The visiting times to $A_k$, rescaled by $\mu(A_k)^{-1}$, converge in law to a Poisson point process of rate 1.*

### The Approach of Two Trajectories for Piecewise-Expanding Maps

Let $f \colon [0,1) \to [0,1)$ be a piecewise-expanding map of the circle $S^1 \equiv [0,1)/\sim$, i.e. identifying 0 and 1 through the covering map $\theta \mapsto e^{2\pi i\theta}$. Suppose $f$ is topologically mixing and let $d\mu(x) = h(x)dx$ be its absolutely continuous invariant measure. We will also assume ess $\inf\{h(x)\} > 0$.

Given $\varepsilon > 0$ and two points $x, y \in S^1$, consider the times $n > 0$ such that

$$|f^n x - f^n y| \leq \varepsilon \,,$$

i.e. the times for which the $f$-orbits of $x$ and $y$ get $\varepsilon$-close. Coelho and Collet ([8], 94) find a scale in which the distribution of these times (after rescaling) converge in law. First we will rephrase this problem in the general setup. Consider the 2-torus $\Omega = \mathcal{T}^2 = S^1 \times S^1$ and the map $T = f \times f \colon \mathcal{T}^2 \to \mathcal{T}^2$ given by $T(x, y) = (fx, fy)$. The map $T$ preserves the product measure $m = \mu \times \mu$. Writing

$$\Delta_\varepsilon = \{(x, y) \in \mathcal{T}^2 \colon |x - y| \leq \varepsilon\} \,,$$

that is, $\Delta_\varepsilon$ is an $\varepsilon$-neighbourhood of the diagonal in $S^1 \times S^1$, we define the visiting times

$$\tau_\varepsilon(\omega) = \tau_\varepsilon^{(1)}(\omega) = \inf\{n > 0 \colon T^n \omega \in \Delta_\varepsilon\} \,,$$

and for $j \geq 2$,

$$\tau_\varepsilon^{(j)}(\omega) = \inf\{n > \tau_\varepsilon^{(j-1)}(\omega) \colon T^n \omega \in \Delta_\varepsilon\} \,.$$

Coelho and Collet show that the distribution of the rescaled random variable $\tau_\varepsilon$, i.e. the function

$$F_\varepsilon(t) = m\{\omega \in \mathcal{T}^2 \colon m(\Delta_\varepsilon)\,\tau_\varepsilon(\omega) \leq t\} \,,$$

satisfies

$$\lim_{\varepsilon \to 0} F_\varepsilon(t) = 1 - e^{-\lambda t} \,,$$

for some $\lambda > 1$. We note that this provides an example where $m(\Delta_\varepsilon)\,\mathbb{E}(\tau_\varepsilon)$ does not tend to 1 when $\varepsilon$ goes to zero. Also, if we consider the inter-occurrence times $\tau_\varepsilon^{(j)} - \tau_\varepsilon^{(j-1)}$, rescaled by $m(\Delta_\varepsilon)^{-1}$, then we do not have convergence of

$$F_\varepsilon^{(j)}(t) = m\{\omega \in \mathcal{T}^2 \colon \mu(\Delta_\varepsilon)\left(\tau_\varepsilon^{(j)}(\omega) - \tau_\varepsilon^{(j-1)}(\omega)\right) \leq t\}$$

to $1 - e^{-\lambda t}$, as $\varepsilon$ tends to zero. However, we know that there is convergence to some non-negative function, due to the convergence of the corresponding point process (see Section 5.5.1). Also, after rescaling, $\tau_\varepsilon^{(1)}, \tau_\varepsilon^{(2)} - \tau_\varepsilon^{(1)}, \cdots,$ $\tau_\varepsilon^{(j)} - \tau_\varepsilon^{(j-1)}, \cdots$ are not asymptotically independent. (Compare this with Theorem 5.2.3.)

Nevertheless, if we consider "entrance times" instead of visiting times, in other words, we define $N_\varepsilon^{(1)}(\omega) = \tau_\varepsilon^{(1)}(\omega)$, and for $j \geq 2$,

$$N_\varepsilon^{(j)}(\omega) = \inf\{n > N_\varepsilon^{(j-1)}(\omega) \colon T^{n-1}\omega \notin \Delta_\varepsilon, \ T^n \omega \in \Delta_\varepsilon\} \,.$$

then the random variables $N_\varepsilon^{(1)}$, $N_\varepsilon^{(2)} - N_\varepsilon^{(1)}$, $N_\varepsilon^{(3)} - N_\varepsilon^{(2)}$, $\cdots$ are asymptotically independent (after rescaling). Moreover, rescaling these by $m(\Delta_\varepsilon)^{-1}$, more precisely, considering $m(\Delta_\varepsilon) N_\varepsilon^{(1)}$, $m(\Delta_\varepsilon) (N_\varepsilon^{(2)} - N_\varepsilon^{(1)})$, $\cdots$, then these random variables are asymptotically identically distributed with exponential limit law of parameter $\lambda > 1$. Using the terminology of point processes (to be introduced in Section 5.5.1), what Coelho and Collet prove are the following results.

**Theorem 5.3.7** *The rescaled entrance times to an $\varepsilon$-neighbourhood of the diagonal converge in law to a Poisson point process of constant parameter $\lambda > 1$.*

**Theorem 5.3.8** *The rescaled visiting times to an $\varepsilon$-neighbourhood of the diagonal converge in law to a marked Poisson point process of constant parameter measure.*

This problem translated into symbolic dynamics reads: given $\varepsilon > 0$ and two independent sequences $x = (x_0, x_1, x_2, \cdots) \in \mathcal{A}^{\mathbb{N}} = \{1, \cdots, \ell\}^{\mathbb{N}}$ and $y = (y_0, y_1, y_2, \cdots) \in \mathcal{A}^{\mathbb{N}}$, record the times $n > 0$ such that $\sigma^n x$, $\sigma^n y$ are $\varepsilon$-close. Here, to be $\varepsilon$-close means that $x_0 = y_0$, $x_1 = y_1$, $\cdots$, $x_k = y_k$ for some integer $k = k(\varepsilon) > 0$. Therefore, this means to wait until observing $k$ matchings of the sequences $x$ and $y$. When $x$ and $y$ are given randomly according to different Markov measures $m_1$, $m_2$ on $\mathcal{A}^{\mathbb{N}}$, then considering the shift space $(\mathcal{A} \times \mathcal{A})^{\mathbb{N}}$ and letting the pair $(x, y)$ denote

$$(x, y) = \big((x_0, y_0), (x_1, y_1), (x_2, y_2), \cdots \big),$$

we see that the product measure $m_1 \times m_2$ is Markov with respect to the shift transformation on $(\mathcal{A} \times \mathcal{A})^{\mathbb{N}}$. Hence this problem may be studied by renewal theory (as outlined in Section 5.2.2).

### Homeomorphisms of the Circle

Let $f \colon S^1 \to S^1$ be an orientation-preserving homeomorphism with irrational rotation number $\rho \in (0, 1)$. We recall that $f$ is semi-conjugate to the rotation by $\rho$, i.e. there exists an orientation-preserving continuous surjective map $h \colon S^1 \to S^1$ such that $f \circ h = h \circ R_\rho$, where $R_\rho(x) = x + \rho \,(\mathrm{mod}\,\mathbb{Z})$ is the rotation by $\rho$. Here we are thinking of $S^1 = \mathbb{R}/\mathbb{Z}$ with an additive group structure. The map $f$ has a unique invariant probability measure $\mu$ satisfying $\mu[0, x) = h(x)$ for all $x \in [0, 1)$. This is equivalent to the fact that $h$ carries $\mu$ to Lebesgue measure on $S^1$. Therefore, from the point of view of Ergodic Theory, the pair $(f, \mu)$ is isomorphic to $(R_\rho, dx)$.

We will mention the main result of Coelho and de Faria ([10], 96) on the subject of visiting times of an $f$-orbit to a particular decreasing sequence of intervals. From the isomorphism mentioned above, the result should be read as if it is for rotations. (However, in [10] there is also an extension of the result for the case when $f$ is a diffeomorphism of the circle which is $C^1$-conjugate to $R_\rho$, and the visiting times are studied using a random point with respect to Lebesgue measure.)

In what follows we will assume $f = R_\rho$ and $\mu$ is Lebesgue measure on $S^1$. Consider the continued fraction expansion of $\rho$:

$$\rho = \cfrac{1}{a_1 + \cfrac{1}{a_2 + \cfrac{1}{\cdots}}},$$

where the $a_i$'s are positive integers. We will use the notation $\rho = [a_1, a_2, \cdots]$. We recall that $a_n = [G^n(\rho)]$, where $G(\rho) = 1/\rho \,(\bmod\, \mathbf{Z})$ is the Gauss map on $[0, 1)$, and $[t]$ denotes the integer part of $t$. In order to describe the result we will also need to define the so-called double Gauss map $\Gamma \colon \mathcal{T}^2 \to \mathcal{T}^2$ given by

$$\Gamma(\rho, y) = (G\rho, [a_1, b_1, b_2, \cdots]),$$

where $y = [b_1, b_2, \cdots]$.

Fix a base point $z \in S^1$. Consider the sequence of best approximations of $z$ by its $f$-orbit $(fz, f^2z, \cdots)$, i.e. find the sequence of times $k$ such that $|f^k z - z| < |f^j z - z|$ for all $j = 1, 2, \cdots, k - 1$. It can be shown that this sequence is given by $q_n$, where

$$\frac{p_n}{q_n} = [a_1, a_2, \cdots, a_n] = \cfrac{1}{a_1 + \cfrac{1}{a_2 + \cfrac{1}{\cdots + \cfrac{1}{a_n}}}}$$

denotes the truncated continued fraction expansion of $\rho$ in its irreducible form. The points $z$ and $\bar z = z + 1/2 \,(\bmod \mathbf{Z})$ divide the circle into two connected components, and for every $n > 1$ the points $f^{q_n-1}z$ and $f^{q_n}z$ lie on opposite components relative to $z$. Let $I_n$ denote the interval with endpoints $\{f^{q_n}z, z\}$ contained in the same component and define $J_n = I_n \cup I_{n-1}$. Coelho and de Faria [10] prove the following

**Theorem 5.3.9** *The visiting times to $J_n$, rescaled by $\mu(J_n)^{-1}$, converge in law under a subsequence $\{n_i\}$ of $n$, if and only if, either $G^{n_i}(\rho)$ converges to zero or $\Gamma^{n_i}(\rho, \rho)$ converges to $(\theta, \beta)$, for some $0 < \theta < 1$ and $\beta \in [0, 1)$. Moreover, for each choice of $(\theta, \beta)$, the corresponding limit point process are pairwise different.*

The above result has the consequence

**Corollary 5.3.10** *The visiting times to $J_n$, rescaled by $\mu(J_n)^{-1}$, do not converge in law for Lebesgue a.e. rotation number $\rho \in [0,1)$.*

In a way, for the purpose of building a general theory of asymptotic limit laws, this result is disappointing, since if we consider visiting times to neighbourhoods of length $\varepsilon$ around a base point, it shows that there is no universal limit law for rotations, i.e. no analog to Hirata's type of result. Nevertheless, this gives incentive to find out what are the ergodic reasons preventing the existence of universal laws in this context. We will return to these questions in the Section 5.5.4.

## 5.4 Shifts of Finite Type and Equilibrium States

Let $\mathcal{A} = \{1, \cdots, \ell\}$ be a finite set of symbols (represented by the number of each state). Let $M$ be an $\ell \times \ell$ matrix of 0's and 1's representing the allowable transitions in a directed graph $G$ of $l$-vertices. There exists an edge of $G$

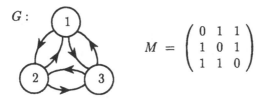

$$M = \begin{pmatrix} 0 & 1 & 1 \\ 1 & 0 & 1 \\ 1 & 1 & 0 \end{pmatrix}$$

Figure 5.4: An example of $G$ and $M$.

joining $i$ to $j$ if and only if $M(i,j) = 1$. The matrix $M$ is called the *incidence matrix* of the directed graph $G$. We consider the subset $\Omega = \Omega_G \subseteq \mathcal{A}^{\mathbb{N}}$ of infinite sequences corresponding to forward infinite paths in the graph $G$, i.e.

$$\Omega = \left\{ w = (w_0, w_1, \cdots) \in \mathcal{A}^{\mathbb{N}} : M(w_{n-1}, w_n) = 1, \ \forall n > 0 \right\}.$$

If $\sigma \colon \mathcal{A}^{\mathbb{N}} \to \mathcal{A}^{\mathbb{N}}$ denotes the shift map

$$\sigma(w_0, w_1, \cdots) = (w_1, w_2, \cdots),$$

then we will also denote by $\sigma$ its restriction to $\Omega$. We note that $\sigma(\Omega) \subseteq \Omega$, and with respect to the product topology on $\mathcal{A}^{\mathbb{N}}$ induced by the discrete topology on $\mathcal{A}$, $\Omega$ is closed. The pair $(\Omega, \sigma)$ is a (one-sided) *shift of finite type.* (Considering $\mathcal{A}^{\mathbb{Z}}$ and doubly infinite allowable paths in $G$, we obtain the *two-sided shift of finite type*). We will assume throughout that the incidence matrix $M$ is irreducible and aperiodic, in other words, it is the support of an irreducible and aperiodic Markov process on the states $\mathcal{A}$.

Let the cylinders in $\Omega$ be the intersection of the cylinders of $\mathcal{A}^{\mathbb{N}}$ with $\Omega$,

$$C[i_0, \cdots, i_m]_s = \{ x = (x_0, x_1, \cdots) \in \Omega \colon x_{j+s} = i_s, \ \forall j = 0, \cdots, m \} \ .$$

Recall that the Markov measures with support on $\Omega$ are the probability measures of the type:

$$\mu(C[i_0, \cdots, i_m]_s) = p_{i_0} P(i_0, i_1) \cdots P(i_{m-1}, i_m) \ ,$$

where $P$ is the transition probability matrix respecting the allowable transitions, in short

$$M(i, j) = 0 \iff P(i, j) = 0 \ .$$

Here $p = (p_1, \cdots, p_\ell)$ is the stationary vector representing the stationary distribution associated to $P$ (i.e. $pP = p$).

## 5.4.1   Hölder Potentials

The consideration of equilibrium states gives a richer class of shift-invariant probability measures which strictly contains the Markov measures. These are defined by a "driving" potential $\varphi$, which we will now describe.

Let $\varphi \colon \Omega \to \mathbb{R}$ be a bounded function. Define the $n$-th variation of $\varphi$ by

$$\mathrm{var}_n(\varphi) = \sup\{ |\varphi(x) - \varphi(y)| \colon x_0 = y_0, \cdots, x_n = y_n \} \ .$$

The function $\varphi$ is continuous if and only if we have $\mathrm{var}_n(\varphi) \to 0$, as $n \to 0$. We say that $\varphi$ has *summable variation* if

$$\sum_{n>0} \mathrm{var}_n(\varphi) < \infty \ ,$$

and $\varphi$ is called *Hölder continuous* if there exists $0 < \theta < 1$ such that $\{\mathrm{var}_n(\varphi)\, \theta^{-n}\}_{n>0}$ is a bounded sequence; that is, $\mathrm{var}_n(\varphi)$ goes to zero exponentially fast with $n$.

We refer to $\varphi$ as a *potential* on $\Omega$, which could be interpreted as a source of interaction of the paths in the graph $G$. For instance, suppose $\varphi$ satisfies

$$\sum_{\{i \colon M(i, x_0)=1\}} e^{\varphi(i, x_0, x_1, \cdots)} = 1, \quad \forall x = (x_0, x_1, \cdots) \in \Omega \ .$$

Then $\varphi$ is said to be *normalised* in this case. We may think of $e^{\varphi(i, x_0, x_1, \cdots)}$ as the transition probability to move from $(x_0, x_1, \cdots)$ to $(i, x_0, x_1, \cdots)$.

This defines a Markov process on $\Omega$ and the stationary distributions of this process are the equilibrium states of $\varphi$ (which will be defined later in another manner).

Consider also the case when $\varphi$ depends only on the first $k+1$ coordinates, i.e. if $x = (x_0, x_1, \cdots)$ and $y = (y_0, y_1, \cdots)$ satisfy $x_0 = y_0, \ldots, x_k = y_k$ then $\varphi(x) = \varphi(y)$. Let $k$ be smallest integer satisfying this property. In this case, a normalised potential $\varphi$ defines a $k$-step Markov process on the symbols $\{1, \cdots, \ell\}$, since the transition probabilities in $\Omega$ would depend only on $(x_0, x_1, \cdots, x_k)$.

When $\varphi$ is not normalised we will use the Ruelle operator and its spectral properties to study equilibrium states, which will be the stationary distributions of the Markov process defined above associated to a normalised $\varphi'$ canonically associated to $\varphi$.

## 5.4.2 Entropy and Pressure

Let $\mu$ be a shift-invariant probability measure on $\Omega$. Recall that the entropy of $\mu$ is given by the limit

$$h(\mu) = \lim_{n \to \infty} -\frac{1}{n} \sum_{C \in B_n(\Omega)} \mu(C) \log \mu(C) ,$$

where $B_n(\Omega)$ denotes the cylinders in $\Omega$ of type $C[x_0, \cdots, x_n]_0$. The pressure of the potential $\varphi \colon \Omega \to \mathbb{R}$ is defined by

$$\mathcal{P}(\varphi) = \sup \left\{ h(\mu) + \int \varphi \, d\mu \right\} ,$$

where the supremum is taken over all shift-invariant probability measures $\mu$ on $\Omega$. We say that a shift-invariant probability measure $m$ is an *equilibrium state* of $\varphi$ if $m$ satisfies

$$\mathcal{P}(\varphi) = h(m) + \int \varphi \, dm .$$

For every continuous potential $\varphi$, there exists at least one equilibrium state $m$. If $\varphi$ has summable variation then there exists a unique equilibrium state (see [4]). Note that when $\varphi = 0$ its equilibrium state $m$ is the measure of maximal entropy, this is a one-step Markov measure known as the Parry measure on the shift of finite type $(\Omega, \sigma)$ (cf. [24]). For general properties of the pressure for continuous dynamical systems see Walters [27].

We will be mostly interested when $\varphi$ is Hölder continuous because of the nice spectral properties of the Ruelle operator. Here it is appropriate to mention that with the hypothesis of $\varphi$ being normalised, there was a belief that $\varphi$ continuous would imply uniqueness of its equilibrium state. This is shown to be false with an example given in [5]. Recently, Coelho and Quas ([12], 97) show that if the variations of $\varphi$ satisfy

$$\sum_{n=r}^{\infty} \prod_{i=r}^{n}(1 - a_i) = \infty \, ,$$

for some $r \geq 1$, where $a_i = \frac{\ell}{2}\mathrm{var}_i(\varphi)$, and $\ell$ is the cardinality of $\mathcal{A}$, then $\varphi$ has a unique equilibrium state $m$ and the natural extension of $(\Omega, \sigma, m)$ is Bernoulli (i.e. metrically isomorphic to a Bernoulli scheme).

In the next section we will drop the condition of $\varphi$ being normalised.

### 5.4.3   Ruelle-Perron-Frobenius Operator

For $\varphi$ continuous, define an operator $L = L_\varphi \colon C(\Omega) \to C(\Omega)$ on the space of continuous functions $C(\Omega)$, $\psi \mapsto L_\varphi(\psi)$, by

$$L_\varphi(\psi)(x) = \sum_{M(i,x_0)=1} e^{\varphi(ix)}\psi(ix) \, ,$$

where $ix = (i, x_0, x_1, \cdots)$ and $x = (x_0, x_1, \cdots)$. The operator $L$ has important spectral properties when restricted to certain subspaces of $C(\Omega)$. For instance, if $\varphi$ is Hölder, $L_\varphi$ acts as an operator on the Hölder continuous functions, and restricted to this subspace, it has a simple and isolated eigenvalue $e^{\mathcal{P}(\varphi)}$. The rest of the spectrum is contained in a disc of radius strictly smaller than $e^{\mathcal{P}(\varphi)}$. There exists a strictly positive eigenfunction associated to $e^{\mathcal{P}(\varphi)}$. Therefore there exists $w$ Hölder continuous such that $w(x) > 0$ everywhere and

$$L_\varphi(w)(x) = e^{\mathcal{P}(\varphi)}w(x) \, ,$$

for all $x \in \Omega$. This implies that

$$\sum_{M(i,x_0)=1} e^{\varphi(ix)}w(ix) = e^{\mathcal{P}(\varphi)}w(x) \, ,$$

and then

$$\sum_{M(i,x_0)=1} e^{\varphi(ix)-\mathcal{P}(\varphi)+\log w(ix)-\log w(x)} = 1 \, .$$

Therefore, defining $\varphi'(x) = \varphi(x) - \mathcal{P}(\varphi) - \log w(\sigma x) + \log w(x)$ we would have $\varphi'$ is Hölder and normalised. The equilibrium state of $\varphi$ is the same equilibrium state of $\varphi'$. In fact, $m$ the equilibrium state of $\varphi$ is the unique shift-invariant probability measure which is fixed by the dual operator $L_\varphi^*$, i.e.

$$\int L_\varphi(\psi)\, dm = \int \psi\, dm \, , \quad \forall \psi \in C(\Omega) \, .$$

Another way to characterise the equilibrium state $m$ is a *homogeneous* property: there exists $K > 1$ such that

$$\frac{1}{K} < \frac{m(C[x_0, x_1, \cdots, x_n]_0)}{\exp\left\{-(n+1)\mathcal{P}(\varphi) + \sum_{j=0}^{n} \varphi(\sigma^j x)\right\}} < K\,,$$

for all $x = (x_0, x_1, \cdots) \in \Omega$. Replacing $\varphi$ by the normalised $\varphi'$, we may assume $\varphi$ is normalised and Hölder. (Recall that this implies $\mathcal{P}(\varphi) = 0$ and the Ruelle operator $L_\varphi$ will have 1 as a simple isolated eigenvalue, with the rest of the spectrum contained in a disc of radius strictly smaller than 1, when operating on the space of Hölder continuous functions.) Therefore, if $\varphi$ is Hölder then there exist $0 < \rho < 1$ and $c > 0$ (independent of $\psi$) such that

$$\left\| L_\varphi^n(\psi) - \int \psi\, dm \right\|_\Omega \leq c\rho^n\,.$$

This implies that, if $\psi$ is continuous then

$$L_\varphi^n(\psi)(x) \xrightarrow[n\to\infty]{} \int \psi\, dm\,,$$

uniformly in $x \in \Omega$ (this is done by approximating continuous functions by Hölder continuous functions).

On what follows we will assume $(\Omega, \sigma)$ is an irreducible and aperiodic shift of finite type with incidence matrix $M$, defining graph $G$, and with equilibrium state $m$ of a Hölder normalised potential $\varphi$.

### 5.4.4   The Central Limit Theorem

Let $\psi\colon \Omega \to \mathbb{R}$ be a Hölder continuous function such that $\int \psi\, dm = 0$ (mean zero). Define the *variance* of $\psi$ by

$$\underline{\sigma}^2 = \underline{\sigma}^2(\psi) = \lim_{n\to\infty} \frac{1}{n} \int \left(\sum_{j=0}^{n-1} \psi(\sigma^j x)\right)^2 dm(x)\,,$$

where we think of $\psi(x), \psi(\sigma x), \psi(\sigma^2 x), \cdots$ being a sequence of random variables defined on the probability space $(\Omega, m)$. This sequence is stationary since $m$ is shift-invariant. If $\psi$ does not have zero mean, define $\underline{\sigma}^2(\psi) = \underline{\sigma}^2(\tilde{\psi})$, where $\tilde{\psi} = \psi - \int \psi\, dm$. The following result shows a connection between the pressure and the above statistical constants (see Ruelle [26] for a proof). Let $\mathcal{H}$ denote the space of Hölder continuous functions defined on $\Omega$.

**Theorem 5.4.1** *Consider the pressure map* $\mathcal{P}\colon \mathcal{H} \to \mathbb{R}$ *given by* $\psi \mapsto \mathcal{P}(\psi)$. *For every fixed function* $\psi \in \mathcal{H}$, *and* $s$ *a real parameter, we have*

(a) $s \mapsto \mathcal{P}(\varphi + s\psi)$ *is a real analytic function;*

(b) $\dfrac{d}{ds}\{\mathcal{P}(\varphi + s\psi)\}\Big|_{s=0} = \displaystyle\int \psi \, dm$ ;

(c) $\dfrac{d^2}{ds^2}\{\mathcal{P}(\varphi + s\psi)\}\Big|_{s=0} = \underline{\sigma}^2(\psi)$ .

The variance $\underline{\sigma}^2(\psi)$ is zero if and only if there exists a Hölder function $h\colon \Omega \to \mathbb{R}$ satisfying $h(\sigma x) - h(x) = \psi(x)$ for all $x \in \Omega$. It can be shown that if there exists a measurable function $h$ satisfying the latter property for $m$-a.e. $x$, then there exists a Hölder continuous function satisfying this property. This is sometimes referred to as *rigidity* of the equilibrium state $m$.

A good exposition of the subject in the present Section, which contains an elegant proof of the next couple of results can be found in [11], see also [25].

**Theorem 5.4.2** *(Central Limit Theorem).  Suppose that $\underline{\sigma}(\psi) \neq 0$ and $\int \psi \, dm = 0$. Define*

$$(S_n\psi)(x) = \sum_{j=0}^{n-1} \psi(\sigma^j x) .$$

*Then $\dfrac{S_n\psi}{\underline{\sigma}\sqrt{n}}$ is asymptotically Normal with zero mean and variance 1, which means that*

$$F_n(t) = m\Big\{x \in \Omega\colon \frac{1}{\underline{\sigma}\sqrt{n}} \sum_{j=0}^{n-1} \psi(\sigma^j x) \le t\Big\} \xrightarrow[n\to\infty]{} \frac{1}{\sqrt{2\pi}} \int_{-\infty}^{t} e^{-s^2/2} \, ds = N(t).$$

*Furthermore, $\|F_n - N\|_\infty = O(1/\sqrt{n})$, i.e. $\{\sqrt{n}\,\|F_n - N\|_\infty\}$ is a bounded sequence.*

This is the Normal asymptotic law for Hölder equilibrium states. When $\psi$ satisfies an extra condition which prevents $S_n\psi$ to be asymptotically lattice distributed, then there is a refinement in the Central Limit Theorem. More precisely, we say that $\psi$ defines a *non-lattice distribution* if, for all $a \in \mathbb{R}$, the values $S_n\psi(x) - na$ for $x$ such that $\sigma^n x = x$, generate a dense additive subgroup of $\mathbb{R}$. This is equivalent to the condition that whenever

$$h(\sigma x) = \alpha \, e^{is\psi(x)} h(x) ,$$

for some complex constant $\alpha$ and measurable function $h$, then necessarily $s = 0$ and $h(x)$ is a constant $m$-a.e. $x$.

**Theorem 5.4.3** *Let $\psi \in \mathcal{H}$ be given such that $\underline{\sigma}(\psi) \neq 0$ and $\int \psi \, dm = 0$. If $\psi$ defines a non-lattice distribution then*

$$F_n(t) - N(t) = \frac{\mathcal{P}^{(3)}(0)}{6 \underline{\sigma} \sqrt{2\pi n}} \left(1 - \frac{t^2}{\underline{\sigma}^2}\right) e^{-t^2/2\underline{\sigma}^2} + o\left(\frac{1}{\sqrt{n}}\right),$$

*where $\mathcal{P}^{(3)}(0)$ denotes the third derivative of the pressure map $s \mapsto \mathcal{P}(\varphi + s\psi)$ at $s = 0$.*

Higher-order asymptotics can also be obtained under stronger conditions on $\psi$ (see [11]).

### 5.4.5 Pianigiani-Yorke Measure

In this section we will find the correct scale to study the convergence of asymptotic visiting times to a subsystem of finite type inside the shift of finite type $(\Omega, \sigma)$.

Suppose for a moment that we have two equilibrium states $m_1$ of $\varphi_1$ and $m_2$ of $\varphi_2$ ($\varphi_1, \varphi_2$ normalised and Hölder continuous on $\Omega$).

Let $x = (x_0, x_1, \cdots)$ be a generic point for $m_1$, i.e. the shift orbit of $x$ determines the statistical behaviour of $m_1$, or more precisely, for every continuous function $g: \Omega \to \mathbb{R}$ we have

$$\frac{1}{n} \sum_{j=0}^{n-1} g(\sigma^j x) \xrightarrow[n \to \infty]{} \int g \, dm_1 .$$

Let $y = (y_0, y_1, \cdots)$ be a generic point for $m_2$. We may think of the pair $(x, y)$ as a point in $(\mathcal{A}^2)^{\mathbb{N}}$ by writing

$$(x, y) = \big((x_0, y_0), (x_1, y_1), \cdots\big) .$$

Defining an incidence matrix on the states $\mathcal{A}^2$ by

$$\widetilde{M}\big((i_1, i_2), (j_1, j_2)\big) = M(i_1, j_1) \, M(i_2, j_2) ,$$

we obtain the product shift of finite type $\widetilde{\Omega} \subseteq (\mathcal{A}^2)^{\mathbb{N}}$. The corresponding shift map $\tilde{\sigma}$ is clearly conjugate to $\sigma \times \sigma$. The product measure $m_1 \times m_2$ is canonically isomorphic to the equilibrium state of the potential $\tilde{\varphi}: \widetilde{\Omega} \to \mathbb{R}$ given by

$$\tilde{\varphi}\big((x_0, y_0), (x_1, y_1), \cdots\big) = \varphi_1(x) + \varphi_2(y) . \tag{5.2}$$

For future reference we will denote $\tilde{\varphi}$ by $\varphi_1 \otimes \varphi_2$.

Now let $\Delta \subseteq \{1, \cdots, \ell\}^2$ be the subset of $\mathcal{A}^2$ consisting of pairs $\{(j, j)\}_{j=1, \cdots, \ell}$. Then $(x, y)$ match on the first $k + 1$ coordinates if

$$(x_i, y_i) \in \Delta, \quad \forall i = 0, 1, \cdots, k .$$

Therefore, studying the asymptotic properties of the waiting time for $k$ matchings in a row for two sequences $x$ and $y$, is equivalent to visiting a subset of states $\Delta$ in the product symbol set, consecutively $k+1$ times, in the product shift of finite type.

This leads us to the consideration of a general case, where $\Delta \subseteq \{1, \cdots, \ell\}$ is a subset of states, $x = (x_0, x_1, \cdots)$ is a random point in $(\Omega, m)$ (i.e. generic with respect to $m$) and we study the times of $k$ successive hits in $\Delta$. (Here we are identifying $\Delta$ with the union of cylinders $\cup_{i \in \Delta} C[i]_0$ in $\Omega$.) Let $\chi_\Delta$ indicate a hit in $\Delta$ at time 0, i.e. the characteristic function of $\Delta$:

$$\chi_\Delta(x) = \begin{cases} 1 & \text{if } x_0 \in \Delta \,; \\ 0 & \text{if } x_0 \notin \Delta \,. \end{cases}$$

Then hitting $k+1$ times in succession means that

$$\prod_{j=r}^{r+k} \chi_\Delta(\sigma^j x) = 1 \,,$$

for some $r \geq 0$. Suppose that the incidence matrix $M$ restricted to the states of $\Delta$ is irreducible and aperiodic. (Therefore, thinking of the states of $\Delta$ as painted states, if we forget the transitions in $\Omega$ which passes through non-painted states, then we obtain an irreducible and aperiodic subsystem of finite type $\Omega_\Delta$ in $(\Omega, \sigma)$. See Figure 5.5 for an example.)

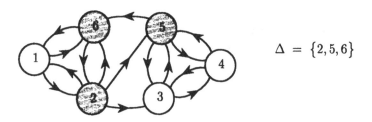

$$\Delta = \{2, 5, 6\}$$

Figure 5.5: An example of $\Delta$.

Let $\Delta_k$ denote the union of cylinders of length $k+1$ with symbols belonging to $\Delta$, that is

$$\Delta_k = \{x = (x_0, x_1, \cdots) \in \Omega \colon x_i \in \Delta, \, \forall i = 0, 1, \cdots, k\} \,.$$

The visiting times to $\Delta_k$ of the shift orbit of a point in $\Omega$, corresponds to the times of $k$-consecutive hits in $\Delta$. Define as before

$$\tau_k(x) = \inf\{n > 0 \colon \sigma^n x \in \Delta_k\}$$

as the first visiting time to $\Delta_k$. (In the next chapter, we will consider the whole process of subsequent visits to $\Delta_k$).

In order to study the asymptotic properties of these times, as mentioned in Section 5.3, we need to find a scale $c_k \to 0$ such that the distribution

$$F_k(t) = m\{x \in \Omega:\ c_k\,\tau_k(x) \le t\}$$

converges as $k \to \infty$. As we saw, the natural scale here is $\mathbb{E}(\tau_k)$. However, as we also noticed, the study of $m(\Delta_k)$ could be a source of inspiration, and indeed we will see that it will provide a good scale.

Recall some properties of the operator $L_\varphi$:

(i) $\displaystyle \int \psi\,dm \ = \ \int L_\varphi(\psi)\,dm\,, \quad \forall\,\psi \in C(\Omega)\,;$

(ii) $\displaystyle \int \psi_1(\psi_2 \circ \sigma)\,dm \ = \ \int \psi_2 L_\varphi(\psi_1)\,dm\,, \quad \forall\,\psi_1, \psi_2 \in C(\Omega)\,.$

First we note that

$$\chi_{\Delta_k} \ = \ \prod_{j=0}^{k} \chi_{\Delta} \circ \sigma^j \ = \ \chi_\Delta \cdot \left[\left(\prod_{j=0}^{k-1} \chi_\Delta \circ \sigma^j\right)\circ\sigma\right]\,.$$

Therefore, using the properties (i) and (ii) we obtain

$$\int_{\Delta_k} \psi\,dm \ = \ \int \psi \cdot \chi_{\Delta_k}\,dm$$

$$= \ \int \psi \cdot \chi_\Delta \cdot (\chi_{\Delta_{k-1}} \circ \sigma)\,dm$$

$$= \ \int L_\varphi(\psi \cdot \chi_\Delta) \cdot \chi_{\Delta_{k-1}}\,dm$$

$$= \ \int_{\Delta_{k-1}} L_\varphi(\psi \cdot \chi_\Delta)\,dm\,.$$

Thus, in short we have

$$\int_{\Delta_k} \psi\,dm \ = \ \int_{\Delta_{k-1}} L_\varphi(\psi \cdot \chi_\Delta)\,dm\,, \quad \forall\,\psi \in C(\Omega)\,. \tag{5.3}$$

Define the restricted Ruelle operator $L_\Delta: C(\Omega) \to C(\Omega)$ by

$$L_\Delta(\psi)(x) \ = \ L_\varphi(\psi \cdot \chi_\Delta)(x)$$

$$= \ \sum_{M(i,x_0)=1} e^{\varphi(ix)}\,\psi(ix)\,\chi_\Delta(ix)$$

$$= \ \sum_{\substack{M(i,x_0)=1 \\ i\in\Delta}} e^{\varphi(ix)}\,\psi(ix)\,.$$

With this definition we obtain from (5.3)

$$\int_{\Delta_k} \psi \, dm \;=\; \int_{\Delta_{k-1}} L_\Delta(\psi) \, dm \;=\; \cdots \;=\; \int L_\Delta^{k+1}(\psi) \, dm \,, \qquad (5.4)$$

for all $\psi \in C(\Omega)$.

Knowing some spectral properties of $L_\Delta$ would be useful here. We recall a recent result by Collet et al. ([14], 97) about Pianigiani-Yorke measures for aperiodic subsystems of finite type. This proves to be the key point here.

Consider the restricted subsystem $\Omega_\Delta$ (suppose that $\Delta \subsetneq \{1, \cdots, \ell\}$). The restriction of $\varphi$ to $\Omega_\Delta$ defines a Hölder potential $\varphi_\Delta$. Consider the pressure $\mathcal{P}_\Delta$ of $\varphi_\Delta$ as a function defined on the subsystem $\Omega_\Delta$. The Ruelle operator $L_{\varphi_\Delta} : C(\Omega_\Delta) \to C(\Omega_\Delta)$ restricted to Hölder continuous functions defined on $\Omega_\Delta$, has a simple isolated eigenvalue $e^{\mathcal{P}_\Delta}$. Thus, there exists a positive Hölder function $\tilde{w} : \Omega_\Delta \to \mathbb{R}^+$ such that

$$\begin{aligned} L_{\varphi_\Delta}(\tilde{w})(x) &= \sum_{\substack{M(i,x_0)=1 \\ i \in \Delta}} e^{\varphi(ix)} \, \tilde{w}(ix) \\ &= e^{\mathcal{P}_\Delta} \, \tilde{w}(x) \,, \end{aligned}$$

for all $x \in \Omega_\Delta$. We note that, since $\Delta \neq \{1, \cdots, \ell\}$ then necessarily $\mathcal{P}_\Delta < 0$ (i.e. $e^{\mathcal{P}_\Delta} < 1$). This is because we are assuming $\varphi$ to be normalised. (In the general case we would have always $\mathcal{P}(\varphi) > \mathcal{P}_\Delta$.)

Collet, Martínez and Schmitt [14] proves

**Theorem 5.4.4** *There exists a strictly positive Hölder continuous function* $w_\Delta : \Omega \to \mathbb{R}^+$ *such that*

$$L_\Delta(w_\Delta) \;=\; e^{\mathcal{P}_\Delta} \, w_\Delta \,,$$

*and* $w_\Delta|_{\Omega_\Delta} \equiv \tilde{w}$. *Furthermore,*

$$\left\| e^{-n\mathcal{P}_\Delta} L_\Delta^n(\psi) - w_\Delta \int_{\Omega_\Delta} \psi \, dm_\Delta \right\|_\Omega \xrightarrow[n \to \infty]{} 0 \,,$$

*for all* $\psi \in C(\Omega)$. *(Here* $m_\Delta$ *is the unique equilibrium state of* $\varphi_\Delta$ *on the subsystem* $\Omega_\Delta$.*)*

Let $\mathcal{B}$ denote the Borel $\sigma$-algebra on $\Omega$. The measure $\mu_{PY}$ on $(\Omega, \mathcal{B})$ defined by

$$\mu_{PY}(B) \;=\; \int_B w_\Delta \, dm \,, \quad \forall B \in \mathcal{B} \,,$$

is called the *Pianigiani-Yorke measure* of the subsystem $(\Omega_\Delta, \sigma)$. This is a quasi-stationary measure related to the asymptotic approach to the subsystem $(\Omega_\Delta, \sigma)$ (i.e. $\mu_{PY} = \alpha\,\mu_{PY}\circ\sigma^{-1}$, for some $\alpha > 1$). In fact, in our context $\alpha = e^{-\mathcal{P}_\Delta}$. This measure satisfies

$$\frac{\mu_{PY}(B)}{\mu_{PY}(\Omega)} \;=\; \lim_{k\to\infty}\, m\!\left(\sigma^{-k}B\,|\,\Delta_{k-1}\right),$$

for all $B \in \mathcal{B}$. The Pianigiani-Yorke measure is not shift-invariant and carries an information of how subsystems of finite type are embedded into larger systems. It would be nice if there would be a non-trivial relation between these measures and certain classification problems of shifts of finite type, but this is subject for further developments.

To finish this section we state the next result, which shows how fast the measure of $\Delta_k$ goes to zero and also gives the implied constant.

**Lemma 5.4.5** *We have*

$$\lim_{k\to\infty}\, e^{-k\mathcal{P}_\Delta}\, m(\Delta_{k-1}) \;=\; \mu_{PY}(\Omega)\,.$$

Coming back to our original problem of visiting times, the above result means that $c_k = e^{k\mathcal{P}_\Delta}$ is a suitable scale for rescaling the visiting times to $\Delta_k$. In the next Section we will recall and give a proof of Lemma 5.4.5. There will be also a characterisation of the asymptotic limit law of visiting times to $\Delta_k$.

## 5.5   Point Processes and Convergence in Law

In this section we discuss some general properties of convergence in law for point processes. This is the natural terminology to treat the problems mentioned earlier concerning visiting times, matching times, etc.

If $(\Omega, \mu)$ is a probability space, a point process with values in $[0, +\infty)$ defined on $\Omega$ is a map $\tau \colon \Omega \to \mathcal{M}_\sigma[0, +\infty)$, where $\mathcal{M}_\sigma[0, +\infty)$ denotes the $\sigma$-finite measures on $[0, +\infty)$ such that the support, $\mathrm{supp}(\tau(w)) \subseteq [0, +\infty)$, of the measure $\tau(w)$ is discrete for $m$-a.e. $w \in \Omega$. For every continuous function $g \colon [0, +\infty) \to \mathbb{R}$ of compact support we may consider the random variable

$$N(g)(w) \;=\; \int_0^\infty g\, d\tau(w)\,.$$

The statistical properties of $\tau$ can then be studied, varying the functions $g$. The typical choice for $g$ is the indicator function of an interval $[a, b)$, and in this case, the above integral defines a random variable corresponding to the number of hits of the point process in the interval $[a, b)$.

An example is the *Poisson point process of rate* $\lambda > 0$, which is the point process defined by

$$\tau(w) = \sum_{n>0} \delta_{X_n(w)} ,$$

where $\delta_t$ denotes the Dirac measure at the point $t$, and the random variables $X_1, X_2 - X_1, X_3 - X_2, \cdots$ are independent identically distributed with exponential law of parameter $\lambda > 0$, i.e. defining $X_0 \equiv 0$ we have

$$\mathbb{P}(X_i - X_{i-1} \le t) = 1 - e^{-\lambda t} ,$$

for all $i \ge 1$, and the joint distributions are products of the marginals,

$$\mathbb{P}(X_1 \le t_1, X_2 - X_1 \le t_2, \cdots, X_j - X_{j-1} \le t_j) = \prod_{i=1}^{j} \left(1 - e^{-\lambda t_i}\right) ,$$

for every $j \ge 1$ and every choice of $t_i \in [0, +\infty)$. Let $g : [0, +\infty) \to \mathbb{R}$ be given by $g(t) = \chi_{[a,b)}(t)$. Then

$$N(g)(w) = \sum_{n>0} \chi_{[a,b)}(X_n(w))$$

i.e. $N(g)$ counts the number of hits at $[a, b)$ for the random impulses given by the Poisson process. Another interesting random variable is $N(g_s)$, where $g_s(t) = \chi_{[0,s)}(t)$, since it has the Poisson distribution of parameter $\lambda s$:

$$\mathbb{P}(N(g_s) = k) = \frac{(\lambda s)^k}{k!} e^{-\lambda s} ,$$

for all $k = 0, 1, \cdots$. We will also consider another naturally defined point process in the next Section.

### 5.5.1   Convergence of Point Processes

We say that a sequence of point processes $\tau_n$ *converges in law* to a point process $\tau$, if for every continuous $g : [0, +\infty) \to \mathbb{R}$ with compact support, the random variables

$$N_n(g)(w) = \int g \, d\tau_n(w)$$

converge in distribution to some random variable $N(g)(w)$. This is equivalent to having convergence of the moment generating function

$$\mathbb{E}(e^{z \, N_n(g)}) \xrightarrow[n \to \infty]{} \mathbb{E}(e^{z \, N(g)}) ,$$

for $z$ in some open neighbourhood of the origin in the complex plane $\mathbb{C}$. In the case $\tau \colon \Omega \to \mathcal{M}_\sigma[0, +\infty)$ is a Poisson point process of parameter $\lambda > 0$, it is not difficult to prove that (cf. [17])

$$\mathbb{E}(e^{z\,N(g)}) = \exp\left\{\lambda \int_0^\infty (e^{zg(y)} - 1)\,dy\right\}. \tag{5.5}$$

Therefore, to show that a sequence of point processes, representing the visits to an asymptotically rare sequence of events, converges to a Poisson point process, it suffices to show that the moment generating functions converge to the expression in (5.5).

Actually, we will show convergence of a sequence of point processes to a marked Poisson point process, which we now describe. This consists of a point process with

$$\tau(w) = \sum_{n>0} K_n(w)\,\delta_{X_n(w)}$$

where $K_n(w) \in \mathbb{N}$ are i.i.d. random variables, $X_1, X_2 - X_1, X_3 - X_2, \cdots$ are real positive i.i.d. random variables and $K_i, X_j$ are independent for any collection of $i$'s and $j$'s. Moreover, the sequence $X_1, X_2 - X_1, X_3 - X_2, \cdots$ have exponential distribution of parameter $\lambda > 0$. If $\pi$ denotes the distribution of each $K_n$, i.e. $\pi = (\pi_k)_{k\geq 1}$, and $\pi_k = \mathbb{P}(K_n = k)$, then $\tau$ is called a *marked Poisson point process with constant parameter measure* $\lambda\pi$.

The moment generating function of such a process is given by

$$\mathbb{E}(e^{z\,N(g)}) = \exp\left\{\lambda \sum_{k=1}^\infty \pi_k \int_0^\infty (e^{z\,k\,g(y)} - 1)\,dy\right\}.$$

Thus, one way to prove the convergence to a marked Poisson point process is to prove convergence of the moment generating function and identifying the limit with the above expression in a neighbourhood of the origin in $\mathbb{C}$. This was the method used by Coelho and Collet [8] for the process of visits to the diagonal for piecewise expanding maps of the circle (see Section 5.3.3).

## 5.5.2 Entrance Times and Visiting Times

What will be done here on what follows is also joint work with P. Collet [9] (under construction). We recall the setup done in Section 5.4 regarding the construction of a particular sequence of asymptotically rare events $\Delta_k$ in an irreducible and aperiodic shift of finite type. For the sake of clarity in the exposition, we will repeat some of the definitions.

Let $\mathcal{A} = \{1, \cdots, \ell\}$ be the states of an irreducible and aperiodic shift of finite type $(\Omega, \sigma)$, $\Omega \subseteq \mathcal{A}^{\mathbb{N}}$. Let $\varphi \in \mathcal{H}$ be a Hölder potential on $\Omega$ defining an equilibrium state $m$ on $\Omega$. We consider the painting of some states of $\Omega$

by choosing $\Delta \subseteq \{1, \cdots, \ell\}$, where $\Delta \neq \{1, \cdots, \ell\}$. As we mentioned before, $\Delta$ defines a subsystem of finite type in $\Omega$, which we called $\Omega_\Delta$. We assume that this subsystem is irreducible and aperiodic.

By changing the potential $\varphi$ to another Hölder potential defining the same equilibrium state $m$, we may suppose $\varphi$ is normalised, i.e.

$$\sum_{M(i,x_0)=1} e^{\varphi(ix)} = 1 ,$$

for all $x \in \Omega$, where $M$ denotes the incidence matrix of $\Omega$. We also defined the restricted Ruelle operator

$$L_\Delta(\psi)(x) = L_\varphi(\psi \cdot \chi_\Delta)(x) = \sum_{\substack{M(i,x_0)=1 \\ i \in \Delta}} e^{\varphi(ix)} \psi(ix) .$$

If $e^{\mathcal{P}_\Delta}$ is the simple isolated eigenvalue of the Ruelle operator $L_{\varphi_\Delta}$ corresponding to the restriction $\varphi_\Delta$ of $\varphi$ to the subsystem $\Omega_\Delta$, then by Theorem 5.4.4, there exists $w_\Delta : \Omega \to \mathbb{R}$ Hölder continuous and strictly positive such that

$$L_\Delta(w_\Delta) = e^{\mathcal{P}_\Delta} w_\Delta, \text{ and } \left\| e^{-n\mathcal{P}_\Delta} L_\Delta^n(\psi) - w_\Delta \int_{\Omega_\Delta} \psi \, dm_\Delta \right\|_\Omega \xrightarrow[n \to \infty]{} 0, \quad (5.6)$$

where $m_\Delta$ denotes the equilibrium state of $\varphi_\Delta$ with support on the subsystem $\Omega_\Delta$. We defined the Pianigiani-Yorke measure

$$\mu_{PY}(B) = \int_B w_\Delta \, dm ,$$

for any Borel set $B \subseteq \Omega$. Let $\Delta_k = \{x \in \Omega : x_i \in \Delta, \ i = 1, \cdots, k\}$.

**Proof of Lemma 5.4.5.** We want to show that

$$\lim_{k \to \infty} e^{-k\mathcal{P}_\Delta} m(\Delta_{k-1}) = \mu_{PY}(\Omega) = \int w_\Delta \, dm .$$

First we recall from (5.4) that

$$\int_{\Delta_{k-1}} \psi \, dm = \int L_\Delta^k(\psi) \, dm ,$$

for all $\psi \in C(\Omega)$. We also have from (5.6)

$$e^{-k\mathcal{P}_\Delta} L_\Delta^k(\psi) = w_\Delta \int_{\Omega_\Delta} \psi \, dm_\Delta + o(1) ,$$

which implies that

$$e^{-k\mathcal{P}_\Delta} \int_{\Delta_{k-1}} \psi \, dm = \left( \int w_\Delta \, dm \right) \left( \int_{\Omega_\Delta} \psi \, dm_\Delta \right) + o(1) .$$

Therefore, for every $\psi \in C(\Omega)$ we obtain

$$\lim_{k \to \infty} e^{-kP_\Delta} \int_{\Delta_{k-1}} \psi \, dm = \left( \int w_\Delta \, dm \right) \left( \int_{\Omega_\Delta} \psi \, dm_\Delta \right).$$

Putting $\psi \equiv 1$ gives the desired result

$$e^{-kP_\Delta} m(\Delta_{k-1}) \xrightarrow[k \to \infty]{} \int w_\Delta \, dm \quad \blacksquare$$

Now we turn our attention to the times when we see blocks of length $k$ consisting of painted symbols. For each fixed $k$, consider the *visiting time point process* $\tau_k \colon \Omega \to \mathcal{M}_\sigma[0, +\infty)$ defined by

$$\tau_k(x) = \sum_{j \geq 0} \chi_{\Delta_{k-1}}(\sigma^j x) \, \delta_{j \cdot c_k} ,$$

where we introduced the scale $c_k = e^{kP_\Delta}$. (Recall that $P_\Delta < 0$.)

Consider also the *entrance time point process* of $\Delta_{k-1}$ rescaled by $c_k$,

$$\tau_k^e(x) = \sum_{j \geq 1} \chi_{\Delta_{k-1}}(\sigma^j x) \, \chi_{\Delta_{k-1}^c}(\sigma^{j-1} x) \, \delta_{j \cdot c_k} .$$

The following results summarise the asymptotic behaviour of these processes.

**Theorem 5.5.1** *The entrance time point process $\tau_k^e$ converges in law to a Poisson point process of parameter $\lambda > 0$.*

**Theorem 5.5.2** *The visiting time point process $\tau_k$ converges in law to a marked Poisson point process with constant parameter measure $\lambda\pi$, where $\pi = (\pi_j)_{j \geq 1}$ is geometrically distributed.*

Intuitively, we may interpret this result by thinking of the subsystem $\Omega_\Delta$ in $\Omega$ as an attractor in probability, in the sense that considering a sequence of open neighbourhoods $\mathcal{U}_k$ in $\Omega$ such that $\Omega_\Delta = \cap_k \mathcal{U}_k$, we must have necessarily $m(\mathcal{U}_k) \to 0$, and for very large $k$, a $m$-generic point $x \in \Omega$ will take (after rescaling by the proper power of the restricted pressure of $\Omega_\Delta$) an exponential time to enter $\mathcal{U}_k$. When it is in there, it takes an asymptotically geometric time to leave this neighbourhood, and as soon as it is out, the process regenerates itself.

Turning to a more technical discussion, we continue to give an idea of how one can prove the above results. First we note that these results are consequences of the following

**Theorem 5.5.3** *There exists $\lambda > 0$ and a probability measure $\pi = (\pi_j)_{j\geq 1}$ on $\{1, 2, \cdots\}$ such that, for every $g: [0, +\infty) \to \mathbb{R}$ continuous with compact support, the random variable*

$$N_k(g)(x) = \sum_{j\geq 0} \chi_{\Delta_{k-1}}(\sigma^j x) g(j \cdot c_k),$$

*satisfies*

$$\mathbb{E}(e^{zN_k(g)}) = \int N_k(g)(x)\, dm(x) \xrightarrow[k\to\infty]{} \exp\Big\{\lambda \sum_{j=1}^{\infty} \pi_j \int_0^{\infty} (e^{z\,j\,g(y)} - 1)dy\Big\}.$$

$$(5.7)$$

In fact the constants $\lambda$ and $\pi_j$ are given by the following expressions

$$\lambda = (1 - e^{P_\Delta}) \int w_\Delta\, dm, \qquad \pi_j = (1 - e^{P_\Delta})\, e^{(j-1)P_\Delta}.$$

**Proof of Theorem 5.5.3 (sketch).** Considering the factorial moments of $N_k(g)$, i.e. $\mathbb{E}(N_k(g)^n)$, we show that the limit $\nu_n = \lim_{k\to\infty} \mathbb{E}(N_k(g)^n)$ exists. Then we identify the function

$$F_g(z) = \sum_{n\geq 0} \frac{z^n}{n!}\nu_n$$

with the complex function in (5.7), for $z$ in a small neighbourhood of the origin in $\mathbb{C}$. This is similar to what is done in [8], so we illustrate only the main steps.

We see that

$$\mathbb{E}(N_k(g)^n) = \mathbb{E}\Big[\Big(\sum_{j\geq 0} \chi_{\Delta_{k-1}}(\sigma^j x) g(j \cdot c_k)\Big)^n\Big]$$

$$= \sum_{0\leq j_1, \cdots, 0\leq j_k} \prod_{s=1}^{n} g(j_s \cdot c_k)\, \mathbb{E}\Big(\prod_{s=1}^{n} \chi_{\Delta_{k-1}} \circ \sigma^{j_s}\Big).$$

Let us compute this for $n = 1$. In this case,

$$\mathbb{E}(N_k(g)) = \mathbb{E}\Big(\sum_{j\geq 0} \chi_{\Delta_{k-1}}(\sigma^j x) g(j \cdot c_k)\Big)$$

$$= \sum_{j\geq 0} g(j \cdot c_k)\, \mathbb{E}(\chi_{\Delta_{k-1}} \circ \sigma^j)$$

$$= \frac{m(\Delta_{k-1})}{c_k} \sum_{j\geq 0} c_k\, g(j \cdot c_k)$$

As $c_k = e^{kP_\Delta}$, and $e^{-kP_\Delta} m(\Delta_{k-1}) \to \int w_\Delta \, dm$, we obtain

$$\mathbb{E}(N_k(g)) \xrightarrow[k \to \infty]{} \left( \int w_\Delta \, dm \right) \int_0^\infty g(y) \, dy \ .$$

In the case $n \geq 2$, as $k$ diverges, there is a need to study the asymptotic behaviour of

$$\mathbb{E}\left( \prod_{s=1}^n (\chi_{\Delta_{k-1}} \circ \sigma^{j_s}) \right) = \int \left( \prod_{s=1}^n (\chi_{\Delta_{k-1}} \circ \sigma^{j_s}) \right) dm \ ,$$

for $j_0 = 0 < j_1 < \cdots < j_s$. In order to do this, we are able to show the following result.

**Lemma 5.5.4** *The limit*

$$C_n = \lim_{k \to \infty} m(\Delta_{k-1})^{-1} \sum_{\substack{0 = q_0 < q_1 < \cdots < q_{n-1} \\ q_s - q_{s-1} \leq k}} \mathbb{E}\left( \prod_{s=0}^{n-1} \chi_{\Delta_{k-1}} \circ \sigma^{q_s} \right) \ ,$$

*exists for every* $n \geq 1$. *Furthermore, for* $n \geq 1$,

$$C_n = \left( \frac{1}{e^{-P_\Delta} - 1} \right)^{n-1} \ .$$

We then use the above constants to show the next preliminary result.

**Proposition 5.5.5** *For each fixed* $n$, *the limit* $\nu_n = \lim_{k \to \infty} \mathbb{E}(N_k(g)^n)$ *exists, and it is the* $n$-th *derivative at the origin of the complex function*

$$F_g(z) = \exp \left\{ \sum_{j=1}^\infty C_j \int_0^\infty (e^{zg(y)} - 1)^j \, dy \right\} \ ,$$

*which is an analytic function on a disc around the origin in* $\mathbb{C}$.

Using the above Proposition, Theorem 5.5.3 follows by solving the equation

$$\exp \left\{ \sum_{j=1}^\infty C_j \int_0^\infty (e^{zg(y)} - 1)^j \, dy \right\} = \exp \left\{ \lambda \sum_{j=1}^\infty \pi_j \int_0^\infty (e^{zjg(y)} - 1) \, dy \right\}$$

in the constants $\lambda$ and $\pi_j$, and showing that both functions define an analytic function in a neighbourhood of the origin in $\mathbb{C}$. Therefore the limiting process is a marked Poisson point process with constant parameter measure $\lambda\pi$. ∎

We finish this section with the ingredients involved in the proof of Proposition 5.5.5. In order to prove that $\lim_{k \to \infty} \int N_k(g)^n \, dm$ exists, we use a sort of relativised decay of correlations of the equilibrium state $m$. It can be stated as

**Lemma 5.5.6** *There exist $K > 0$ and $0 < \gamma < 1$ such that*

$$\left| \mathbb{E}(\chi_{\Delta_{k-1}} \cdot \chi_{B} \circ \sigma^{k+r-1}) - m(\Delta_{k-1}) \, m(B) \right| \leq K \gamma^r e^{k \mathcal{P}_\Delta} \, m(B) \, ,$$

*for every $k \geq 1$, $r \geq 1$, and for every Borel set $B \subseteq \Omega$.*

The above Lemma allows us to control the summations involving the expectation of the various products of characteristic functions, which appear as "clusters" of successive visits to $\Delta_{k-1}$, when dealing with the quantities

$$\sum_{0 \leq j_1, \cdots, 0 \leq j_p} \mathbb{E}\left( \prod_{s=1}^{p} (\chi_{\Delta_{k-1}} \circ \sigma^{j_s}) \right) \, .$$

More details will be given in the final version of [9].

## 5.5.3    Final Remarks

Here we discuss how the results of the previous section can be used in the examples given in the Preliminaries, and the various comments made after the motivating examples coming from different ergodic dynamical systems of Section 5.3.

First we note that the consideration of $\Delta$ being a subset of the symbol space $\mathcal{A}$ is general enough to obtain all irreducible and aperiodic subsystems of finite type in $(\Omega, \sigma)$. This is because of the higher-block representation of any subshift of finite type (see M. Boyle's lectures in this volume), which could be represented as $\Omega_\Delta$ for $\Delta$ being the choice of blocks which define the subshift of finite type.

The main result of the last Section, applied to products of shifts of finite type, has a consequence for the waiting time for matchings of two random sequences $x = (x_0, x_1, \cdots)$ and $y = (y_0, y_1, \cdots)$, when say, $x$ is randomly chosen according to an equilibrium state $m_1$, and $y$ randomly chosen according to the equilibrium state $m_2$. In this case, the scale we need to apply in order to study the asymptotic behaviour of these matching times is the exponential of the pressure on the restricted system defined by the diagonal in $\{1, \cdots, \ell\}^2$.

We have already mentioned in (5.2) that, if $\varphi_1$ is the Hölder potential defining $m_1$ (and $\varphi_2$ defining $m_2$), then $m_1 \times m_2$ is the equilibrium state of the potential $\varphi_1 \otimes \varphi_2$. Denote the restriction of $\varphi_1 \otimes \varphi_2$ to the diagonal subsystem by $\varphi_\Delta(x, x) = \varphi_1(x) + \varphi_2(x)$. We may think of $\varphi_\Delta(x, x) = \varphi_\Delta(x)$ as a function defined on $(\Omega, \sigma)$. Therefore the pressure $\mathcal{P}_\Delta$ will be given by

$$\mathcal{P}_\Delta = \sup \left\{ h(\mu) + \int (\varphi_1 + \varphi_2) \, d\mu \right\} \, ,$$

where the supremum is taken over all shift-invariant probability measures $\mu$ on $\Omega$. The latter is true because the the diagonal subsystem is canonically conjugate to $(\Omega, \sigma)$. The above results prove our comments in Section 5.2, when we talked about the inter-occurrence times for matchings.

We also return to Hirata's type of result. Given a generic $x \in \Omega$ consider the cylinders $C[x_0, x_1, \cdots, x_k]_0$ and study the visiting times to these cylinders asymptotically with $k$. This corresponds to defining $\Delta_i = \{x_i\}$ and writing

$$\Delta^{(k)} = \{w \in \Omega : w_i \in \Delta_i, \forall i = 0, 1, \cdots, k\}.$$

If $x$ is not eventually periodic, then it can be shown, using a similar technique sketched in the last Section, that the visiting time process, rescaled by $m(C[x_0, x_1, \cdots, x_k]_0)^{-1}$, converges in law to a Poisson point process with rate 1, which we mentioned earlier. Hopefully, all of these results are going to appear in the forthcoming paper [9].

### 5.5.4 Some Questions

One can also consider the case when $\Delta \subsetneq \{1, \cdots, \ell\}$ is chosen such that the subsystem $\Omega_\Delta$ is not irreducible (or if it is irreducible, but maybe not aperiodic). It is expected that periodicity may imply no convergence of the rescaled process. Something like the example given in Figure 5.6 could also prove to be intriguing.

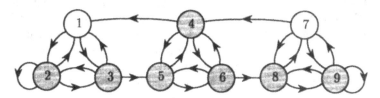

Figure 5.6: Example of a non-irreducible $\Omega_\Delta$.

Intuition may imply that asymptotically, as we approach $\Omega_\Delta$, a generic point $x$ will prefer to see the irreducible subshift on the symbols $\{8,9\}$, since the pair of subsystems, one determined by the symbols $\{2,3\}$, and the other determined by $\{4,5,6\}$, have a drift in the bigger space to the right of Figure 5.6. Finding a scale to study the convergence in law of asymptotic visiting times in this case is not an easy task. (We know the answer in this particular case and it turns out that the exponential of the pressure, relative to the restriction of the potential on the subsystem determined by $\{8,9\}$ is the correct scale.)

How about other subshifts? For instance, substitution systems like the Morse sequence (see B. Host's lectures in this volume)

$$a \mapsto ab, \quad \text{and} \quad b \mapsto ba.$$

The corresponding subshift of $\{a, b\}^{\mathbb{N}}$ is defined as the orbit closure (under the shift) of the infinite word

$$w_\infty = (abbabaabbaababbabaababbaabbabaab\cdots),$$

where $w_\infty$ is obtained as the direct limit of a sequence of substitutions starting with the letter $a$, i.e. $a \mapsto ab \mapsto abba \mapsto abbabaab \mapsto \cdots$. This system is uniquely ergodic, i.e. there exists a unique shift-invariant probability measure $\mu$ and

$$\mu(C[x_0, x_1, \cdots, x_k]_0) = \lim_{n \to \infty} \frac{1}{n} \sum_{j=0}^{n-1} \chi_C(\sigma^j w_\infty),$$

where $C = C[x_0, x_1, \cdots, x_k]_0$. Therefore the measure of $C$ is the relative frequency in which $C$ appears in $w_\infty$. How the times to visit $C$ distribute as $k \to \infty$ if $x = (x_0, x_1, \cdots)$ is a generic point for $\mu$?

Due to some correspondence between certain substitution systems (for instance, the Fibonacci subshift) and irrational rotations of the circle, there may be a way to study these asymptotic processes using the technique in [10]. We mentioned in Section 5.3.3 that for Lebesgue almost every rotation number the process of visits to a neighbourhood of a base point does not converge, if the process is rescaled by any scale. However, when the rotation number is the golden mean, then *there is convergence* and the limiting process is not Poisson and the inter-occurrence times are not asymptotically independent.

It seems to us that the Poisson asymptotic law is associated to Bernoulli systems. The consideration of rotations shows that these processes do not converge in general. Therefore, we are tempted to believe that there is a need of a higher order of mixing to get convergence of these processes. Since some substitution systems are weakly-mixing (although never strongly mixing), it would be interesting to find any analog of the discussions of our lectures in this context.

We should also mention that it is not clear what happens if we consider the same type of problems for irrational rotations of the $n$-torus. In [10] we used renormalisation techniques to prove our results and these are not totally well developed for the $n$-torus.

Finally, we propose a general question. Let $F(t)$ be a distribution function of a positive random variable and consider the full shift on $\ell$ symbols, i.e. $(\{1, \cdots, \ell\}^{\mathbb{N}}, \sigma)$. Are there an ergodic shift-invariant measure $\mu$ on $\{1, \cdots, \ell\}^{\mathbb{N}}$ and a sequence of positive measure sets $A_n$ with $\mu(A_n) \to 0$ as $n \to \infty$, such that

$$F_n(t) = \mu\{x \colon \mu(A_n)\,\tau_n(x) \le t\} \underset{n \to \infty}{\longrightarrow} F(t) \,?$$

(where as usual $\tau_n(x)$ denotes $\inf\{s > 0 : \sigma^s x \in A_n\}$.)

**Acknowledgement.** The author would like to thank François Blanchard for the invitation to give these lectures. Thanks are also due to the organisers and participants of the Summer School for providing an excellent atmosphere during the meeting. The present notes are based on transparencies which were handwritten by the author for the School and copies were given to the participants. The author thanks Rodrigo Davila who did a superb job typing the contents of the transparencies. Rodrigo's source file was edited and modified later by the author to produce the present notes.

# Bibliography

[1] R. Adler and B. Weiss, *Similarity of Automorphisms of the Torus*, Memoirs of the AMS **98** (1970).

[2] D. Aldous, *Probability Approximations via the Poisson Clumping Heuristic*, Springer-Verlag, NY (1989).

[3] R. Arratia, L. Gordon and M.S. Waterman, *The Erdős-Rényi Law in Distribution, for Coin Tossing and Sequence Matching*, Ann. Statist. **18**, pp. 539-570 (1990).

[4] R. Bowen, *Equilibrium States and the Ergodic Theory of Anosov Diffeomorphisms*, Springer Lect. Notes in Maths. **470** (1975).

[5] M. Bramson and S.A. Kalikow, *Non-uniqueness in g-Functions*, Israel J. Math. **84**, pp. 153-160 (1993).

[6] L. Breiman, *Probability Theory*, Addison-Wesley Series in Statistics (1968).

[7] P. Collet, *Some Ergodic Properties of Maps of the Interval*, in Dynamical Systems and Frustrated Systems, eds. R. Bamon, J.-M. Gambaudo and S. Martínez, pp. XX–XX (1996).

[8] Z. Coelho and P. Collet, *Limit Law for the Close Approach of two Trajectories of Expanding Maps of the Circle*, Prob. Th. and Rel. Fields **99**, pp. 237–250 (1994).

[9] Z. Coelho and P. Collet, *Asymptotic Limit Law for Subsystems of Shifts of Finite Type*, in preparation.

[10] Z. Coelho and E. de Faria, *Limit Laws of Entrance Times for Homeomorphisms of the Circle*, Israel J. Math. **93**, pp. 93-112 (1996).

[11] Z. Coelho and W. Parry, *Central Limit Asymptotics for Shifts of Finite Type*, Israel J. Math. **69**, pp. 235-249 (1990).

[12]  Z. Coelho and A. Quas, *Criteria for d̄-Continuity*, Trans. of Amer. Math. Soc. **350**, pp. 223-230 (1998).

[13]  P. Collet and A. Galves, *Asymptotic Distribution of Entrance Times for Expanding Maps of the Interval*, Dynamical Systems and Applications, World Sci. Ser. Appl. Anal., World Sci. Publishing **4**, pp. 139-152 (1995).

[14]  P. Collet, S. Martínez and B. Schmitt, *The Pianigiani-Yorke Measure for Topological Markov Chains*, Israel J. Math. **97**, pp. 61-70 (1997).

[15]  W. Doeblin, *Remarques sur la Théorie Métrique des Fractions Continues*, Compositio Math. **7**, pp. 353-371 (1940).

[16]  W.F. Donoghue Jr, *Monotone Matrix Functions and Analytic Continuation*, Die Grundlehren der Mathematischen Wissenschaften in Einzeldarstellungen Band, Springer-Verlag **207** (1974).

[17]  D.J. Daley and D. Vere-Jones, *An Introduction to the Theory of Point Processes*, Springer Series in Statistics, Springer-Verlag (1988).

[18]  P. Erdös and A. Rényi, *On a New Law of Large Numbers*, J. Analyse Math. **23**, pp. 103-111 (1970).

[19]  G.R. Grimmett and D.R. Stirzaker, *Probability and Random Processes*, Clarendon Press, Oxford (1992).

[20]  M. Hirata, *Poisson Law for Axiom A Diffeomorphisms*, Erg. Th. and Dyn. Sys. **13**, pp. 533-556 (1993).

[21]  M.R. Leadbetter, G. Lindgreen and H. Rootzén, *Extremes and Related Properties of Random Sequences and Processes*, Springer-Verlag, NY (1983).

[22]  J. Neveu, *Processus Pontuels*, in Springer Lecture Notes in Maths **598**, pp. 249-445 (1976).

[23]  B. Pitskel, *Poisson Limit Law for Markov Chains*, Erg. Th. and Dyn. Sys. **11**, pp. 501-513 (1991).

[24]  W. Parry, *Intrinsic Markov Chains*, Trans. of Amer. Math. Soc. **112**, pp. 55-65 (1964).

[25]  W. Parry and M. Pollicott, *Zeta Functions and the Periodic Orbit Structure of Hyperbolic Dynamical Systems*, Astérisque, Soc. Math. de France, pp. 187-188 (1990).

[26]  D. Ruelle, *Thermodynamic Formalism*, Addison-Wesley, Reading, Mass. (1978).

[27]  P. Walters, *An Introduction to Ergodic Theory*, Graduate Texts in Maths, Springer Verlag, New York **79** (1982).

# Chapter 6

# ERGODIC THEORY AND DIOPHANTINE PROBLEMS

*Vitaly BERGELSON*
*The Ohio State University, Department of Mathematics*
*231 West 18th Avenue, Columbus, OH 43210-1174*
*U.S.A.*

## 6.1   Introduction

The topic of these notes is the interplay between ergodic theory, some diophantine problems, and the area of combinatorics called Ramsey Theory.

The first section deals with some classical and well-known diophantine results and their connections with topological and measure-preserving dynamics. Some of the proofs offered in Section 6.2 are very elementary, while some use ergodic-theoretic machinery which is actually much more sophisticated than the results it gives us as applications. This should not in any way discourage the reader since the author's intention was not to produce proofs that are as elementary as possible (see Appendix), but to show how intertwined the different and seemingly remote areas of mathematics may be.

The combinatorial results discussed in Sections 6.3, 6.4, and 6.5 are more recent, but they are as beautiful and, in our opinion, as important as the diophantine facts dealt with in Section 6.2.

It was H. Furstenberg, who, with his publication in 1977 of the ergodic-theoretic proof of Szemerédi's theorem (see Section 6.4), established the link between Ramsey theory and the theory of multiple recurrence. Since then, many open problems of combinatorics and number theory have been solved by the methods of ergodic theory and topological dynamics (see for example, [15], [16], [17], [5], and [6]). As it happens, the developments brought to light many new and intriguing problems which are of interest to both the ergodic theory and combinatorics. (See for example, Section 5 in [3].)

The author hopes that after reading these notes, the reader will develop enough of an interest and curiosity to undertake an in-depth study of Ergodic Ramsey Theory. Such a reader is especially encouraged to look into the book [13] and the recent survey [3].

We conclude this introduction with some general terminological and notational remarks.

An *abstract dynamical system* is a space endowed with some structure and a group or a semigroup of its self-mappings which preserves this structure. *Topological dynamics* concerns itself with compact metric spaces and semigroups of continuous mappings. Measure-preserving dynamics, or *ergodic theory*, works with measure spaces and semigroups of measure-preserving transformations.

Because we are going to be mostly interested in applications to and connections with number theory and combinatorics, the semigroups of structure-preserving mappings we encounter will usually be countable and abelian. However, some material in Section 6.4 will be developed for the general set-up of countable amenable groups, partly because confining ourselves to abelian groups would not make the presentation easier and partly because countable amenable groups seem to be the rightly general object for developing Ergodic Ramsey Theory.

We shall denote topological dynamical systems by $(X, \{T_g\}_{g \in G})$ or $(X, G)$ where $X$ will always mean a compact metric space on which a (countable) semigroup $G$ acts by continuous mappings $T_g$, $g \in G$. In case $G$ is either $\mathbb{N}$ or $\mathbb{Z}$ (i.e. $G$ is generated by a single continuous mapping $T : X \to X$), we shall denote the topological dynamical system by $(X, T)$.

Similarly, a typical notation for a measure-preserving system will be $(X, \mathcal{B}, \mu, \{T_g\}_{g \in G})$, where $(X, \mathcal{B}, \mu)$ is a probability measure space and the transformations $T_g$, $g \in G$ are measure-preserving (i.e. $\forall A \in \mathcal{B}$ and $\forall g \in G$ one has $\mu(T_g^{-1} A) = \mu(A)$). Again, if $G$ is generated by a single, not necessarily invertible, measure-preserving transformation $T$, we shall denote the measure-preserving system by $(X, \mathcal{B}, \mu, T)$.

Given a point $x \in X$, its *orbit* under the action $\{T_g\}_{g \in G}$ is defined by $\{T_g x\}_{g \in G}$. In both the topological and measure-preserving situations, it is important to know how the points of an orbit are distributed in $X$, what can be said about the orbit of a typical point (in this or that sense), how massive is the set of semigroup elements $g$ for which the images $T_g x$ of the point $x$ are close to $x$, etc.

Finally, we want to emphasize the following important point: even in the purely topological set-up, it is often helpful to introduce an invariant measure. For example, the Bogoliouboff-Kryloff theorem (see Theorem 6.2.21 below) tells us that for any topological dynamical system $(X, G)$ with $G$ abelian (or, more generally, amenable), an invariant measure exists. When it is unique, the system $(X, G)$ is called *uniquely ergodic*, and one is then able to make strong statements about the uniform distribution of orbits in $X$. See the discussion at the end of Section 6.2.

Still another setting that warrants the introduction of an invariant measure is discussed in Section 6.4 (see Theorems 6.4.4 and 6.4.17).

## 6.2   Some Diophantine Problems Related to Polynomials and their Connections with Combinatorics and Dynamics

We start this section by giving a simple, "dynamically flavored" proof of the one-dimensional case of Kronecker's theorem on diophantine approximation.

**Theorem 6.2.1** ([24], Theorem 438). *If $\theta$ is irrational, $\alpha$ arbitrary, and $N$ and $\varepsilon$ are positive, then there are integers $n$ and $p$ such that $n > N$ and $|n\theta - p - \alpha| < \varepsilon$.*

**Proof.** Let $\mathbb{T} = \mathbb{R}/\mathbb{Z}$ be the one-dimensional torus, and let $T_\theta \colon x \to x + \theta \pmod 1$ be the "rotation" defined by $\theta$. It is easy to see that Kronecker's theorem is equivalent to the following.

**Statement.** The forward semiorbit of 0, $\{n\theta \pmod 1\}_{n \in \mathbb{N}}$, is dense in $\mathbb{T}$.

Indeed, let $\alpha - [\alpha] = \alpha' \in (0, 1)$, and assume without loss of generality that $\alpha' + \varepsilon < 1$. If for some $n \in \mathbb{N}$ $n\theta \pmod 1 \in (\alpha', \alpha' + \varepsilon)$, then for $p = [n\theta] + [\alpha]$ one has

$$0 < n\theta - p - \alpha < \varepsilon. \tag{6.1}$$

Clearly, if the semiorbit $\{n\theta \pmod 1\}_{n \in \mathbb{N}}$ is dense, then inequality (6.1) is satisfied for infinitely many $n \in \mathbb{N}$ (in particular for some $n > N$).

The proof of the Statement is very short. Note that if for some $n_0 \in \mathbb{N}$ one has either $0 < n_0\theta \pmod 1 < \varepsilon$ or $1 - \varepsilon < n_0\theta \pmod 1 < 1$, then the set of multiples $\{nn_0\theta \pmod 1\}_{n \in \mathbb{N}}$ is $\varepsilon$-dense in $[0, 1]$. But, any set of $M > [\frac{1}{\varepsilon}] + 1$ points in $[0, 1]$ contains a pair of points at a distance $< \varepsilon$. (Just use the pigeon hole principle!). Applying this remark to the set $\{n\theta \pmod 1\}_{n=1}^{M}$, we find $1 \le n_1 < n_2 \le M$ so that either $0 < (n_2 - n_1)\theta \pmod 1 < \varepsilon$ or $1 - \varepsilon < (n_2 - n_1)\theta \pmod 1 < 1$. $\blacksquare$

**Question.** Where exactly was the irrationality of $\theta$ used in the proof?

**Definition 6.2.2** A set $S \subseteq \mathbb{R}$ is called relatively dense, or *syndetic*, if there exists an $L > 0$ such that any interval of length $L$ contains at least one element from $S$.

**Exercise 6.2.3** Derive from the proof of Kronecker's theorem above that for any irrational $\theta$ and any $0 \leq a < b \leq 1$ the set $\{n \in \mathbb{Z} : a < n\theta \ (\mathrm{mod}\ 1) < b\}$ is syndetic.

We now formulate Kronecker's theorem in its general form. Recall that the numbers $x_1, ..., x_m \in \mathbb{R}$ are called *rationally independent* if the relation $\sum_{i=1}^{m} n_i x_i = 0$ with $n_i \in \mathbb{Q}$ is possible only if $n_i = 0$ for all $1 \leq i \leq m$.

**Theorem 6.2.4** ([24], Theorem 442). *Suppose $\theta_1, \theta_2, ..., \theta_k$ are rationally independent, $\alpha_1, \alpha_2, ..., \alpha_k$ are arbitrary, and $N$ and $\varepsilon$ are positive, then there are integers $n > N$ and $p_1, p_2, ..., p_k$ such that $|n\theta_m - p_m - \alpha_m| < \varepsilon$ $(m = 1, 2, ..., k)$.*

**Exercise 6.2.5** Let $f(x) = \sin x + \sin \sqrt{2}x + \sin \sqrt{3}x$, $x \in \mathbb{R}$.
(i) Prove that $f$ is not periodic.
(ii) Prove that for any $\varepsilon > 0$ the set

$$\{\tau \in \mathbb{R} : |f(x + \tau) - f(x)| < \varepsilon \ \forall x \in \mathbb{R}\}$$

is syndetic.
(iii) Functions satisfying (ii) are called *almost periodic*. Prove that if $f$ and $g$ are almost periodic, then $f + g$ is also almost periodic.

**Exercise 6.2.6** Prove that Theorem 6.2.4 is equivalent to the following:

**Statement.** Let $\mathbb{T}^k = \mathbb{R}^k / \mathbb{Z}^k$, and for $\theta = (\theta_1, ..., \theta_k) \in \mathbb{R}^k$ define

$$T_\theta : \mathbb{T}^k \to \mathbb{T}^k \ \text{by} \ T_\theta(x_1, ..., x_k) = (x_1 + \theta_1, ..., x_k + \theta_k) \ (\mathrm{mod}\ 1).$$

If $\theta_1, ..., \theta_k$, and 1 are rationally independent, then for any $x = (x_1, ..., x_k) \in \mathbb{T}^k$ the forward semiorbit $\{T_\theta^n x\}_{n \in \mathbb{N}}$ is dense in $\mathbb{T}^k$.

One can give a proof of Theorem 6.2.4 by refining the argument of the proof of Theorem 6.2.1 above. We prefer to indicate a different and, in a sense, more fruitful approach.

Let $X$ be a compact metric space and $\mu$ a probability measure on $\mathcal{B}_X$ (the $\sigma$-algebra generated by open sets in $X$). We say that a sequence $\{x_n\}_{n \in \mathbb{N}}$ of points in $X$ is *uniformly distributed* with respect to $\mu$ if for any $f \in C(X)$ one has

$$\frac{1}{N} \sum_{n=1}^{N} f(x_n) \xrightarrow[N \to \infty]{} \int_X f d\mu. \tag{6.2}$$

A useful observation is that if $\Phi$ is a countable family of functions in $C(X)$, such that linear combinations of elements of $\Phi$ are dense in $C(X)$, then in order to verify the uniform distribution of a sequence $\{x_n\}_{n\in\mathbb{N}}$ it suffices to check that (6.2) holds for any $f \in \Phi$.

Specifying $X = \mathbb{T}^k$, $\mu = m$ (the Lebesgue measure), and taking into account the fact that finite linear combinations (with complex coefficients) of functions of the form $e^{2\pi i\langle h,t\rangle}$, where $h = (h_1, ..., h_k) \in \mathbb{Z}^k$ and $t = (t_1, ..., t_k) \in \mathbb{T}^k$, are dense in $C(\mathbb{T}^k)$, we obtain the following.

**Theorem 6.2.7** (Weyl criterion). *A sequence $\{x_n\}_{n\in\mathbb{N}} \subset \mathbb{T}^k$ is uniformly distributed if and only if for any nonzero $h = (h_1, ..., h_k) \in \mathbb{Z}^k$ one has*

$$\frac{1}{N}\sum_{n=1}^{N} e^{2\pi i\langle h,x_n\rangle} \xrightarrow[N\to\infty]{} 0.$$

**Exercise 6.2.8** Give a detailed proof of Theorem 6.2.7.

**Exercise 6.2.9** Prove that a sequence $\{x_n\}_{n\in\mathbb{N}} \subset \mathbb{T}^k$ is uniformly distributed if and only if for any nonzero $h \in \mathbb{Z}^k$, the sequence $\{\langle h, x_n\rangle(\text{mod }1)\}_{n\in\mathbb{N}}$ is uniformly distributed in $\mathbb{T}$.

**Exercise 6.2.10** Prove that a sequence $\{x_n\}_{n\in\mathbb{N}}$ is uniformly distributed in $\mathbb{T}$ if and only if for any $0 \le a < b \le 1$ one has

$$\lim_{N\to\infty} \frac{\#\{1 \le n \le N : a \le x_n \le b\}}{N} = b - a.$$

As Hardy and Wright ([24], 23.10) put it, $\{x_n\}_{n\in\mathbb{N}}$ is uniformly distributed in $\mathbb{T}$ "if every subinterval contains its proper quota of points."

Note that even for $k = 1$ Theorem 6.2.7 gives significantly more than Theorem 6.2.1. Indeed, since for any irrational $\theta$ and $h \in \mathbb{Z}$, $h \ne 0$,

$$\frac{1}{N}\sum_{n=1}^{N} e^{2\pi ihn\theta} = \frac{1}{N}e^{2\pi ih\theta}\frac{e^{2\pi ihN\theta} - 1}{e^{2\pi ih\theta} - 1} \xrightarrow[N\to\infty]{} 0,$$

we have for any $f \in C(\mathbb{T})$

$$\frac{1}{N}\sum_{n=1}^{N} f(n\theta\ (\text{mod }1)) \xrightarrow[N\to\infty]{} \int_{\mathbb{T}} f\,dm .$$

This implies (cf. Exercise 6.2.10) that not only does the sequence $\{n\theta\ (\text{mod }1)\}_{n\in\mathbb{N}}$ visit any subinterval $[a,b] \subseteq [0,1]$, but it does so with the right frequency. In other words, the sequence $\{n\theta\ (\text{mod }1)\}$ is *uniformly dense* in $[0,1]$.

Now let $\theta_1, \theta_2, ..., \theta_k$, and 1 be rationally independent. To see that the sequence $x_n = (n\theta_1, ..., n\theta_k)$ (mod 1), $n = 1, 2, ...$ is uniformly distributed in $\mathbb{T}^k$, it is enough (in accordance with Exercise 6.2.9) to show that for any nonzero $h = (h_1, h_2, ..., h_k) \in \mathbb{Z}^k$ the sequence $\{\langle h, x_n \rangle \pmod 1\}_{n \in \mathbb{N}}$ is uniformly distributed in $\mathbb{T}$. Observe that if $(h_1, h_2, ..., h_k) \neq (0, 0, ..., 0)$, then the number $\gamma = \sum_{i=1}^{k} h_i \theta_i$ is irrational. Hence, the sequence $\{n\gamma \pmod 1\}_{n \in \mathbb{N}}$ is uniformly distributed in $\mathbb{T}$. The identity

$$n\gamma(\text{mod } 1) = \left(n \sum_{i=1}^{k} h_i \theta_i\right)(\text{mod } 1) = \left(\sum_{i=1}^{k} h_i n \theta_i\right)(\text{mod } 1) = \langle h, x_n \rangle(\text{mod } 1)$$

implies that we are done.

**Exercise 6.2.11** Let $T_\alpha$, where $\alpha = (\alpha_1, ..., \alpha_k) \in \mathbb{R}^k$ be the rotation of $\mathbb{T}^k$ defined by $T_\alpha(x_1, ..., x_k) = (x_1 + \alpha_1, ..., x_k + \alpha_k)$ (mod 1). For a fixed $x \in \mathbb{T}^k$, let $X = \overline{\{T_\alpha^n x, n \in \mathbb{Z}\}}$. Prove that there exists a unique, $T_\alpha$-invariant Borel measure on $X$. ($T_\alpha$-invariance means that for any Borel set $A \subseteq X$, $\mu(T_\alpha^{-1}A) = \mu(A)$.) Show that the sequence $x_n = T_\alpha^n x, n = 1, 2, ...$, is uniformly distributed with respect to $\mu$.

We will now turn our attention to another refinement of Kronecker's theorem.

**Theorem 6.2.12** ([33]). *If $p(x) = \alpha_s x^s + \alpha_{s-1}x^{s-1} + ... + \alpha_1 x + \alpha_0$ is a polynomial with at least one of its coefficients other than the constant term irrational, then the sequence $\{p(n) \ (mod \ 1)\}_{n \in \mathbb{N}}$ is dense in $\mathbb{T}$.*

We shall discuss several different approaches to the proof of Theorem 6.2.12.

The first possibility is to polynomialize the simple combinatorial proof of Theorem 6.2.1 above. We illustrate the ideas involved on the special case of $p(x) = \theta x^k$, where $\theta$ is irrational, and indicate how to extend the proof to the general case. First notice that to show that the sequence $\{n^k\theta \pmod 1\}_{n \in \mathbb{N}}$ is dense in $\mathbb{T}$, it is enough to prove that for any $\varepsilon > 0$ there exists $n_0$ satisfying either $0 < n_0^k\theta \pmod 1 < \varepsilon$ or $1 - \varepsilon < n_0^k\theta \pmod 1 < 1$.

Indeed, let $0 < a < \varepsilon$. Assuming that $\varepsilon$ is close enough to $a$, let $N = \max\{n \in \mathbb{N} : n^k a < 1\}$ (so that, under our assumption, $N < \varepsilon^{-\frac{1}{k}}$). The largest possible distance between consecutive numbers of the form $n^k a$, $1 \leq n \leq N$, is not greater than $N^k a - (N-1)^k a < kN^{k-1}\varepsilon < k\varepsilon^{\frac{1}{k}}$. Also, it follows from the definition of $N$ that $1 - N^k a < (N+1)^k a - N^k a < 2^k N^{k-1}\varepsilon < 2^k \varepsilon^{\frac{1}{k}}$.

We see that if for some $n_0 \in \mathbb{N}$, $0 < n_0^k\theta \pmod 1 = a < \varepsilon$, then the multiples $n^k n_0^k\theta \pmod 1 = (nn_0)^k\theta \pmod 1$, $n = 1, 2, ..., N$, with $N = [\varepsilon^{-\frac{1}{k}}]$, are $2^k \varepsilon^{\frac{1}{k}}$-dense in $[0,1]$. Since $k$ is fixed and $\varepsilon$ is arbitrary, it gives the desired density of $\{n^k \theta \pmod 1\}_{n \in \mathbb{N}}$. Now, to show that for any $\varepsilon > 0$ there exists an $n_0$, satisfying either $0 < n_0^k \theta \pmod 1 < \varepsilon$ or $1-\varepsilon < n_0^k \theta \pmod 1 < 1$, we shall employ the multidimensional version of the celebrated Van der Waerden's theorem on arithmetic progressions, which will be proved in the next section. (An alternative, more elementary approach is offered in the Appendix.)

**Theorem 6.2.13** (Gallai). *If $k \in \mathbb{N}$ and $\mathbb{N}^k = \bigcup\limits_{i=1}^{r} C_i$ is an arbitrary finite partition of $\mathbb{N}^k$, then one of the $C_i$ contains an "affine k-cube" of the form*

$$Q(n_1, n_2, ..., n_k; h) = \{(n_1 + \delta_1 h, n_2 + \delta_2 h, ..., n_k + \delta_k h) : \\ \delta_i \in \{0,1\}, \ i = 1, 2, ..., k\}.$$

To apply Theorem 6.2.13 to our situation, let us, given $\varepsilon > 0$, induce a partition of $\mathbb{N}^k$ into $r = [\frac{2^k}{\varepsilon}] + 1$ subsets by the following rule:

$$(n_1, n_2, ..., n_k) \in C_i \Leftrightarrow n_1 \cdot n_2 \cdots n_k \theta \pmod 1 \in \left[\frac{i-1}{r}, \frac{i}{r}\right), \ i = 1, 2, ..., r.$$

We shall need the following identity:

$$h^k = \sum_{d=0}^{k} \sum_{\substack{\mathcal{D} \subseteq \{1,2,...,k\} \\ |\mathcal{D}|=d}} (-1)^d \prod_{i \in \mathcal{D}} n_i \prod_{i \notin \mathcal{D}} (n_i + h).$$

(The identity looks frightening, but it is just a concise form of writing down a simple counting procedure. The reader is invited to gain some insight by reflecting on the following special case:

$$h^3 = (n_1 + h)(n_2 + h)(n_3 + h) - n_1(n_2 + h)(n_3 + h) \\ -(n_1 + h)n_2(n_3 + h) - (n_1 + h)(n_2 + h)n_3 + n_1 n_2(n_3 + h) \\ +n_1(n_2 + h)n_3 + (n_1 + h)n_2 n_3 - n_1 n_2 n_3).$$

Let $Q(n_1, n_2, ..., n_k; h)$ be the cube generated by Theorem 6.2.13. For any vertex of this cube of the form $(n_1 + \delta_1 h, n_2 + \delta_2 h, ..., n_k + \delta_k h)$, where $\delta_i \in \{0,1\}$, $i = 1, 2, ...., k$, one has $(\prod\limits_{i=1}^{k}(n_i + \delta_i h))\theta \pmod 1 \in [\frac{i-1}{r}, \frac{i}{r})$. Since the sum of the coefficients $\sum\limits_{d=0}^{k} \sum\limits_{\substack{\mathcal{D} \subseteq \{1,2,...,k\} \\ |\mathcal{D}|=d}} (-1)^d$ equals zero (there are precisely $2^k$ coefficients), one has either $0 < h^k \theta \pmod 1 < \frac{2^k}{r} < \varepsilon$ or $1-\varepsilon < h^k \theta \pmod 1 < 1$. This finishes the proof.

**Exercise 6.2.14** Extend the previous proof to general polynomials by using the following strengthened form of Theorem 6.2.13: Given $k_1, k_2, ..., k_s \in \mathbb{N}$,

suppose $N^{k_j} = \bigcup_{i=1}^{r_j} C_i^{(j)}$, $1 \leq i \leq s$, are arbitrary finite partitions of the lat-
tices. There exists $h \in N$ so that the cubes $Q_i$, $i = 1, 2, ..., s$, whose existence
is promised by Theorem 6.2.13, all have $h$ as the same "edge length".

We now show how one can replace the use of Theorem 6.2.13 by a simple
dynamical argument due to H. Furstenberg (see [13]).

**Definition 6.2.15** Let $(X, \{T_g\}_{g \in G})$ be a topological dynamical system. A
point $x \in X$ is called *recurrent* if for any neighborhood $V$ containing $x$ there
exists $g \in G$, $g \neq e$ so that $T_g x \in V$.

**Exercise 6.2.16** Show that every point $x \in T$ is recurrent with respect to
the transformation $T_\alpha : x \to x + \alpha$ (mod 1). (This follows from the discussion
in the beginning of this section. Note that it is immaterial whether $\alpha$ is
rational or irrational.)

Now, fix an irrational number $\theta$ and consider the dynamical system on $T^2$
defined by $T : (x, y) \to (x + \theta, y + 2x + \theta)$ (mod 1). Following Furstenberg,
we show that *every* point in $T^2$ is recurrent under T. It is easy to verify
by induction that for any integer $n$, $T^n(0,0) = (n\theta, n^2\theta)$ (mod 1), and so
the fact that $(0,0)$ is a recurrent point will imply, in particular, that for
any $\varepsilon$, there is an integer $n \neq 0$ such that $|n^2\theta - m| < \varepsilon$, or, equivalently,
$|\theta - \frac{m}{n^2}| < \varepsilon$. As it was explained above, this implies, in its turn, that the
sequence $\{n^2\theta \pmod 1\}_{n=1}^\infty$ is dense in $[0, 1]$.

We remark that the argument below applies actually to a much wider
class of dynamical systems; namely, the so-called *group extensions* (see [13]).

To show that a point $(x, y) \in T^2$ is recurrent, it is enough to show that
$(x, 0)$ is recurrent. Indeed, denoting the transformation $(x, y) \to (x, y + t)$
(mod 1) by $S_t$, we see that if $T^{n_k}(x, 0) \to (x, 0)$ as $k \to \infty$, then

$$T^{n_k}(x, y) = T^{n_k} S_y(x, 0) = S_y T^{n_k}(x, 0) \to S_y(x, 0) = (x, y).$$

Now, to prove that the point $(x, 0)$ is recurrent, one argues as follows:
Let $O((x, 0))$ denote the (forward) orbit closure of $(x, 0)$: $O((x, 0)) = \overline{\{T^n(x, 0), n \geq 1\}}$. Since every point $x \in T$ is a recurrent point for the transformation $T_\theta : x \to x + \theta$ (mod 1) (see Exercise 6.2.16), there exists $y_0 \in T$
so that $(x, y_0) \in O((x, 0))$. Using the fact that for any $z \in T$, $O(S_y z) = S_y(O(z))$, one gets $(x, 2y_0) \in O((x, y_0))$ (we are suppressing the (mod 1)
sign), which implies $(x, 2y_0) \in O((x, 0))$. Repeating this argument, we get
that for every $n \geq 1$, $(x, ny_0) \in O((x, 0))$. Applying again the result of
Exercise 6.2.16 to $y_0$ and using the fact that $O((x, 0))$ is closed, we obtain
$(x, 0) \in O((x, 0))$. ∎

**Exercise 6.2.17** (i) Define $T : \mathbb{T}^3 \to \mathbb{T}^3$ by

$$T(x, y, z) = (x + \theta, y + 2x + \theta, z + 3y + 3x + \theta) \pmod 1.$$

Check that for any $n \in \mathbb{Z}$, $T^n(0,0,0) = (n\theta, n^2\theta, n^3\theta) \pmod 1$. Generalize this to arbitrarily many dimensions.
(ii) Using (i), show that $\{n^k\theta \pmod 1\}_{n \in \mathbb{N}}$ is dense in $[0,1]$ for any $k \in \mathbb{N}$ and irrational $\theta$.

It turns out that the sequence $\{p(n) \pmod 1\}_{n \in \mathbb{N}}$, where the polynomial $p(t) \in \mathbb{R}[t]$ has at least one "non-constant" coefficient irrational, is actually uniformly distributed in $[0, 1]$.

We shall now describe two approaches to the proof of this fact. The first one, due to Weyl, is based on the fact that if $p(n)$ is a polynomial of degree $d \geq 1$, then for any $h \neq 0$, the polynomial $p(n + h) - p(n)$ has degree $d - 1$. The principle behind the Weyl proof was succinctly formulated by van der Corput in the form of the following proposition (cf. [27]; see also Proposition 6.4.27 and Exercise 6.5.5 (i), (ii) below).

**Theorem 6.2.18** (van der Corput's Difference Theorem). *Let $\{x_n\}_{n \in \mathbb{N}}$ be a sequence of real numbers. If for any $h \in \mathbb{N}$, $h \neq 0$, the sequence $\{x_n - x_{n+h}\}_{n \in \mathbb{N}}$ is uniformly distributed* (mod 1), *then the sequence $\{x_n\}_{n \in \mathbb{N}}$ is uniformly distributed* (mod 1).

**Exercise 6.2.19** Use Theorem 6.2.18 to prove that if the polynomial $p(n)$ has at least one "non-constant" coefficient irrational, then the sequence $\{p(n)\}_{n \in \mathbb{N}}$ is uniformly distributed (mod 1).

The ergodic approach to equidistribution of polynomials, due to Furstenberg, relies on the notion of unique ergodicity. We give here only some general explanations (see [13], Chapter 3 for the details).

**Definition 6.2.20** Let $X$ be a compact metric space and $T : X \to X$ a continuous mapping. The dynamical system $(X, T)$ is called *uniquely ergodic* if there exists only one $T$-invariant Borel probability measure on $X$.

To be able to talk about the uniqueness of invariant measures, one should first make sure that the set of invariant probability measures is non-empty. This is guaranteed by the following:

**Theorem 6.2.21** (Bogoliouboff-Kryloff, [26]). *For any continuous self-map $T : X \to X$ of a compact metric space, there exists a $T$-invariant Borel probability measure.*

The following exercise indicates a way of proving Theorem 6.2.21. Recall that given a compact metric space $X$, the set $M(X)$ of Borel probability measures on $X$ is non-empty, in particular, it contains the point masses (namely, the measures $\mu_x, x \in X$, defined for any Borel set $A$ by $\mu_x(A) = 1$ if $x \in A$ and 0 otherwise), and is compact in the weak*-topology.

**Exercise 6.2.22** Given a measure $\nu \in M(X)$, define $\mu_n = \frac{1}{n} \sum_{k=0}^{n-1} T^k \nu$. Let $\mu$ be any weak* limit point of the sequence $\{\mu_n\}_{n=1}^{\infty}$. Show that $\mu$ is $T$-invariant.

**Exercise 6.2.23** Let $T(x) = ax$ be a rotation of a compact metrizable group $(G, \cdot)$. Prove that the Haar measure is the unique invariant measure for $(G, T)$ if and only if the sequence $\{a^n\}_{n \in \mathbb{N}}$ is dense in $G$.

The uniquely ergodic systems are characterized by the following theorem (for a proof see for example [32], Theorem 6.19).

**Theorem 6.2.24** *Let $T : X \to X$ be a continuous self-mapping of a compact metric space $X$. The dynamical system $(X, T)$ is uniquely ergodic if and only if $\frac{1}{N} \sum_{n=0}^{N-1} f(T^n x)$ converges uniformly to a constant.*

Return now to the transformation on $\mathbb{T}^2$ defined by $T(x, y) = (x + \alpha, y + 2x + \theta)$ (mod 1) which we dealt with above. One can show that the normalized Lebesgue measure $m$ on $\mathbb{T}^2$ is the unique invariant measure with respect to $T$. (See, for example, Proposition 3.10 in [13].) It follows from Theorem 6.2.24 that for any continuous function $f \in C(\mathbb{T}^2)$

$$\frac{1}{N} \sum_{n=0}^{N-1} f(T^n(x,y)) \xrightarrow[N \to \infty]{} \int_{\mathbb{T}^2} f \, dm \qquad \text{uniformly.}$$

In particular, taking $(x, y) = (0, 0)$ and remembering that $T^n(0,0) = (n\alpha, n^2\alpha)$ (mod 1), one obtains the following result:

**Proposition 6.2.25** *The sequence $\left\{(n\alpha, n^2\alpha) \ (\text{mod } 1)\right\}_{n=1}^{\infty}$ is uniformly distributed in $\mathbb{T}^2$.*

**Exercise 6.2.26** Taking for granted the unique ergodicity of the transformations suggested in Exercise 6.2.17 (i) and (ii), prove, with the help of Exercise 6.2.9, the Weyl's theorem on equidistribution of polynomials.

## 6.3    Ramsey Theory and Topological Dynamics

We start this section by formulating several combinatorial results (cf. [21]). The first of them, van der Waerden's theorem, is actually a 1-dimensional special case of Theorem 6.2.13.

**Theorem 6.3.1** *If* $\mathbf{Z} = \bigcup_{i=1}^{r} C_i$ *is an arbitrary finite partitioning of* $\mathbf{Z}$, *then one of the* $C_i$ *contains arbitrarily long arithmetic progressions.*

**Exercise 6.3.2** Call a set $S \subset \mathbf{Z}$ *AP-rich* if $S$ contains arbitrarily long arithmetic progressions. Prove that Theorem 6.3.1 is equivalent to the following statement: If $S \subset \mathbf{Z}$ is AP-rich, then for any finite partition of $S$, $S = \bigcup_{i=1}^{r} C_i$, one of the $C_i$ is also AP-rich.

Now let $F$ be a finite field and $V_F$ an infinite vector space over $F$. (Example: $F = \mathbf{Z}_p$ and $V_F = \mathbf{Z}_p^{\infty} = \{(x_1, x_2, ...) : x_i \in \mathbf{Z}_p, i \in \mathbf{N}$, and all but finitely many $x_i = 0\}$.) A set $A \subset V_F$ is a $d$-dimensional *affine subspace* of $V_F$ if for some $v, x_1, ..., x_d \in V$, where $x_1, ..., x_d$ are linearly independent, $A = v + \mathrm{Span}\{x_1, ..., x_d\}$.

**Theorem 6.3.3** (Geometric Ramsey Theorem, [20]). *If* $r \in \mathbf{N}$ *and* $V_F = \bigcup_{i=1}^{r} C_i$, *then one of the* $C_i$ *contains affine subspaces of arbitrarily large (finite) dimension.*

**Exercise 6.3.4** Call a subset $S \subset V_F$ *AS-rich* (*AS* stands for affine subspace) if $S$ contains affine subspaces of arbitrarily high dimension. Prove that Theorem 6.3.3 is equivalent to the following statement: If $S \subset V_F$ is AS-rich and, for some $r \in \mathbf{N}, S = \bigcup_{i=1}^{r} C_i$ is a finite partitioning, then one of the $C_i$ is also AS-rich.

**Exercise 6.3.5** Give an example of a finite partition of $\mathbf{Z}_p^{\infty}$ (where $p$ is a prime bigger than 2) so that none of the cells of the partition contains an infinite affine subspace. (When $p = 2$ the situation is different. See Exercise 6.3.7 (ii) below.)

The last combinatorial result we want to discuss is the celebrated Hindman's theorem. To formulate it we first introduce some notation that will be used throughout this section.

Let $\mathcal{F}$ denote the set of all non-empty finite sets in $\mathbf{N}$. If $\alpha, \beta \in \mathcal{F}$, then we write $\alpha < \beta$ if the maximal element in $\alpha$ is smaller than the minimal element of $\beta$. We shall say that $\{\alpha_i\}_{i=1}^\infty$ is an increasing sequence in $\mathcal{F}$ if $\alpha_1 < \alpha_2 < \dots$ .

Given an increasing sequence $\{\alpha_i\}_{i=1}^\infty$ in $\mathcal{F}$, we shall denote by $FU(\{\alpha_i\}_{i=1}^\infty)$ the set of all finite unions of "atoms" $\alpha_i, i \in \mathbb{N}$. Note that the set $FU(\{\alpha_i\}_{i=1}^\infty)$ has, in a sense, the same structure as $\mathcal{F}$ (the atoms $\alpha_i$ play the same role as the singletons $\{i\}$ in $\mathbb{N}$).

**Theorem 6.3.6** ([22]). *If $\mathcal{F} = \bigcup_{i=1}^r C_i$ is an arbitrary finite partition of $\mathcal{F}$, then one of the $C_i$ contains $FU(\{\alpha_i\}_{i=1}^\infty)$ for some increasing sequence $\{\alpha_i\}_{i=1}^\infty$ in $\mathcal{F}$.*

**Exercise 6.3.7** For an infinite subset $\{x_1, x_2, \dots\} \subset \mathbb{N}$, let

$$FS(\{x_i\}_{i=1}^\infty) = \{x_{i_1} + x_{i_2} + \dots + x_{i_k} \ : \ i_1 < i_2 < \dots < i_k, \ k \in \mathbb{N}\}.$$

In other words, $FS(\{x_i\}_{i=1}^\infty)$ is the set of all finite sums of elements of the set $\{x_1, x_2, \dots\}$ having distinct indices.
(i) Prove that Theorem 6.3.6 is equivalent to the following statement: If $r \in \mathbb{N}$ and $\mathbb{N} = \bigcup_{i=1}^r C_i$, then one of the $C_i$ contains $FS(\{x_i\}_{i=1}^\infty)$ for some infinite subset $\{x_1, x_2, \dots\}$.
(ii) Prove that for any finite partition of the vector space $\mathbb{Z}_2^\infty$, one of the cells must contain an infinite subspace with the possible exception of the zero vector.

The combinatorial results above have a common feature: they all state that certain structures are undestroyable by finite partitioning. Theorems 6.3.1, and 6.3.3 belong to a vast variety of results which form the body of Ramsey Theory and which have the following general form: if a highly organized structure (complete graph, ordered set, vector space, etc.) is finitely partitioned (or as they say, finitely colored), then one of the pieces will still be highly organized.

We are not going to give a proof of Hindman's theorem here. There are many interesting proofs of this theorem in the literature, each of them lending some new insights (see for example [23], [18], [13], or [3]).

We are now going to formulate and prove a dynamical theorem which has Theorems 6.3.1 and 6.3.3 as corollaries. Before formulating it, we introduce a few more definitions and some notation.

An $\mathcal{F}$-*sequence* in an arbitrary space $Y$ is a sequence $\{y_\alpha\}_{\alpha \in \mathcal{F}}$ indexed by the set $\mathcal{F}$ of the finite non-empty subsets of $\mathbb{N}$. If $Y$ is a (multiplicative) semigroup, one says that an $\mathcal{F}$-sequence defines an *IP-system* if for any $\alpha = \{i_1, i_2, \dots, i_k\} \in \mathcal{F}$, one has $y_\alpha = y_{i_1} y_{i_2} \dots y_{i_k}$. IP-systems should be viewed as generalized semigroups. Indeed, if $\alpha \cap \beta = \emptyset$, then $y_{\alpha \cup \beta} = y_\alpha y_\beta$. We shall often use this formula for sets $\alpha, \beta$ satisfying $\alpha < \beta$.

We will be working with IP-systems generated by homeomorphisms belonging to a commutative group $G$ acting minimally on a compact space $X$. Recall that $(X, G)$ is a minimal dynamical system if for each non-empty open set $V \subset X$ there exist $S_1, ..., S_r \in G$ so that $\bigcup_{i=1}^{r} S_i V = X$.

The following theorem was first proved in [18]; the proof that we give here is based on a proof of its special case in [10].

**Theorem 6.3.8** *Let $X$ be a compact topological space and $G$ a commutative group of its homeomorphisms such that the dynamical system $(X, G)$ is minimal. For any non-empty open set $V \subseteq X$, $k \in \mathbb{N}$, any IP-systems $\{T_\alpha^{(1)}\}_{\alpha \in \mathcal{F}}, ..., \{T_\alpha^{(k)}\}_{\alpha \in \mathcal{F}}$ in $G$ and any $\alpha_0 \in \mathcal{F}$, there exists $\alpha \in \mathcal{F}, \alpha > \alpha_0$ such that $V \cap T_\alpha^{(1)} V \cap ... \cap T_\alpha^{(k)} V \neq \emptyset$.*

**Proof.** We fix a non-empty open $V \subseteq X$ and $S_1, ..., S_r \in G$ with the property that $S_1 V \cup S_2 V \cup ... \cup S_r V = X$. (The existence of $S_1, ..., S_k$ is guaranteed by the minimality of $(X, G)$.) The proof proceeds by induction on $k$. The case $k = 1$ is almost trivial, but we shall do it in detail to set up the notation in a way that indicates the general idea.

So, let $\{T_i\}_{i=1}^{\infty}$ be a fixed sequence of elements in $G$ and $\{T_\alpha\}_{\alpha \in \mathcal{F}}$ the IP-system generated by $\{T_i\}_{i=1}^{\infty}$. (This means of course that for any finite non-empty set $\alpha = \{i_1, i_2, ..., i_m\} \subset \mathbb{N}$, one has $T_\alpha = T_{i_1} T_{i_2} \cdots T_{i_m}$.)

Now we construct a sequence $W_0, W_1, ...$ of non-empty open sets in $X$ so that:

(i) $W_0 = V$;

(ii) $T_n^{-1} W_n \subseteq W_{n-1}, \forall n \geq 1$;

(iii) each $W_n, n \geq 1$, is contained in one of the sets $S_1 V, S_2 V, ..., S_r V$. (We recall that $S_1 V \cup S_2 V \cup ... \cup S_r V = X$.)

To define $W_1$, let $t_1, 1 \leq t_1 \leq r$, be such that $T_1 V \cap S_{t_1} V = T_1 W_0 \cap S_{t_1} V \neq \emptyset$; let $W_1 = T_1 W_0 \cap S_{t_1} V$. If $W_n$ was already defined, then let $t_{n+1}$ be such that $1 \leq t_{n+1} \leq r$ and $T_{n+1} W_n \cap S_{t_{n+1}} V \neq \emptyset$, and let $W_{n+1} = T_{n+1} W_n \cap S_{t_{n+1}} V$. By the construction, each $W_n$ is contained in one of the $S_1 V, ..., S_r V$, so there will necessarily be two natural numbers $i < j$ and $1 \leq t \leq r$ such that $W_i \cup W_j \subseteq S_t V$ (pigeon hole principle!). Let $U = S_t^{-1} W_j$ and $\alpha = \{i+1, i+2, ..., j\}$. We have

$$T_\alpha^{-1} U = T_{i+1}^{-1} T_{i+2}^{-1} \cdots T_j^{-1} S_t^{-1} W_j = S_t^{-1} T_{i+1}^{-1} T_{i+2}^{-1} \cdots T_j^{-1} W_j \subseteq$$
$$\subseteq S_t^{-1} T_{i+1}^{-1} T_{i+2}^{-1} \cdots T_{j-1}^{-1} W_{j-1} \subseteq ... \subseteq S_t^{-1} T_{i+1}^{-1} W_{i+1} \subseteq S_t^{-1} W_i \subseteq V.$$

So, $U \subseteq T_\alpha V$ and $U \subseteq V$ which implies $V \cap T_\alpha V \neq \emptyset$.

Notice that since the pair $i < j$ for which there exists $t$ with the property $W_i \cup W_j \subseteq S_t V$ could be chosen with arbitrarily large $i$, it follows that the set $\alpha = \{i+1, ..., j\}$ for which $V \cap T_\alpha V \neq \emptyset$ could be chosen so that $\alpha > \alpha_0$.

Assume now that the theorem holds for any $k$ IP-systems in $G$. Fix a non-empty set $V$ and $k+1$ IP-systems $\{T_\alpha^{(1)}\}_{\alpha \in \mathcal{F}}, ..., \{T_\alpha^{(k+1)}\}_{\alpha \in \mathcal{F}}$. We shall also fix the homeomorphisms $S_1, ..., S_r \in G$ (whose existence is guaranteed by minimality) satisfying $S_1 V \cup ... \cup S_r V = G$. We shall inductively construct a sequence $W_0, W_1, ...$ of non-empty open sets in $X$ and an increasing sequence $\alpha_1 < \alpha_2 < ...$ in $\mathcal{F}$ so that

(a) $W_0 = V$,

(b) $(T_{\alpha_n}^{(1)})^{-1} W_n \cup (T_{\alpha_n}^{(2)})^{-1} W_n \cup ... \cup (T_{\alpha_n}^{(k+1)})^{-1} W_n \cup \subseteq W_{n-1}$ for all $n \geq 1$, and

(c) each $W_n, n \geq 1$ is contained in one of the sets $S_1 V, ..., S_r V$.

To define $W_1$, apply the induction assumption to the non-empty open set $W_0 = V$ and IP-systems

$$\{(T_\alpha^{(k+1)})^{-1} T_\alpha^{(1)}\}_{\alpha \in \mathcal{F}}, ..., \{(T_\alpha^{(k+1)})^{-1} T_\alpha^{(k)}\}_{\alpha \in \mathcal{F}}.$$

There exists $\alpha_1 \in \mathcal{F}$ such that

$$V \cap (T_{\alpha_1}^{(k+1)})^{-1} T_{\alpha_1}^{(1)} V \cap ... \cap (T_{\alpha_1}^{(k+1)})^{-1} T_{\alpha_1}^{(k)} V$$
$$= W_0 \cap (T_{\alpha_1}^{(k+1)})^{-1} T_{\alpha_1}^{(1)} W_0 \cap ... \cap (T_{\alpha_1}^{(k+1)})^{-1} T_{\alpha_1}^{(k)} W_0 \neq \emptyset.$$

Applying $T_{\alpha_1}^{(k+1)}$, we get

$$T_{\alpha_1}^{(k+1)} W_0 \cap T_{\alpha_1}^{(1)} W_0 \cap ... \cap T_{\alpha_1}^{(k)} W_0 \neq \emptyset.$$

It follows that for some $1 \leq t_1 \leq r$

$$W_1 := T_{\alpha_1}^{(1)} W_0 \cap T_{\alpha_1}^{(2)} W_0 \cap ... \cap T_{\alpha_1}^{(k+1)} W_0 \cap S_{t_1} V \neq \emptyset.$$

Clearly, $W_0$ and $W_1$ satisfy (b) and (c) above for $n = 1$

If $W_{n-1}$ and $\alpha_{n-1} \in \mathcal{F}$ have already been defined, apply the induction assumption to the non-empty open set $W_{n-1}$ (and the IP-systems $\{(T_\alpha^{(k+1)})^{-1} T_\alpha^{(1)}\}_{\alpha \in \mathcal{F}}, ..., \{(T_\alpha^{(k+1)})^{-1} T_\alpha^{(k)}\}_{\alpha \in \mathcal{F}}$) to get $\alpha_n > \alpha_{n-1}$ such that

$$W_{n-1} \cap (T_{\alpha_n}^{(k+1)})^{-1} T_{\alpha_n}^{(1)} W_{n-1} \cap ... \cap (T_{\alpha_n}^{(k+1)})^{-1} T_{\alpha_n}^{(1)} W_{n-1} \neq \emptyset,$$

and hence, for some $1 \leq t_n \leq r$,

$$W_n := T_{\alpha_n}^{(1)} W_{n-1} \cap ... \cap T_{\alpha_n}^{(k+1)} W_{n-1} \cap S_{t_n} V \neq \emptyset.$$

Again, this $W_n$ clearly satisfies the conditions (b) and (c).

Since, by the construction, each $W_n$ is contained in one of the sets $S_1 V, ..., S_r V$, there is $1 \leq t \leq r$ such that infinitely many of the $W_n$ are contained in $S_t V$. In particular, there exists $i$ as large as we please and $j > i$ so that $W_i \cup W_j \subseteq S_t V$. Let $U = S_t^{-1} W_j$ and $\alpha = \alpha_{i+1} \cup ... \cup \alpha_j$.

Notice that $U \subseteq V$, and for any $1 \leq m \leq k+1$, $(T_\alpha^{(m)})^{-1} U \subseteq V$. Indeed,

$$(T_\alpha^{(m)})^{-1} U = (T_{\alpha_{i+1} \cup ... \cup \alpha_j}^{(m)})^{-1} S_t^{-1} W_j = S_t^{-1} (T_{\alpha_{i+1}}^{(m)})^{-1} \cdots (T_{\alpha_j}^{(m)})^{-1} W_j$$

$$\subseteq S_t^{-1} (T_{\alpha_{i+1}}^{(m)})^{-1} \cdots (T_{\alpha_{j-1}}^{(m)})^{-1} W_{j-1} \subseteq \cdots \subseteq S_t^{-1} (T_{\alpha_{i+1}}^{(m)})^{-1} W_{i+1} \subseteq S_t^{-1} W_i \subseteq V$$

It follows that $U \cup (T_\alpha^{(1)})^{-1} U \cup ... \cup (T_\alpha^{(n+1)})^{-1} U \subseteq V$, and this, in turn, implies $V \cap T_\alpha^{(1)} V \cap ... \cap T_\alpha^{(k+1)} V \neq \emptyset$. ∎

**Corollary 6.3.9** *If $X$ is a compact metric space and $G$ a group of its home-omorphisms, then for any $k$ IP-systems $\{T_\alpha^{(1)}\}_{\alpha \in \mathcal{F}}, ..., \{T_\alpha^{(k)}\}_{\alpha \in \mathcal{F}}$ in $G$, any $\alpha_0 \in \mathcal{F}$, and any $\varepsilon > 0$ there exists $\alpha > \alpha_0$ and $x \in X$ such that the diameter of the set $\{x, T_\alpha^{(1)} x, ..., T_\alpha^{(k)} x\}$ is smaller than $\varepsilon$.*

**Proof.** If $(X, G)$ is minimal, then the claim follows immediately from Theorem 6.3.8. If not, then pass to a minimal, non-empty, closed $G$-invariant subset of $X$. (It always exists by Zorn's lemma.)

**Exercise 6.3.10** Under the conditions of Corollary 6.3.9, show that for any $m \in \mathbb{N}$ one can always find $\alpha_1 < \alpha_2 < ... < \alpha_m$ and $x$ such that the set

$$\{T_{\alpha_1}^{(i_1)} T_{\alpha_2}^{(i_2)} \cdots T_{\alpha_m}^{(i_m)} x : i_1, ..., i_m \in \{1, ..., k\}\}$$

has diameter smaller than $\varepsilon$.

Let us now show how to derive Theorems 6.3.1 and 6.3.3 from Corollary 6.3.9. We start with Theorem 6.3.1. Let $r \in \mathbb{N}, r \geq 2$, and let $\Omega = \{1, 2, ..., r\}^{\mathbb{Z}}$ be the (compact) space of all bilateral sequences with entries from the set $\{1, 2, ..., r\}$ with the product topology. We shall use the standard metric on $\Omega$ defined by

$$d(x, y) = \begin{cases} 0, & \text{if } x = y \\ 1, & \text{if } x(0) \neq y(0) \\ \frac{1}{k+1} & \text{otheriwise, where } k \text{ is the maximal natural number with} \\ & \text{the property } x(i) = y(i) \text{ for all } |i| < k. \end{cases}$$

The points of $\Omega$ are in a natural one-to-one correspondence with the partitions of $\mathbb{Z}$ into $r$ sets: if $x = x(n) \in \Omega$, set $C_i = \{n \in \mathbb{Z} : x(n) = i\}$, $i = 1, 2, ..., r$. Let $T : \Omega \to \Omega$ be the shift homeomorphism defined by

$(Tx)(n) = x(n + 1)$. Fix a partition $\mathbf{Z} = \bigcup_{i=1}^{r} C_i$, and let $\xi = \xi(n)$ be the corresponding sequence in $\Omega$. Finally, let $X \subseteq \Omega$ be the orbital closure of $\xi$:

$$X = \overline{\{T^n \xi, n \in \mathbf{Z}\}}$$

Let $\{n_i\}_{i=1}^{\infty}$ be an arbitrary sequence in $\mathbf{Z}$. For $\alpha \in \mathcal{F}$ define $n_\alpha := \sum_{i \in \alpha} n_i$, and consider IP-systems

$$T_\alpha^{(m)} := T^{mn_\alpha}, \quad m = 1, 2, ..., k.$$

By Corollary 6.3.9, for any $\varepsilon > 0$ there exist $\alpha \in \mathcal{F}$ and $x \in X$ such that the diameter of $\{x, T^{n_\alpha} x, ..., T^{kn_\alpha} x\}$ is less than $\varepsilon$. Let $\eta > 0$ be such that $d(x, y) < \eta$ implies that the sequences $x(n)$ and $y(n)$ coincide for $|n| \leq k|n_\alpha|$. Since the orbit $\{T^n \xi, n \in \mathbf{Z}\}$ is dense in $X$, there exists $m_0 \in \mathbf{Z}$ such that $T^{m_0}\xi$ and $x$ agree on the interval $\left[-k|n_\alpha|, k|n_\alpha|\right]$.

It follows that $\xi(m_0) = \xi(m_0 + n_\alpha) = ... = \xi(m_0 + kn_\alpha)$. If this common value is $j$, then clearly $C_j$ contains an arithmetic progression of length $k$. Note that as a by-product, we showed that the difference of the length $k$ arithmetic progression always to be found in one cell of the partition can be chosen from any prescribed IP-set.

**Exercise 6.3.11** Prove that van der Waerden's theorem is equivalent to the following:

**Statement.** Let $\mathbf{Z} = \bigcup_{i=1}^{r} C_i$ be an arbitrary finite partition of the integers and $F \in \mathcal{F}$. One of the $C_i$ necessarily contains an affine image of $F$. In other words, one of the $C_i$ contains a set of the form $a + nF = \{a + nx : x \in F\}$ for some $a \in \mathbf{Z}$ and $n \in \mathbf{N}$.

The above statement is a special case of the following multidimensional version of van der Waerden's theorem which also follows from Theorem 6.3.8.

**Theorem 6.3.12** *For any $d \in \mathbf{N}$ and finite subset $F \subset \mathbf{Z}^d$, if $\mathbf{Z}^d = \bigcup_{i=1}^{r} C_i$ is a finite partition of $\mathbf{Z}^d$, then one of the $C_i$ contains a set of the form $a + nF = \{a + nx : x \in F\}$ for some $a \in \mathbf{Z}^d, n \in \mathbf{N}$.*

**Exercise 6.3.13** Prove that Theorems 6.3.12 and 6.2.13 are equivalent.

We move now to the derivation of Theorem 6.3.3. Let $V_F$ be a vector space over a finite field $F$ of characteristic $p$. Without loss of generality, we shall assume that $V_F$ is countable. As an abelian group, $V_F$ has a natural representation as the direct sum of countably many copies of $F$:

$$F^\infty = \{g = (a_1, a_2, ..., ) : a_i \in F \text{ and all but finitely many } a_i = 0\}.$$

Fix an IP-system $\{g_\alpha\}_{\alpha\in\mathcal{F}}$ such that $\mathrm{Span}\{g_\alpha, \alpha \in \mathcal{F}\}$ is an infinite sub-set in $V_F$. We will show a stronger fact that if $V_F$ is partitioned into $r$ subsets $C_1, C_2, ..., C_r$, then one of the $C_i$ contains an affine subspace of arbitrarily large dimension which is generated by elements of $\{g_\alpha\}_{\alpha\in\mathcal{F}}$. Let $\Omega = \{1, 2, ...r\}^{V_F}$. In other words, $\Omega$ is the set of all functions defined on $V_F$ which take values in $\{1, 2, ..., r\}$. With its product topology, $\Omega$ is a compact topological space. Introduce a metric on $\Omega$ analogous to the one used for $\{1, ..., r\}^{\mathbf{Z}}$ above. For $g = (a_1, a_2, ...) \in V_F$, let $|g|$ be the minimal natural number such that $a_i = 0$ for all $i > |g|$. For $x = x(g)$ and $y = y(g)$ in $\Omega$, define

$$
d(x,y) = \begin{cases} 0, & \text{if } x = y \\ 1, & \text{if } x(0) \neq y(0) \\ \frac{1}{k+1} & \text{otheriwse, where } k \text{ is the maximal natural number with} \\ & \text{the property } x(g) = y(g) \text{ for all } |g| < k \end{cases}
$$

(where $\mathbf{0}$ denotes the element $(0, 0, ...) \in V_F$). Let $V_F = \bigcup_{i=1}^{r} C_i$ be a partition of $V_F$, and define $\xi \in \Omega$ by $\xi(g) = i \Leftrightarrow g \in C_i$.

We show first that one of the $C_i$ contains an affine line (i.e. a one-dimensional affine subspace). For $h \in V_F$, define $T_h : \Omega \to \Omega$ by $(T_h x)(g) = x(gh)$. Clearly $T_h$ is a homeomorphism of $\Omega$ for every $h \in V_F$. Let $X \subseteq \Omega$ be the orbital closure of $\xi(g)$: $X = \overline{\{T_h\xi, h \in V_F\}}$.

Use now the IP-system $\{g_\alpha\}_{\alpha\in\mathcal{F}}$ to define an IP-system of homeomor-phisms of $X$. Put $T_\alpha := T_{g_\alpha}, \alpha \in \mathcal{F}$, and for each $c \in F, c \neq 0$, define an IP-system by $T_\alpha^{(c)} = T_{cg_\alpha}, \alpha \in \mathcal{F}$. This way we get $q - 1$ (where $q = |F|$) IP-systems of commuting homeomorphisms of $\Omega$ (and of $X$). Applying Corol-lary 6.3.9 to the space $X$ and these IP-systems and taking $\varepsilon < 1$, we get a point $x \in X$ and $\alpha_1 \in \mathcal{F}$ such that the diameter of $\{T_{c\alpha_1}x, c \in \mathcal{F}\}$ is less than 1. This implies that $x(0) = x(cg_{\alpha_1})$ for every $c \in F$. Since the orbit $\{T_h\xi, h \in V_F\}$ is dense in $X$, there exists $h_0 \in V_F$ such that $(T_{h_0}\xi)(g)$ and $x(g)$ agree on all $g$ satisfying $|g| \leq |g_{\alpha_1}|$. If $\xi(h_0) = i$ then $C_i$ contains the affine line $\{h_0 + cg_{\alpha_1}, c \in F\}$ (in view of our assumptions on $\{g_\alpha\}_{\alpha\in\mathcal{F}}$, we, of course, took care to choose $\alpha_1$ so that $g_{\alpha_1} \neq 0$.) The statement about affine spaces of arbitrary dimension follows by iteration (cf. Exercise 6.3.10) and is left to the reader.

# 6.4 Density Ramsey Theory and Ergodic Theory of Multiple Recurrence

In this section, we concern ourselves with *density* Ramsey Theory and its links to ergodic theory. While the main theme of *partition* Ramsey theory is to look for nontrivial patterns in one cell of an arbitrary finite partition, the typical density Ramsey theory statement concerns an appropriately defined notion of largeness: any large subset of a highly organized structure contains large, highly organized substructures. Two basic properties are usually required from the notion of largeness:

(i) if $A$ is large and $A = \bigcup_{i=1}^{r} C_i$, then at least one of the $C_i$ is large;

(ii) the family of large subsets of a given set with a particular structure is invariant under some natural semigroup of structure preserving transformations.

We shall discuss now the density version of Theorems 6.3.1 and 6.3.3. (As for the density version of Theorem 6.3.6, see the discussion in [3], Section 6.4.) As Graham, Rothschild and Spencer put it in [21], "for all Ramsey theorems, one can express (but not always prove) the corresponding density statements." For a set $E \subseteq \mathbf{Z}$ define its *upper density* by

$$\bar{d}(E) = \limsup_{N \to \infty} \frac{|E \cap [-N, N]|}{2N + 1}.$$

Clearly the property of a set of integers to have positive upper density satisfies property (i) above. It is also invariant with respect to the shift: for any $k \in \mathbf{Z}, \bar{d}(E + k) = \bar{d}(E)$, where $E + k = \{x + k, x \in E\}$.

We formulate now a density version of van der Waerden's theorem. We say *a* density rather than *the* density version since $\bar{d}$ is not the only notion of largeness which leads to a density generalization of van der Waerden's theorem. See Exercise 6.4.2 below.

**Theorem 6.4.1** (Szemerédi, [31]). *If a set $E \subseteq \mathbf{Z}$ has positive upper density, then it contains arbitrarily long arithmetic progressions.*

**Exercise 6.4.2** (i) Derive from Theorem 6.4.1 the following finitistic version of it: For any $\varepsilon > 0$ and $k \in \mathbf{N}$ there exists $N = N(\varepsilon, k) \in \mathbf{N}$ such that if $I = [a, b] \subset \mathbf{Z}$ is an interval with $|b - a| \geq N$ and $E \subseteq I$ satisfies $\frac{|E|}{|I|} > \varepsilon$, then $E$ contains a $k$-term arithmetic progression.

(ii) For any $E \subseteq \mathbf{Z}$, let the quantity $d^*(E) = \limsup_{N-M \to \infty} \frac{|E \cap \{M, M+1, \ldots, N\}|}{N - M + 1}$ denote its *upper Banach density*. Call a set $E \subseteq \mathbf{Z}$ $d^*$-*large* if $d^*(E) > 0$. Clearly, $E$ is $d^*$-large if and only if for some sequence of intervals

$$I_N = [a_N, b_N] \subseteq \mathbf{Z} \text{ with } |b_N - a_N| \to \infty, \ \limsup_{N \to \infty} \frac{|E \cap I_N|}{|I_N|} > 0.$$

(In other words, each such sequence $I_N, N = 1, 2, \ldots$, defines a notion of largeness (check!), and to be $d^*$-large means to be large with respect to some sequence of intervals of increasing length). Prove that Theorem 6.4.1 implies that a $d^*$-large set contains arbitrarily long arithmetic progressions.

The original proof of Theorem 6.4.1 in [31] is a brilliant and highly non-trivial piece of combinatorial reasoning. A different, ergodic theoretical proof was given by Furstenberg in [12], thereby starting a new branch of mathematics, Ergodic Ramsey Theory. Soon the methods of ergodic theory turned out to be very useful in proving some natural density conjectures and have led to some strong results, which so far have no conventional combinatorial proofs (cf. [15], [16], [17], [4], [5], [6], [29], [7], [8]).

Furstenberg derived Szemerédi's theorem from a beautiful, far-reaching extension of the classical Poincaré recurrence theorem which corresponds to the case $k = 1$ in the following:

**Theorem 6.4.3** (Furstenberg, [12]). *Let $(X, \mathcal{B}, \mu, T)$ be an invertible probability measure preserving system. For any $k \in \mathbb{N}$ and $A \in \mathcal{B}$ with $\mu(A) > 0$ there exists $n \in \mathbb{N}$ such that*

$$\mu(A \cap T^{-n}A \cap \ldots \cap T^{-kn}A) > 0.$$

To see that Theorem 6.4.1 follows from Theorem 6.4.3, one needs the following:

**Theorem 6.4.4** *(Furstenberg's Correspondence Principle.) Let $E \subseteq \mathbb{Z}$ with $d^*(E) > 0$. Then there exist an invertible probability measure preserving system $(X, \mathcal{B}, \mu, T)$ and a set $A \in \mathcal{B}$ with $\mu(A) = d^*(E)$ such that for any $k \in \mathbb{N}$ and $n_1, n_2, \ldots, n_k \in \mathbb{Z}$ one has:*

$$d^*(E \cap (E - n_1) \cap \ldots \cap (E - n_k)) \geq \mu(A \cap T^{-n_1}A \cap \ldots \cap T^{-n_k}A).$$

**Exercise 6.4.5** Prove that Theorem 6.4.1, in its turn, implies Theorem 6.4.3.

We shall formulate now a density version of Theorem 6.3.3. Let $V_F$ be a countable vector space over a finite field $F$. As before, it will be convenient to work with the realization of $V_F$ as a direct sum of countably many copies of $F$:

$$F^\infty = \{g = (a_1, a_2, \ldots) : a_i \in F, i \in \mathbb{N} \text{ and } a_i = 0 \text{ for all but finitely many } i\}$$

Let $F_n = \{g = (a_1, a_2, \ldots) : a_i = 0 \ \forall i > n\}$. For a set $S \subseteq V_F$ define its *upper density*, $\bar{d}(s)$ by

$$\bar{d}(S) = \limsup_{n \to \infty} \frac{|S \cap F_n|}{|F_n|}.$$

**Theorem 6.4.6** *Let $S \subseteq V_F$ with $\bar{d}(S) > 0$. Then $S$ contains affine subspaces of arbitrarily high dimension.*

**Exercise 6.4.7** Define in $V_F = F^\infty$ the notion of upper Banach density similar to that defined in Exercise 6.4.2 (ii) for subsets of $\mathbb{Z}$. Derive from 6.4.6 a finitistic statement similar to that of Exercise 6.4.2 (i) and a statement obtained by replacing in the formulation of Theorem 6.4.6 the upper density by the upper Banach density.

Similarly to the situation with Szemerédi's theorem, Theorem 6.4.6 follows from a measure theoretical theorem dealing with multiple recurrence:

**Theorem 6.4.8** Let $\{T_g\}_{g \in V_F}$ be a measure preserving action of $V_F = F^\infty$ on a probability measure space $(X, \mathcal{B}, \mu)$. Let $A \in \mathcal{B}, \mu(A) > 0$. Then for some $g \in V_F, g \neq e$ one has $\mu(\bigcap_{c \in F} T_{cg} A) > 0$.

**Corollary 6.4.9** Under the assumptions of Theorem 6.4.8, for any $k$ there exist $g_1, \ldots, g_k$ in $V_F$ such that $\dim\left(\mathrm{Span}\{g_1, \ldots, g_k\}\right) = k$ and

$$\mu\left(\bigcap_{i=1}^{k} \bigcap_{c \in F} T_{cg_i} A\right) > 0.$$

**Exercise 6.4.10** Derive Corollary 6.4.9 from Theorem 6.4.8.
*Hint: if $g_1, \ldots, g_k$ have already been found, apply Theorem 6.4.8 to the set $A_k = \bigcap_{i=1}^{k} \bigcap_{c \in F} T_{cg_i} A$ and to the (sub) action $\{T_g\}_{g \in V_F^{(k)}}$, where, for some fixed $m$ satisfying $m \geq \max_{1 \leq i \leq k} |g_i|$,*

$$V_F^{(k)} = \left\{g = (a_1, a_2, \ldots) \in V_F \ : \ a_1 = a_2 = \ldots = a_m = 0\right\}$$

To derive Theorem 6.4.6 from Theorem 6.4.8 (or, rather, from Corollary 6.4.9) one utilizes the Furstenberg correspondence principle for $V_F$:

**Theorem 6.4.11** Let $S \subseteq V_F$ with $\bar{d}(S) > 0$. Then there exist a measure preserving system with "time" $V_F, (X, \mathcal{B}, \mu, \{T_g\}_{g \in V_F})$ and a set $A \in \mathcal{B}$ with $\mu(A) = \bar{d}(S)$ such that for any $k \in \mathbb{N}$ and any $g_1, g_2, ..., g_k \in V_F$ one has:

$$\bar{d}(S \cap (S - g_1) \cap \ldots \cap (S - g_k)) \geq \mu(A \cap T_{g_1}^{-1} A \cap \ldots \cap T_{g_k}^{-1} A).$$

The apparent similarity of Theorems 6.4.4 and 6.4.11 hints that there is a general Furstenberg correspondence principle which encompasses both theorems. This is indeed so and the "right" class of groups to which it applies are countable *amenable* groups. Among the many equivalent definitions of amenability the one of major importance to ergodic theory is the following one.

**Definition 6.4.12** A countable group $G$ is called amenable, if it has a (left) Følner sequence, namely, a sequence of finite sets $\Phi_n \subset G, n \in \mathbb{N}$ with $|\Phi_n| \to \infty$ and such that $\frac{|\Phi_n \cap g\Phi_n|}{|\Phi_n|} \to 1$ for all $g \in G$.

**Remark 6.4.13** Strictly speaking, Definition 6.4.12 is a definition of *left* amenability. *Right* amenability is defined as the property of group $G$ of possessing a *right Følner sequence*, i.e. a sequence of finite sets $\Phi_n \subset G, n \in \mathbb{N}$ with $|\Phi_n| \to \infty$ such that $\frac{|\Phi_n \cap \Phi_n g|}{|\Phi_n|} \to 1$ for all $g \in G$. It is known, however, that the notions of right and left amenability coincide. At the same time one should be aware of the fact that in non-abelian groups not every right Følner sequence is necessarily a left Følner sequence and vice versa. Since in these notes we are mostly concerned with abelian groups, these subtleties will not bother us too much.

**Exercise 6.4.14** (i) Let $M_n = [a_n^{(1)}, b_n^{(1)}] \times \ldots \times [a_n^{(d)}, b_n^{(d)}] \subset \mathbb{Z}^d, n \in \mathbb{N}$, and assume that $|b_n^{(i)} - a_n^{(i)}| \to \infty, i = 1, 2, ..., d$. Show that $M_n, n \in \mathbb{N}$ is a Følner sequence in $\mathbb{Z}^d$.
(ii) Check that the sets $F_n, n \in \mathbb{N}$ defined before the formulation of Theorem 6.4.6 form a Følner sequence in $V_F$.
(iii) Give an example of a left Følner sequence in a non-abelian group which is not a right Følner sequence.

The set of countable amenable groups is quite rich and includes all countable abelian groups and many classes of non-abelian ones, for example, solvable, locally finite, etc. On the other hand, the free group on two or more generators and the groups $SL(n, \mathbb{Z}), n \geq 2$ are not amenable.

The following version of the classical von Neumann ergodic theorem for amenable groups is merely an illustration of the principle that many results of conventional ergodic theory of one-parameter actions extend naturally to amenable groups.

**Theorem 6.4.15** *Let $H$ be a Hilbert space and let $\{U_g\}_{g \in G}$ be an antirepresentation of a countable amenable group $G$ as a group of unitary operators on $H$ (i.e. $U_{g_1} U_{g_2} = U_{g_2 g_1}$ for all $g_1, g_2 \in G$). Let $H_c = \{f \in H : U_g f = f \; \forall g \in G\}$ and let $P$ be the orthogonal projection on $H_{\text{inv}}$. For any left Følner sequence $\Phi_n, n \in \mathbb{N}$ in $G$ one has:*

$$\lim_{n \to \infty} \left\| \frac{1}{|\Phi_n|} \sum_{g \in \Phi_n} U_g f - Pf \right\| = 0.$$

**Proof.** In complete analogy with the case of a single operator (or rather the $\mathbb{Z}$-action generated by a single unitary operator) one checks that the orthogonal complement of $H_{\text{inv}} \in H$, call it $H_{\text{erg}}$, coincides with the space $\overline{\text{Span}\{f - U_g f : f \in H, g \in G\}}$. It remains to show that on $H_{\text{erg}}$ the limit in question is zero. It is enough to prove it for the elements of the form $f - U_{g_0} f$.

We have:

$$\left\| \frac{1}{|\Phi_n|} \sum_{g \in \Phi_n} U_g(f - U_{g_0} f) \right\| = \left\| \frac{1}{|\Phi_n|} \sum_{g \in \Phi_n} U_g f - \frac{1}{|\Phi_n|} \sum_{g \in \Phi_n} U_{g_0 g} f \right\|$$

$$= \left\| \frac{1}{|\Phi_n|} \sum_{g \in \Phi_n} U_g f - \frac{1}{|\Phi_n|} \sum_{g \in g_0 \Phi_n} U_g f \right\| \le \frac{|\Phi_n \triangle g_0 \Phi_n|}{|\Phi_n|} \|f\|.$$

Since the definition of left Følner sequence clearly implies $\dfrac{|\Phi_n \triangle g_0 \Phi_n|}{|\Phi_n|} \to 0$, we are done.

**Exercise 6.4.16** Recall that a measure preserving action $\{T_g\}_{g \in G}$ on a probability space $(X, \mathcal{B}, \mu)$ is *ergodic* if any set $A \in \mathcal{B}$ which satisfies $\mu(T_g A \triangle A) = 0$ for all $g \in G$ has measure zero or measure one. Derive from Theorem 6.4.15 the following statement:

If $\{T_g\}_{g \in G}$ is an ergodic measure preserving action of an amenable group $G$, then for any (left or left) Følner sequence $\Phi_n, n \in \mathbb{N}$ and any $A_1, A_2 \in \mathcal{B}$ one has:

$$\frac{1}{|\Phi_n|} \sum_{g \in \Phi_n} \mu(A_1 \cap T_g A_2) \to \mu(A_1)\mu(A_2).$$

We shall formulate and prove now the Furstenberg correspondence principle for countable amenable groups.

**Theorem 6.4.17** *Let $G$ be a countable amenable group and $\Phi_n, n \in \mathbb{N}$ a left Følner sequence in $G$. Let $E \subseteq G$ have positive upper density with respect to $\{\Phi_n\}_{n=1}^{\infty}$: $\bar{d}(E) = \limsup\limits_{n \to \infty} \frac{|E \cap \Phi_n|}{|\Phi_n|} > 0$ (for convenience we are suppressing in the notation $\bar{d}(E)$ the dependence on $\{\Phi_n\}$). Then there exists a probability measure preserving system with "time" $G$, $(X, \mathcal{B}, \mu, \{T_g\}_{g \in G})$ and a set $A \in \mathcal{B}$ with $\mu(A) = \bar{d}(E)$ such that for any $k \in \mathbb{N}$ and $g_1, g_2, ..., g_k \in G$ one has*

$$\bar{d}(E \cap g_1^{-1} E \cap ... \cap g_k^{-1} E) \ge \mu(A \cap T_{g_1}^{-1} A \cap ... \cap T_{g_k}^{-1} A).$$

Before giving a proof of Theorem 6.4.17 we shall need a crucial fact linking sets of positive upper density in $G$ with invariant linear functionals on the space $B(G)$ of all complex-valued bounded functions on $G$. We recall first some definitions. Let $G$ be a group and $Y$ a closed subset of the space $B(G)$ of all bounded complex valued functions equipped with the uniform norm $\| \cdot \|_\infty$. Assume additionally that $Y$ is closed under complex conjugation and contains all the constants. A linear functional $L : B(G) \to \mathbb{C}$ is a *left-invariant mean*, if it has the following additional properties:

(i) $L(\bar{f}) = \overline{L(f)} \quad \forall f \in Y$.

(ii) if $f \geq 0$ then $L(f) \geq 0$ and $L(1) = 1$.

(iii) For all $g \in G$ and $f \in Y$  $L(_g f) = L(f)$ where $_g f(t) = f(gt)$.

**Exercise 6.4.18** For $f \in Y$ define $\check{f}(t) = f(t^{-1})$. Show that, if $L$ is a left-invariant mean on $Y$, the functional $\check{L}$ defined on $\check{Y} = \{\check{f} : f \in Y\}$ by $\check{L}(\check{f}) = L(f)$ is a *right-invariant* mean, i.e. for all $g \in G$ and $f \in \check{Y}$, $\check{L}(f^g) = \check{L}(f)$, where $f^g(t) = f(tg)$.

One can show that a countable group $G$ is amenable if and only if the space $B(G)$ admits a left-invariant mean. We shall need this fact only in one direction; it follows from the following result.

**Proposition 6.4.19** *Let $G$ be a countable group and $\Phi_n, n \in \mathbb{N}$ a left Følner sequence in $G$. Assume that a set $E \subseteq G$ has positive upper density with respect to $\{\Phi_n\}$ : $\bar{d}(E) = \limsup\limits_{n \to \infty} \frac{|E \cap \Phi_n|}{|\Phi_n|} > 0$. There exists a left-invariant mean $L$ on $B(G)$ satisfying the following conditions:*

(i) $L(1_E) = \bar{d}(E)$

(ii) *for any $k \in \mathbb{N}$ and any $g_1, g_2, .., g_k \in G$*

$$\bar{d}(E \cap g_1^{-1} E \cap ... \cap g_k^{-1} E) \geq L(1_E \cdot 1_{g_1^{-1} E} \cdot .... \cdot 1_{g_k^{-1} E}).$$

**Proof.** Let $P$ be the (countable!) family of subsets of $G$ having the form $\bigcap\limits_{i=1}^{k} g_i^{-1} E$, where $g_i \in G$, $i = 1, \ldots, k$. By using the diagonal procedure we arrive at a subsequence $\{\Phi_{n_i}\}_{i=1}^{\infty}$ of our Følner sequence such that for our fixed set $E$ we have $\bar{d}(E) = \lim\limits_{i \to \infty} \frac{|E \cap \Phi_{n_i}|}{|\Phi_{n_i}|}$ and for any $S \in P$ the limit $L(S) = \lim\limits_{i \to \infty} \frac{|S \cap \Phi_{n_i}|}{|\Phi_{n_i}|} = \lim\limits_{i \to \infty} \frac{1}{|\Phi_{n_i}|} \sum\limits_{y \in \Phi_{n_i}} 1_S(g)$ exists. Notice that for any $g_1, ..., g_k \in G$ this gives

$$\bar{d}(\bigcap_{j=1}^{k} g_j^{-1} E) = \limsup_{n \to \infty} \frac{\left|(\bigcap\limits_{j=1}^{k} g_j^{-1} E) \cap \Phi_n\right|}{|\Phi_n|} \geq \lim_{i \to \infty} \frac{\left|(\bigcap\limits_{j=1}^{k} g_j^{-1} E) \cap \Phi_{n_i}\right|}{|\Phi_{n_i}|}$$

$$= L(\bigcap_{j=1}^{k} 1_{g_j^{-1} E}).$$

Extending out by linearity we get a linear functional $L$ on the subspace $Y_0 \subset B_{\mathbb{R}}(G)$ of finite linear combinations of characteristic functions of sets in $P$. To extend $L$ from $Y_0$ to $B_{\mathbb{R}}(G)$ define Minkowski functional $p(\varphi)$ by $p(\varphi) =$

$\limsup_{i\to\infty} \frac{1}{|\Phi_{n_i}|} \sum_{g\in\Phi_{n_i}} \varphi(g)$. Clearly, for all $\varphi_1, \varphi_2 \in B_{\mathbb{R}}(G)$, $p(\varphi_1 + \varphi_2) \le p(\varphi_1) +$
$p(\varphi_2)$ and for any non-negative $t$, $p(t\varphi) = tp(\varphi)$. Also, on $Y_0$, $L(\varphi) = p(\varphi)$.
By Hahn-Banach theorem there is an extension of $L$ (which we denote by $L$ as
well) to $B_{\mathbb{R}}(G)$ satisfying $L(\varphi) \le p(\varphi) \; \forall \varphi \in B_{\mathbb{R}}(G)$. This $L$ naturally extends
to a functional on the space $B(G)$ of complex-valued bounded functions and
we are done.

**Remark 6.4.20** For the proof of the Furstenberg correspondence principle
we shall need only the existence of linear functional satisfying the conditions
(i) and (ii) and defined on the uniformly closed and closed under conjugation
algebra of functions on $G$ which is generated by the characteristic function
$1_E$ and its shifts. As the first half of the proof above shows this could be
achieved without involving the Hahn-Banach theorem.

**Proof of Theorem 6.4.17.** Let $f(h) = 1_E(h)$ be the characteristic function
of $E$. Let $\mathcal{A}$ be the uniformly closed and closed under the conjugation algebra generated by the function $f$ and its shifts of the form $_g f(h) = f(gh)$. $\mathcal{A}$
is a separable (linear combinations with rational coefficients are dense in $\mathcal{A}$),
commutative $C^*$-algebra with respect to sup norm. By Gelfand representation theorem, $\mathcal{A}$ is isomorphic to the space $C(X)$ of continuous functions on
a compact metric space $X$. Let $L$ be a right invariant mean on $B(G)$ satisfying the condition (i) and (ii) of the Proposition 6.4.19 above. The linear
functional $L$ induces a positive linear functional $\tilde{L}$ on $C(X)$. By the Riesz
representation theorem there exists a regular Borel measure $\mu$ on the Borel
$\sigma$-algebra $\mathcal{B}$ of $X$ such that for any $\varphi \in \mathcal{A}$

$$L(\varphi) = \tilde{L}(\tilde{\varphi}) = \int_X \tilde{\varphi} d\mu,$$

where $\tilde{\varphi}$ denotes the image of $\varphi$ in $C(X)$. Now, since the Gelfand transform
establishing the isomorphism between $\mathcal{A}$ and $C(X)$ preserves the algebraic
operations and since the characteristic functions of sets are the only idempotents in $C(X)$, it follows that the function $\tilde{f}$ in $C(X)$ which corresponds to
$f(h) = 1_E(h)$ is the characteristic function of a set $A \subset X : \tilde{f}(x) = 1_A(x)$.
This gives

$$\bar{d}(E) = L(1_E) = \tilde{L}(1_A) = \int_X 1_A d\mu = \mu(A)$$

Finally, notice that the shift operators $\varphi(h) \to \varphi(gh)$, $\varphi \in \mathcal{A}$, $g \in G$,
form an anti-action of $G$ on $\mathcal{A}$, which induces an anti-action $\{T_g\}_{g\in G}$ on
$C(X)$ defined for $\varphi \in \mathcal{A}$ by $(T_g)\tilde{\varphi} = \widetilde{_g\varphi}$, where $_g\varphi(h) = \varphi(gh)$.

Now, the transformations $T_g, g \in G$ are $C^*$-isomorphisms of $C(X)$ (since they are induced by $C^*$-isomorphisms $\varphi \to {}_g\varphi$ of $\mathcal{A}$). It is known that algebra isomorphisms of $C(X)$ are induced by homeomorphisms of $X$ which we will, by slight abuse of notation, also denote by $T_g, g \in G$. The homeomorphisms $T_g \colon X \to X$ form an action of $G$ and preserve the measure $\mu$. Indeed, let $C \in \mathcal{B}$ and let $\varphi \in \mathcal{A}$ be the preimage of $1_C$ (so that $\tilde\varphi = 1_C$). Then we have:

$$\mu(C) = \int_X 1_C(x) d\mu(x) = \tilde{L}(\tilde\varphi) = L(\varphi) = L({}_g\varphi) = \tilde{L}(\widetilde{{}_g\varphi})$$

$$= \tilde{L}(\tilde\varphi(T_g x)) = \int_X 1_C(T_g x) d\mu(x) = \int_X 1_{T_g^{-1}C}(x) d\mu(x) = \mu(T_g^{-1}C)$$

Notice also that since $L(1) = 1, \mu(X) = L(1_X) = 1$. It follows that $\{T_g\}_{g \in G}$ is a measure preserving action on the probability space $(X, \mathcal{B}, \mu)$. Taking into account that the functional $L$ satisfies the conditions of the Proposition 6.4.19, we have for $f = 1_E, g_0 = e$ and any $g_1, ..., g_k \in G$:

$$\bar{d}\left(\bigcap_{i=0}^{k} g_i^{-1}E\right) \geq L\left(\prod_{i=0}^{k} {}_{g_i}f\right) = \tilde{L}\left(\prod_{i=0}^{k} \widetilde{{}_{g_i}f}\right) = \tilde{L}\left(\prod_{i=0}^{k} ((T_{g_i})\check{f})\right)$$

$$= \int_X \prod_{i=0}^{k} 1_{T_{g_i}^{-1}A} d\mu = \mu\left(\bigcap_{i=0}^{k} T_{g_i}^{-1}A\right).$$

We are done. ∎

**Remark 6.4.21** Let us describe an alternative way of proving Theorem 6.4.17 which avoids the use of Gelfand transform (cf. [14]). Given a Følner sequence $\Phi_n$, $n \in \mathbb{N}$ and $E \subseteq G$ having positive upper density $\bar{d}(E)$ with respect to $\{\Phi_n\}$, identify $E$ with $1_E = \xi \in \Omega = \{0,1\}^G$ and take the orbital closure $X = \overline{\{T_g\xi, g \in G\}}$, where the shift transformations $T_g \colon \Omega \to \Omega$, $g \in G$, are defined by $T_g f(h) = f(gh)$. By utilizing a procedure similar to that employed in the proof of Proposition 6.4.19, one constructs a functional $L$ on $C(X)$ which, in addition to $L(1_A) = \bar{d}(E)$ (where $A = \{\varphi \in X : \varphi(e) = 1\}$; notice that $1_A \in C(X)$), satisfies the conditions (i) $L(F) \geq 0$ for $F \geq 0$, (ii) $L(1) = 1$, (iii) $L(F \circ T_g) = L(F)$ $\forall g \in G$. It is admittedly somewhat confusing to deal with functions $F \in C(X)$ which are defined on the space $X$ which itself consists of functions, mapping $G$ into $\{0,1\}$, but one has to get used to that! Now, by Riesz representation theorem such a (positive, normalized) linear functional is given by a probability measure $\mu$ on Borel sets of $X$. The condition $L(1_A) = d(E)$ implies $\mu(A) = \bar{d}(E)$ and condition (iii) implies that $\mu$ is $T_g$ invariant for every $g \in G$.

Unfortunately, the scope of these notes does not allow us to present full proofs of Theorems 6.4.3 and 6.4.8. The reader is referred to Furstenberg's original paper [12] which contains the proof of Theorem 6.4.3 and to [16] where both theorems are derived from a very general ergodic *IP-Szemerédi theorem*. We shall, however, be able to give a proof of the following result which contains among its corollaries some nontrivial cases of Theorems 6.4.3 and 6.4.8.

**Theorem 6.4.22** *Let $(G,+)$ be a countable abelian group and $(X,\mathcal{B},\mu,\{T_g\}_{g\in G})$ a probability measure preserving system. For any Følner sequence $\{\Phi_n\}_{n=1}^{\infty}$ in $G$ and any $A \in \mathcal{B}$ with $\mu(A) > 0$ one has*

$$\lim_{n\to\infty} \frac{1}{|\Phi_n|} \sum_{g\in\Phi_n} \mu(A \cap T_{-g}A \cap T_gA) > 0.$$

**Corollary 6.4.23** *Under the conditions and notation of Theorem 6.4.22, the set*

$$R_A = \{g \in G : \mu(A \cap T_{-g}A \cap T_gA) > 0\}$$

*is syndetic. In other words, for some finite set $F \subset G$ one has:*

$$F + R_A = \{x + y : x \in F, y \in R_A\} = G.$$

**Corollary 6.4.24** *Let $E \subseteq \mathbb{Z}$ be $d^*$-large, and let $A \subseteq \mathbb{Z}^2$ be defined by*

$$A = \{(a,d) : \{a, a+d, a+2d\} \subset E\}.$$

*A is a large subset of $\mathbb{Z}^2$ in the sense that for some sequence of rectangles $M_n = [a_n^{(1)}, b_n^{(1)}] \times [a_n^{(2)}, b_n^{(2)}]$, $n \in \mathbb{N}$ with $|b_n^{(i)} - a_n^{(i)}| \to \infty$, $i = 1, 2$ one has $\lim_{n\to\infty} \frac{|A\cap M_n|}{|M_n|} > 0$.*

**Corollary 6.4.25** (cf. [1]). *Let $E \subseteq \mathbb{Z}_3^{\infty}$ with $\bar{d}(E) > 0$. Then $E$ is AS-rich, i.e. contains affine subspaces of arbitrarily high dimension.*

**Exercise 6.4.26** Derive Corollaries 6.4.23, 6.4.24 and 6.4.25 from Theorem 6.4.22.

We shall preface the proof of Theorem 6.4.22 with some remarks regarding the machinery that one needs for the proof.

First of all, we are going to take for granted the theorem about *ergodic decomposition*. Recall that a probability measure preserving system $(X, \mathcal{B}, \mu, \{T_g\}_{g\in G})$ is called *ergodic*, if there are no nontrivial invariant sets: $\mu(T_gA\triangle A) = 0$ for all $g \in G$ implies $\mu(A) = 0$ or $\mu(A) = 1$.

Under some mild regularity assumptions (which are satisfied for the spaces we are working with like those featured in the proof(s) of Theorem 6.4.17) one can assume that if our measure preserving system $(X, \mathcal{B}, \mu, \{T_g\}_{g\in G})$ is not ergodic, then there exists a family of invariant probability measures $\{\mu_\omega\}_{\omega\in\Omega}$

indexed by another probability space $(\Omega, \mathcal{D}, \nu)$ such that with respect to each $\mu_\omega$ the measure preserving system $(X, \mathcal{B}, \mu_\omega, \{T_g\}_{g \in G})$ is ergodic and such that for any $f \in L^1(X, \mathcal{B}, \mu)$ one has:

$$\int_X f d\mu = \int_\Omega (\int_X f d\mu_\omega) d\nu(\omega)$$

To illustrate the usefulness of the theorem about the ergodic decomposition, let us show that Theorem 6.4.22 follows from its special, ergodic case.

Fix $A \in \mathcal{B}$ with $\mu(A) > 0$ and a Følner sequence $\Phi_n, n \in \mathbb{N}$. Let $\{\mu_\omega\}_{\omega \in \Omega}$ be the ergodic decomposition of the measure $\mu$ and assume that Theorem 6.4.22 is valid for each $\mu_\omega, \omega \in \Omega$. Since

$$\mu(A) = \int_X 1_A d\mu = \int_\Omega (\int_X 1_A d\mu_\omega) d\nu(\omega) = \int_\Omega \mu_\omega(A) d\nu(\omega)$$

there exist $\delta > 0$ and a measurable set $C \subset \Omega$, $\nu(C) > 0$, such that for any $\omega \in C$ one has $\mu_\omega(A) > \delta$.

We have:

$$\lim_{n \to \infty} \frac{1}{|\Phi_n|} \sum_{g \in \Phi_n} \mu(A \cap T_{-g} A \cap T_g A)$$

$$= \lim_{n \to \infty} \frac{1}{|\Phi_n|} \sum_{g \in \Phi_n} \int_\Omega \mu_\omega(A \cap T_{-g} A \cap T_g A) d\nu(\omega)$$

$$= \int_\Omega \lim_{n \to \infty} \frac{1}{|\Phi_n|} \sum_{g \in \Phi_n} \mu_\omega(A \cap T_{-g} A \cap T_g A) d\nu(\omega)$$

$$\geq \int_C \lim_{n \to \infty} \frac{1}{|\Phi_n|} \sum_{g \in \Phi_n} \mu_\omega(A \cap T_{-g} A \cap T_g A) d\nu(\omega) > 0.$$

Another useful tool that we shall employ in the proof is a version of *van der Corput trick*:

**Proposition 6.4.27** ([9]). *Let $H$ be a Hilbert space and let $\{v_g\}$ be a bounded family of elements of $H$ indexed by a countable abelian group $G$. Let $\{\Phi_n\}_{n \in \mathbb{N}}$ be a Følner sequence in $G$. If*

$$\lim_{m \to \infty} \frac{1}{|\Phi_m|^2} \sum_{h,k \in \Phi_m} \left| \lim_{n \to \infty} \frac{1}{|\Phi_n|} \sum_{g \in \Phi_n} \langle v_{g+h}, v_{g+k} \rangle \right| = 0,$$

*then*

$$\lim_{n \to \infty} \left\| \frac{1}{|\Phi_n|} \sum_{g \in \Phi_n} v_g \right\| = 0.$$

Finally, we shall need the following splitting theorem (see, for example, [25]).

**Proposition 6.4.28** *Let $H$ be a Hilbert space and $U_g : H \to H$, $g \in G$, a unitary action of a countable abelian group $G$. Then $H = H_c \bigoplus H_{wm}$, where the orthogonal, $\{U_g\}$-invariant spaces $H_c$ and $H_{wm}$ are characterized in the following way:*

$$H_c = \{v \in H : \text{the orbit } \{U_g v, g \in G\} \text{ is precompact in the norm topology}\}$$
$$= \overline{\text{Span}}\{v \in H : \text{there exist } \lambda_g \in \mathbb{C} \text{ with } U_g v = \lambda_g v, g \in G\},$$

$$H_{wm} = \{v \in H : \text{for any Følner sequence } \{\Phi_n\} \text{ and any } v' \in H,$$
$$\lim_{n \to \infty} \frac{1}{|\Phi_n|} \sum_{g \in \Phi_n} |\langle U_g v, v' \rangle| = 0\}.$$

**Exercise 6.4.29** Prove that the following statements are equivalent:
(i) $v \in H_{wm}$,
(ii) for any $v' \in H$ one has $\displaystyle\lim_{n \to \infty} \frac{1}{|\Phi_n|^2} \sum_{h,k \in \Phi_n} |\langle U_h v, U_k v' \rangle|^2 = 0$.

**Exercise 6.4.30** Prove that a non-zero $v \in H_c$ if and only if for any $\varepsilon > 0$ the set $\{g \in G : \|U_g v - v\| < \varepsilon\|\}$ is syndetic in $G$.

**Proof of Theorem 6.4.22.** We shall give the detailed proof for the special case when $2G = \{2g : g \in G\} = G$. The same proof works with only minor changes for the case when $2G$ is a subgroup of finite index in $G$ (this condition is clearly satisfied by the groups $\mathbb{Z}$ and $\mathbb{Z}_3^\infty$, dealt with in Corollaries 6.4.24 and 6.4.25). We leave the treatment of the case when $2G$ has infinite index in $G$ to the reader.

In light of the remarks above, we may and will assume that the action $\{T_{2g}\}_{g \in G}$ is ergodic (remembering that we work under the assumption $2G = G$). Let $H = L^2(X, \mathcal{B}, \mu)$ and $f = 1_A$. We shall show first that the limit

$$\lim_{n \to \infty} \frac{1}{|\Phi_n|} \sum_{g \in \Phi_n} f(T_{-g}x) f(T_g x)$$

exists in the norm of $H = L^2(X, \mathcal{B}, \mu)$. It will follow then that the limit

$$\lim_{n \to \infty} \frac{1}{|\Phi_n|} \sum_{g \in \Phi_n} \mu(A \cap T_{-g}A \cap T_g A) = \lim_{n \to \infty} \frac{1}{|\Phi_n|} \int_X f(x) f(T_{-g}x) f(T_g x) d\mu$$

also exists and we will be left only with showing that it is positive.

Denote by $U_g$, $g \in G$ the unitary action on $H$ induced by $T_g$, $g \in G$ : $(U_g \varphi)(x) := \varphi(T_g x)$, $\varphi \in H$, and utilize the splitting $H = H_c \bigoplus H_{wm}$ described in the Proposition 6.4.28 above.

Let $f = f_c + f_{wm}$ be the decomposition of $f = 1_A$ corresponding to this splitting. We shall show first that $f_c$, and hence $f_{wm}$, are bounded functions. First of all, notice that since $f$ is a real-valued function, and since the operators $U_g$, $g \in G$ send real-valued functions into real-valued functions, the components $f_c$ and $f_{wm}$ are also real-valued. We claim that $f_c \geq 0$. Indeed, $f_c$ minimizes the distance from $H_c$ to $f$ and the function $\bar{f} = \max\{f_c, 0\}$ (which also has precompact orbit and hence belongs to $H_c$) would do as well in minimizing this distance. Similarly one shows that $f_c \leq 1$. Consider now the average

$$\frac{1}{|\Phi_n|} \sum_{g \in \Phi_n} U_{-g} f \, U_g f = \frac{1}{|\Phi_n|} \sum_{g \in \Phi_n} U_{-g} f_c U_g f_c + \frac{1}{|\Phi_n|} \sum_{g \in \Phi_n} U_{-g} f_c U_g f_{wm}$$
$$+ \frac{1}{|\Phi_n|} \sum_{g \in \Phi_n} U_{-g} f_{wm} U_g f_c + \frac{1}{|\Phi_n|} \sum_{g \in \Phi_n} U_{-g} f_{wm} U_g f_{wm}$$

Note that the limits of the first three expressions (namely, those involving $f_c$) exist in view of Theorem 6.4.15. Consider, for example, the expression

$$\frac{1}{|\Phi_n|} \sum_{g \in \Phi_n} U_{-g} f_c U_g f_{wm}.$$

Since $f_c$ is a linear combination (potentially infinite) of eigenfunctions, namely, functions $\varphi$, satisfying, for some $\lambda_g \in \mathbb{C}$, $g \in G$, the equation $U_g \varphi = \lambda_g \varphi, g \in G$, it is enough to prove that for each such $\varphi$ the limit

$$\lim_{n \to \infty} \frac{1}{|\Phi_n|} \sum_{g \in \Phi_n} U_{-g} \varphi U_g f_{wm}$$

exists in norm. But

$$\lim_{n \to \infty} \frac{1}{|\Phi_n|} \sum_{g \in \Phi_n} U_{-g} \varphi U_g f_{wm} = \varphi \cdot \lim_{n \to \infty} \frac{1}{|\Phi_n|} \sum_{y \in \Phi_n} \lambda_{-g} U_g f_{wm}$$
$$= \varphi \cdot \lim_{n \to \infty} \frac{1}{|\Phi_n|} \sum_{g \in \Phi_n} \tilde{U}_g f_{wm} = \varphi \cdot P f_{wm}$$

where $P$ is the orthogonal projection on the space of invariant functions of the action $\tilde{U}_g = \lambda_{-g} U_g, g \in G$.

It is easy to see that since $\tilde{U}_g \psi = \psi$ implies $U_g \psi = \lambda_g \psi$, the projection $P f_{wm}$ belongs to $H_c$. But $f_{wm} \perp H_c$ and we get $P f_{wm} = 0$. In complete analogy one shows that

$$\lim_{n \to \infty} \left\| \frac{1}{|\Phi_n|} \sum_{g \in \Phi_n} U_{-g} f_{wm} U_g f_c \right\| = 0.$$

We shall show now that

$$\lim_{n\to\infty}\left\|\frac{1}{|\Phi_n|}\sum_{g\in\Phi_n}U_{-g}f_{\mathrm{wm}}U_g f_{\mathrm{wm}}\right\|=0.$$

This will follow from the van der Corput trick (Proposition 6.4.27) and from Exercise 6.4.29.

Let $v_g = U_{-g}f_{\mathrm{wm}}U_g f_{\mathrm{wm}}$. We have, using the ergodicity of $\{U_{2g}\}$:

$$
\begin{aligned}
\frac{1}{|\Phi_n|}\sum_{g\in\Phi_n}\langle v_{g+h}, v_{g+k}\rangle &= \frac{1}{|\Phi_n|}\sum_{g\in\Phi_n}\int U_{-g-h}f_{\mathrm{wm}}U_{g+h}f_{\mathrm{wm}}U_{-g-k}f_{\mathrm{wm}}U_{g+k}f_{\mathrm{wm}}\,d\mu \\
&= \frac{1}{|\Phi_n|}\sum_{g\in\Phi_n}\int U_{-g}(U_{-h}f_{\mathrm{wm}}U_{-k}f_{\mathrm{wm}})U_g(U_h f_{\mathrm{wm}}U_k f_{\mathrm{wm}})\,d\mu \\
&= \frac{1}{|\Phi_n|}\sum_{g\in\Phi_n}\int (U_{-h}f_{\mathrm{wm}}U_{-k}f_{\mathrm{wm}})U_{2g}(U_h f_{\mathrm{wm}}U_k f_{\mathrm{wm}})\,d\mu \\
&\xrightarrow[n\to\infty]{} \int U_{-h}f_{\mathrm{wm}}U_{-k}f_{\mathrm{wm}}\,d\mu \int U_h f_{\mathrm{wm}}U_k f_{\mathrm{wm}}\,d\mu \\
&= \left(\int U_h f_{\mathrm{wm}}U_k f_{\mathrm{wm}}\,d\mu\right)^2 = |\langle U_h f_{\mathrm{wm}}, U_k f_{\mathrm{wm}}\rangle|^2.
\end{aligned}
$$

Now use the characterization of $H_{\mathrm{wm}}$ (and Exercise 6.4.29):

$$\lim_{n\to\infty}\frac{1}{|\Phi_n|^2}\sum_{h,k\in\Phi_n}|\langle U_h f_{\mathrm{wm}}, U_k f_{\mathrm{wm}}\rangle|^2=0.$$

So, it follows that

$$\lim_{n\to\infty}\left\|\frac{1}{|\Phi_n|}\sum_{g\in\Phi_n}U_{-g}f_{\mathrm{wm}}U_g f_{\mathrm{wm}}\right\|=0,$$

and that

$$\lim_{n\to\infty}\frac{1}{|\Phi_n|}\sum_{g\in\Phi_n}U_{-g}fU_g f = \lim_{n\to\infty}\frac{1}{|\Phi_n|}\sum_{g\in\Phi_n}U_{-g}f_c U_g f_c.$$

Noticing that the products of bounded functions from $H_c$ belong to $H_c$ and that $f_{\mathrm{wm}} \perp H_c$, we have:

$$
\begin{aligned}
&\lim_{n\to\infty} \frac{1}{|\Phi_n|} \sum_{g\in\Phi_n} \mu(A \cap T_{-g}A \cap T_g A) \\
&= \lim_{n\to\infty} \frac{1}{|\Phi_n|} \sum_{g\in\Phi_n} \int_X f(x) f(T_g x) f(T_{-g} x) d\mu \\
&= \int_X f \cdot \Big( \lim_{n\to\infty} \frac{1}{|\Phi_n|} \sum_{g\in\Phi_n} U_g f U_{-g} f \Big) d\mu \\
&= \int_X (f_c + f_{\mathrm{wm}}) \Big( \lim_{n\to\infty} \frac{1}{|\Phi_n|} \sum_{g\in\Phi_n} U_g f_c U_{-g} f_c \Big) d\mu \\
&= \lim_{n\to\infty} \frac{1}{|\Phi_n|} \sum_{g\in\Phi_n} \int_X f_c U_g f_c U_{-g} f_c \, d\mu.
\end{aligned}
$$

Now, the positivity of the last expression follows from Exercise 6.4.30. ∎

## 6.5 Polynomial Ergodic Theorems and Ramsey Theory

This section's intention is to give a glimpse of a relatively new subject: ergodic theorems along polynomials. The natural limitations allow us only to discuss a few results and applications, but the interested reader is referred, for the details, to the recent survey [3], as well as to the papers [2], [4], [5], [6], [7], [8], [11], [19], [28], [29].

The following theorem, due to Furstenberg (see [12], [13], [14]), shows that measure preserving systems exhibit regular behavior along polynomials.

**Theorem 6.5.1** *If $p(t) \in \mathbb{Q}[t]$, $p(\mathbb{Z}) \subseteq \mathbb{Z}$, and $p(0) = 0$, then for any invertible measure preserving system $(X, \mathcal{B}, \mu, T)$ and $A \in \mathcal{B}$, with $\mu(A) > 0$, one has:*

$$
\lim_{N\to\infty} \frac{1}{N} \sum_{n=0}^{N-1} \mu(A \cap T^{p(n)} A) > 0.
$$

Applying Furstenberg's correspondence principle, one gets the following corollary, independently obtained by Sárközy using methods of analytic number theory.

**Theorem 6.5.2** *([12], [13], [14], [30]). If $p(t) \in \mathbb{Q}[t]$, $p(\mathbb{Z}) \in \mathbb{Z}$, $p(0) = 0$, and $E \subseteq \mathbb{Z}$ has positive upper Banach density, then for some $x, y \in E$ and $n \in \mathbb{N}$ one has $x - y = p(n)$.*

Both Theorems 6.5.1 and 6.5.2 are quite striking if one takes into account that the set of values of $p(n)$ is a "small" subset of $\mathbf{Z}$ for any polynomial $p(n)$ whose degree is larger than 1. One can view Theorem 6.5.1 as a polynomial refinement of the classical Poincaré recurrence theorem.

The next natural question is whether Theorem 6.4.3 has a polynomial generalization. The answer is *yes* (see [5]). We shall formulate here a special case of the main result from [5].

**Theorem 6.5.3** *Assume that $(X, \mathcal{B}, \mu, T)$ is an invertible probability measure-preserving system, $k \in \mathbf{N}$, $A \in \mathcal{B}$ with $\mu(A) > 0$, and $p_i(t) \in \mathbb{Q}[t]$ are polynomials satisfying $p_i(\mathbf{Z}) \subseteq \mathbf{Z}$ and $p_i(0) = 0$, $1 \le i \le k$. Then*

$$\liminf_{N\to\infty} \frac{1}{N} \sum_{n=0}^{N-1} \mu(A \cap T^{p_1(n)}A \cap \ldots \cap T^{p_k(n)}A) > 0.$$

**Remark 6.5.4** Actually, one can show ([7]) that:

$$\liminf_{N-M\to\infty} \frac{1}{N-M} \sum_{n=M}^{N-1} \mu(A \cap T^{p_1(n)}A \cap \ldots \cap T^{p_k(n)}A) > 0.$$

This result is itself a special case of two different and far-reaching extensions. See [8] and [29].

We now return to Theorem 6.5.1. The original proof of Furstenberg was based on the spectral theorem and Weyl's theorem on the equidistribution of polynomials (see Section 6.2). Later, a few more proofs appeared based on different ideas. See [2], [4] and [3].

The following sequence of exercises, supplied with hints, leads to a proof of Theorem 6.5.1 in stages.

**Exercise 6.5.5** (i) Let $H$ be a Hilbert space and $U: H \to H$ a unitary operator. Let

$$H_{\mathrm{rat}} = \overline{\mathrm{Span}\{f \in H : \text{there exists } i \in \mathbf{N} \text{ with } U^i f = f\}},$$
$$H_{\mathrm{tot.erg}} = \Big\{f \in H : \text{for all } i \in \mathbf{N}, \Big\|\tfrac{1}{N}\sum_{n=0}^{N-1} U^{in} f\Big\| \to 0\Big\}.$$

Show that $H = H_{\mathrm{rat}} \oplus H_{\mathrm{tot.erg}}$.
(ii) Prove the following version of the van der Corput trick: If $\{v_n\}_{n\in\mathbf{N}}$ is a bounded sequence in $H$ such that for any $h \in \mathbf{N}$ $\lim_{N\to\infty} \frac{1}{N}\sum_{n=1}^{N}\langle v_{n+h}, v_n\rangle = 0$, then

$$\Big\|\frac{1}{N}\sum_{n=1}^{N} v_n\Big\| \to 0.$$

*Hint*: for any $\varepsilon > 0$ and $M \in \mathbb{N}$, if $N$ is large enough, then one has

$$\left\| \frac{1}{N} \sum_{n=1}^{N} v_n - \frac{1}{NM} \sum_{n=1}^{N} \sum_{h=1}^{M} v_{n+h} \right\| < \varepsilon.$$

(iii) (Aside.) Derive from (ii) the van der Corput difference theorem (Theorem 6.2.18).
*Hint*: take $H = \mathbb{C}$ and use the Weyl criterion (6.2.7).
(iv) Prove that for any polynomial $p(t) \in \mathbb{Q}[t]$ with $p(\mathbb{Z}) \subseteq \mathbb{Z}$ and any $f \in H$,
the limit $\lim\limits_{N\to\infty} \left\| \dfrac{1}{N} \sum\limits_{n=1}^{N} U^{p(n)} f \right\|$ exists.
*Hint*: use (ii) to show that if $\deg p(n) > 0$, and $f \in H_{\text{tot.erg}}$, then the limit in question is 0. If $f \in H_{\text{rat}}$, verify that it is enough to check the existence of the limit for $f$ satisfying $U^i f = f$ for some $i$.
(v) Let $H = L^2(X, \mathcal{B}, \mu)$, $(U\varphi)(x) := \varphi(Tx)$, $\varphi \in H$. Let $f = 1_A$ (where $A \in \mathcal{B}$ with $\mu(A) > 0$). Let $f_a$, $a \in \mathbb{N}$, be the orthogonal projection of $f$ onto the subspace

$$H_a = \{g \in H : U^a g = g\} \subseteq H_{\text{rat}}$$

Notice that each $H_a$ contains the constants and show that $f_a \geq 0$, $f_a \neq 0$.
(vi) Assume that $p(0) = 0$ in addition to the assumptions of (iv). Conclude the proof by showing that for any $a \in \mathbb{N}$

$$\lim_{N\to\infty} \frac{1}{N} \sum_{n=1}^{N} \langle U^{p(n)} f_a, f_a \rangle > 0,$$

and that it implies

$$\lim_{N\to\infty} \frac{1}{N} \sum_{n=1}^{N} \mu(A \cap T^{p(n)} A) = \lim_{N\to\infty} \frac{1}{N} \sum_{n=1}^{N} \langle U^{p(n)} f, f \rangle > 0$$

(vii) Prove that Theorem 6.5.1 remains true if one replaces the condition $p(0) = 0$ by the following, weaker one: $\{p(n) : n \in \mathbb{Z}\} \cap a\mathbb{Z} \neq \emptyset$ for all $a \in \mathbb{N}$.

We conclude this section with an application of the polynomial theorem to partition Ramsey theory (cf. [3], p. 53 and [7], Theorem 0.4).

**Theorem 6.5.6** *Let $p(t) \in \mathbb{Q}[t]$, $p(\mathbb{Z}) \subseteq \mathbb{Z}$, $p(0) = 0$. For any finite partition of $\mathbb{N}$, $\mathbb{N} = \bigcup\limits_{i=1}^{r} C_i$, one can find $i$, $1 \leq i \leq r$, and $x, y, z \in C_i$ so that $x - y = p(z)$.*

**Remark 6.5.7** By Theorem 6.5.2 any set of positive upper Banach density in $\mathbf{N}$ contains $x, y$ with $x - y = p(n)$ for *some* $n$. The crux of Theorem 6.5.6 is that for any finite partition of $\mathbf{N}$ one cell has the additional property that in the equation $x - y = p(z)$, all three parameters are from the same set.

**Proof of Theorem 6.5.6.** Let $\mathbf{N} = \bigcup_{i=1} C_i$ be a given partition. Reindexing if necessary, we may assume that the first $k$ sets $C_1, C_2, \ldots, C_k$ are such that $\bar{d}(C_i) > 0$, $i = 1, 2, \ldots, k$, and $\bar{d}(\bigcup_{i=1}^{k} C_i) = 1$. Let $(X_i, \mathcal{B}_i, \mu_i, T_i)$ and $A_i \in \mathcal{B}_i$ with $\mu(A_i) = \bar{d}(C_i)$, $i = 1, 2, \ldots, k$, be the measure preserving systems and sets guaranteed by the Furstenberg correspondence principle (Theorem 6.4.4). Form the product system $(X, \mathcal{B}, \mu, T)$ where $X = X_1 \times \ldots \times X_k$, $\mathcal{B}$ is the $\sigma$-algebra generated by $\mathcal{B}_1 \times \ldots \times \mathcal{B}_k$, $T = T_1 \times \ldots \times T_k$, and $\mu = \mu_1 \otimes \ldots \otimes \mu_k$. Finally, let $A = A_1 \times \ldots \times A_k \in \mathcal{B}$. Applying Theorem 6.5.1 to the system $(X, \mathcal{B}, \mu, T)$ and the set $A$, we obtain that the set $R = \{n \in \mathbf{N} : \mu(A \cap T^{p(n)} A) > 0\}$ has positive lower density (i.e. $\liminf_{n \to \infty} \frac{|R \cap [-n,n]|}{2n+1} > 0$).
It follows that $R \cap (\bigcup_{i=1}^{k} C_i) \neq \emptyset$. Let $i_0$, $1 \le i_0 \le k$, be such that for some $n \in C_{i_0}$

$$\mu(A \cap T^{p(n)} A) = \mu\Big((A_1 \times \ldots \times A_k) \cap (T_1^{p(n)} \times \ldots \times T_k^{p(n)})(A_1 \times \ldots \times A_k)\Big)$$
$$= \mu\Big((A_1 \cap T_1^{p(n)} A_1) \times \ldots \times (A_k \cap T_k^{p(n)} A_k)\Big) > 0.$$

Applying again Furstenberg's correspondence principle, we get

$$\bar{d}\big(C_{i_0} \cap (C_{i_0} - p(n))\big) > 0.$$

If $y \in C_{i_0} \cap \big(C_{i_0} - p(n)\big)$, then $x = y + p(n) \in C_{i_0}$, and this establishes the partition regularity of the equation $x - y = p(z)$. $\blacksquare$

## 6.6 Appendix

In this appendix, we shall give an elementary proof of Theorem 6.2.12 which makes the claim to be, in Erdös' terminology, from THE BOOK, or at least from THE BOOK OF ELEMENTARY PROOFS. Actually, since the readers of these notes surely have no problem with the method of mathematical induction, we shall confine ourself to the following special case, whose proof has all the ingredients of the proof of the general case.

**Theorem 6.6.1** *For any irrational $\alpha$, the sequence $\{n^2\alpha \pmod 1\}_{n \in \mathbf{N}}$ is dense in $[0, 1]$.*

In light of the (elementary!) discussion in Section 6.2, Theorem 6.6.1 clearly follows from the following.

**Theorem 6.6.2** *For any $\alpha \in \mathbb{R}$ there is a sequence $\{n_k\} \subseteq \mathbb{N}$ such that $n_k^2 \alpha \pmod 1 \underset{k \to \infty}{\longrightarrow} 0$.*

Before giving the promised elementary proof of Theorem 6.6.2, we shall reveal the source of our inspiration. It is $\beta \mathbb{N}$, the Stone-Čech compactification of $\mathbb{N}$, and, specifically, those elements of $\beta \mathbb{N}$ which are called *idempotents*. Section 6.4 of [3] contains all information about $\beta \mathbb{N}$ and idempotent ultrafilters that we are going to need. (Readers who for this or that reason do not like the ultrafilters can skip the following ultrafilter proof of Theorem 6.6.2 and go straight to the elementary proof below.) Let $p \in \beta \mathbb{N}$ be any idempotent ultrafilter. The only property of $p$ that we are interested in is given by the following easy proposition.

**Theorem 6.6.3** (Theorem 3.8 in [3]). *Let $X$ be a compact Hausdorff space and let $\{x_n\}_{n \in \mathbb{N}}$ be a sequence in $X$. Let $p \in \beta \mathbb{N}$ be an idempotent ultrafilter. Then*

$$p\text{-}\lim_{r \in \mathbb{N}} x_r = p\text{-}\lim_{t \in \mathbb{N}} p\text{-}\lim_{s \in \mathbb{N}} x_{s+t}.$$

**Ultrafilter Proof of Theorem 6.6.2.** Let $X = [0, 1]$ with the conventional metric. (Another possibility would be to work with $X = \mathbb{T}$.) Fix an idempotent $p \in \beta \mathbb{N}$. All we have to do is to show that either $p\text{-}\lim_{n \in \mathbb{N}}(n^2 \alpha \pmod 1) = 0$ or $p\text{-}\lim_{n \in \mathbb{N}}(n^2 \alpha \pmod 1) = 1$. (The reader is invited to check that the latter case is possible and is as good for our purposes.) Let us show first that for any $\gamma \in \mathbb{R}$ one has either $p\text{-}\lim_{n \in \mathbb{N}}(n \gamma \pmod 1) = 0$ or $p\text{-}\lim_{n \in \mathbb{N}}(n \gamma \pmod 1) = 1$. Indeed, if $p\text{-}\lim_{n \in \mathbb{N}} n \gamma \pmod 1 = c$, then we have

$$c = p\text{-}\lim_{n \in \mathbb{N}}(n \gamma \pmod 1) = p\text{-}\lim_{n \in \mathbb{N}} p\text{-}\lim_{k \in \mathbb{N}}((n + k)\gamma \pmod 1)$$
$$= p\text{-}\lim_{n \in \mathbb{N}}((n\gamma + c) \pmod 1) = 2c \pmod 1.$$

So, $c = 2c \pmod 1$ and hence $c \in \{0, 1\}$. But the same proof works for any polynomial! Indeed, let $p\text{-}\lim_{n \in \mathbb{N}} n^2 \alpha \pmod 1 = c$. We have:

$$c = p\text{-}\lim_{n \in \mathbb{N}}(n^2 \alpha \pmod 1) = p\text{-}\lim_{n \in \mathbb{N}} p\text{-}\lim_{k \in \mathbb{N}}((n + k)^2 \alpha \pmod 1)$$
$$= p\text{-}\lim_{n \in \mathbb{N}} p\text{-}\lim_{k \in \mathbb{N}}((n^2 \alpha + k(2n\alpha) + k^2 \alpha) \pmod 1)$$
$$= p\text{-}\lim_{n \in \mathbb{N}}((n^2 \alpha + c) \pmod 1) = 2c \pmod 1.$$

So, again, $c = 2c \pmod 1$ which implies $c \in \{0, 1\}$ and we are done.

Now we shall show how this proof may be elementarized. The most important hint which the perspicacious reader may extract from this proof is that if for some sequence $\{n_k\} \subseteq \mathbf{N}$ one has $n_k^2\alpha \pmod 1 \xrightarrow[k\to\infty]{} c$, then, along the finite sums of elements from $\{n_k\}$, one can approach $lc \pmod 1$ for any $l \in \mathbf{N}$. This is all that one needs since the sequence $\{lc \pmod 1\}_{l\in\mathbf{N}}$ has 0 as a limit point.

**Elementary Proof of Theorem 6.6.2.** Let $\{n_k\}$ be an increasing sequence of positive integers satisfying $n_k\alpha \pmod 1 \xrightarrow[k\to\infty]{} 0$. Passing, if needed, to a subsequence, assume that simultaneously $n_k^2\alpha \pmod 1 \xrightarrow[k\to\infty]{} c \in [0,1]$. If $c = 0$, we are done. If $c = 1$, we are also done, since it is easy to see that for appropriately chosen $m_k$ one will have $(m_k n_k)^2\alpha \pmod 1 \xrightarrow[k\to\infty]{} 0$. So assume that $c \in (0,1)$. Again, if $c = r/s$ is a rational number, then replacing $n_k$ by $s n_k$ we are done, so assume without loss of generality that $c$ is irrational, and let us show how, for any $l \in \mathbf{N}$ and any $\varepsilon > 0$, to find $m \in \mathbf{N}$ with $lc \pmod 1 - \varepsilon < m^2\alpha \pmod 1 < lc \pmod 1 + \varepsilon$. Assuming, as we may, that $\varepsilon$ is so small that $\varepsilon < lc \pmod 1 < 1 - \varepsilon$, let us show that such an $m$ can be found among the numbers of the form $n_{k_1} + n_{k_2} + \ldots + n_{k_l}$ with $k_1 \leq k_2 \leq \ldots \leq k_l$. Let us do it, for simplicity, for $l = 3$. Let $k_1 \in \mathbf{N}$ be such that for any $k \geq k_1$ one has $c - \frac{\varepsilon}{6} < n_k^2\alpha \pmod 1 < c + \frac{\varepsilon}{6}$. Choose now $k_3 \geq k_2 \geq k_1$ so that, in addition, $0 < n_{k_2}\alpha \pmod 1 < \frac{\varepsilon}{12 n_{k_1}}$ and $0 < n_{k_3}\alpha \pmod 1 < \frac{\varepsilon}{12 n_{k_2}}$. One trivially checks that such a choice of $k_1$, $k_2$, and $k_3$ guarantees that

$$3c \pmod 1 - \frac{\varepsilon}{2} < (n_{k_1} + n_{k_2} + n_{k_3})^2\alpha \pmod 1 < 3c \pmod 1 + \varepsilon. \quad \blacksquare$$

**Acknowledgment.** I wish to express my sincere appreciation to Mrs. Gladys Cavallone, Steve Cook, Paul Larick and Sasha Leibman for their help in preparing these notes. I also would like to thank Francois Blanchard, Alejandro Maass and Arnaldo Nogueira for organizing, under the aegis of CIMPA and UNESCO, the wonderful workshop in Temuco.

This work was partially supported by NSF under grant DMS-970605.

# Bibliography

[1] V. Bergelson, Ergodic Ramsey theory, logic and combinatorics, edited by S. Simpson, *Contemporary Mathematics* **65**, pp. 63-87 (1987).

[2] V. Bergelson, Weakly mixing PET, *Ergodic Theory and Dynamical Systems* **7**, pp. 337-349 (1987).

[3]  V. Bergelson, Ergodic Ramsey theory – an update, *Ergodic Theory of* $\mathbf{Z}^d$-*actions*, edited by M. Pollicott and K. Schmidt, London Math. Soc. Lecture Note Series **228**, pp. 1-61 (1996).

[4]  V. Bergelson, H. Furstenberg and R. McCutcheon, IP-sets and polynomial recurrence, *Ergodic Theory and Dynamical Systems* **16**, pp. 963-974 (1996).

[5]  V. Bergelson and A. Leibman, Polynomial extensions of van der Waerden's and Szemerédi's theorems, *Journal of AMS* **9**, pp. 725-753 (1996).

[6]  V. Bergelson and A. Leibman, Set-polynomials and a polynomial extension of Hales-Jewett theorem, *Annals of Math.*, to appear.

[7]  V. Bergelson and R. McCutcheon, Uniformity in polynomial Szemerédi theorem, *Ergodic Theory of* $\mathbf{Z}^d$-*actions*, edited by M. Pollicott and K. Schmidt, London Math. Soc. Lecture Note Series **228**, pp. 273-296 (1996).

[8]  V. Bergelson and R. McCutcheon, An IP polynomial Szemerédi theorem for finite families of commuting transformations, *Memoirs of AMS*, to appear.

[9]  V. Bergelson, R. McCutcheon and Q. Zhang, A Roth theorem for amenable groups, *American Journal of Mathematics* **119**, pp. 1173-1211 (1997).

[10]  A. Blaszczyk, S. Plewik and S. Turek, Topological multidimensional van der Waerden theorem, *Comment. Math. Univ. Carolinae* **30**, pp. 783-787 (1989).

[11]  J. Bourgain, Pointwise ergodic theorems for arithmetic sets, *Publ. Math. IHES* **69**, pp. 5-45 (1989).

[12]  H. Furstenberg, Ergodic behavior of diagonal measures and a theorem of Szemerédi on arithmetic progressions, *J. d'Analyse Math.* **31**, pp. 204-256 (1977).

[13]  H. Furstenberg, *Recurrence in Ergodic Theory and Combinatorial Number Theory*, Princeton University Press (1981).

[14]  H. Furstenberg, Poincaré recurrence and number theory, *Bull. AMS* (New Series) **5**, pp. 211-234 (1981).

[15]  H. Furstenberg and Y. Katznelson, An ergodic Szemerédi theorem for commuting transformations, *J. d'Analyse Math.* **34**, pp. 275-291 (1978).

[16] H. Furstenberg and Y. Katznelson, An ergodic Szemerédi theorem for IP-systems and combinatorial theory, *J. d'Analyse Math.* **45**, pp. 117-168 (1985).

[17] H. Furstenberg and Y. Katznelson, A density version of the Hales-Jewett theorem, *J. d'Analyse Math.* **57**, pp. 64-119 (1991).

[18] H. Furstenberg and B. Weiss, Topological dynamics and combinatorial number theory, *J. d'Analyse Math.* **34**, pp. 61-85 (1978).

[19] H. Furstenberg and B. Weiss, A mean ergodic theorem for $\frac{1}{N}\sum_{n=1}^{N} f(T^n x)g(T^{n^2} x)$, *Convergence in Ergodic Theory and Probability*, edited by V. Bergelson, P. March and J. Rosenblatt, Ohio State University Math. Research Institute Publications, de Gruyter (1995).

[20] R. Graham, K.Leeb and B. Rothschild, Ramsey's theorem for a class of categories, *Advances in Math.* **8**, pp. 417-433 (1972). Errata **10**, pp. 326-327 (1973).

[21] R. Graham, B. Rothschild and J. Spencer, *Ramsey Theory*, Wiley, New York (1980).

[22] D. Hilbert, Über die Irreduzibilität Ganzer Rationaler Funktionen mit Ganzzahligen Koeffizienten, *J. Math.* **110**, pp. 104-129 (1892).

[23] N. Hindman, Ultrafilters and combinatorial number theory, *Number Theory Carbondale, M. Nathanson, ed., Lecture Notes in Math.* **751**, pp. 119-184 (1979).

[24] G. Hardy and E. Wright, *An Introduction to the Theory of Numbers*, fifth edition, Clarendon Press, Oxford (1979).

[25] V. Krengel, *Ergodic Theorems*, de Gruyter (1985).

[26] N. Kryloff and N. Bogoliouboff, La théorie générale de la measure dans son application à l'étude des systémes dynamiques de la mécanique non linéaire, *Ann. of Math.* **38**, pp. 65-113 (1937).

[27] L. Kuipers and H. Niederreiter, *Uniform Distribution of Sequences*, John Wiley (1974).

[28] A. Leibman, Multiple recurrence theorem for nilpotent group actions, *Geom. and Funct. Anal.* **4**, pp. 648-659 (1994).

[29] A. Leibman, Multiple recurrence theorem for measure preserving actions of a nilpotent group, *Geom. and Funct. Anal.* **8** pp. 853-931 (1998).

[30] A. Sárközy, On difference sets of integers III, *Acta. Math. Acad. Sci. Hungar.* **31**, pp. 125-149 (1978).

[31] E. Szemerédi, On sets of integers containing no four elements in arithmetic progression, *Acta Math. Acad. Sci. Hungar.* **20**, pp. 89-104 (1969).

[32] P. Walters, *An Introduction to Ergodic Theory*, Springer (1982).

[33] H. Weyl, Über die Gleichverteilung von Zahlen mod1, *Math. Ann.* **77**, pp. 313-352 (1916).

# Chapter 7

# NUMBER REPRESENTATION AND FINITE AUTOMATA

*Christiane FROUGNY*
*Université Paris 8 and*
*CNRS, Laboratoire d'Informatique Algorithmique:*
*Fondements et Applications*
*Case 7014, 2 place Jussieu, 75251 Paris Cedex 05*
*France*

In positional numeration systems, numbers are represented as finite or infinite strings of digits, and thus Symbolic Dynamics is a useful conceptual framework for the study of number representation. We use finite automata to modelize simple arithmetic operations like addition. In this chapter, we first recall some results on the usual representation of numbers to an integer base. Then we consider the case where the base is a real number but not an integer, the so-called beta-expansions. Finally, we treat of the representation of integers with respect to a sequence of integers, for instance the Fibonacci numbers.

## 7.1  Introduction

In this survey, numbers are seen as finite or infinite strings of digits from a finite set. This implies of course that several concepts from Symbolic Dynamics find an illustration in number representation. For instance, the set of base 2 expansions of real numbers from the interval $[0, 1]$ is the one-sided full 2-shift. Number representation has a long and fascinating history; some of its developments can be found in the book of Knuth [28]. In particular, non-classical numeration systems like the Fibonacci numeration system are quite well known. In Computer Arithmetics, it is recognized that algorithmic possibilities depend on the representation of numbers. For instance, addition of two integers represented in the usual binary system with the digits 0 and 1, takes a time which is proportional to the size of the data. But if these numbers are represented with the digits 0, 1, and −1, then addition can be realized in a time that is independent of the size of the data.

For any definition or result from Symbolic Dynamics the reader is referred to the book of Lind and Marcus [30]. For the links between Symbolic Dynamics and Finite Automata Theory, see [1]. We begin by quickly introducing the model of finite state automata. Since numbers are strings of digits, these tools are useful to describe sets of number representations, and also to characterize the complexity of arithmetic operations. For instance, addition in integral base is computable by a finite automaton, but multiplication is not.

Section 7.4 is devoted to the study of the so-called beta-expansions, introduced by Rényi [36]. It consists in taking a real number $\beta > 1$ as a base. When $\beta$ is actually an integer, we obtain the standard numeration. When $\beta$ is not an integer, a number may have several different $\beta$-representations. One of them is obtained by the greedy algorithm of Rényi, and is called the $\beta$-expansion; it is the greatest in the lexicographic ordering. The set of $\beta$-expansions of numbers of $[0, 1]$ is shift-invariant, and its closure, called the $\beta$-shift, is a symbolic dynamical system. We give several results on these topics. We do not cover the whole field, which is very lively and still growing. It has interesting connexions with Number Theory.

We then consider representation of integers with respect to a sequence of integers $U$. The basic example is that of Fibonacci numbers. Every positive integer can be represented in such a system. This field is closely related to the theory of beta-expansions.

Representation of complex numbers is not treated here. Nevertheless, the domain is quite well studied. Taking for base a complex number allows to represent complex numbers as strings of digits, and to handle them without separating real and imaginary parts. Interested readers can find practical results in [28], [27] and [22], and a theoretical approach more related to ours in [41].

## 7.2   Words and Finite Automata

We recall some definitions. More details can be found in [15] or [25]. An *alphabet* $A$ is a finite set. A finite sequence of elements of $A$ is called a *word*, and the set of words on $A$ is the free monoid $A^*$. The *empty word* is denoted by $\varepsilon$. A *factor* or a *block* of a word $w$ is a word $f$ such that there exist words

$w'$ and $w''$ with $w = w'fw''$. When $w' = \varepsilon$, $f$ is said to be a *prefix* of $w$, and when $w'' = \varepsilon$, $f$ is said to be a *suffix* of $w$. If $H$ is a subset of $A^*$ we denote by $F(H)$ (resp. $PF(H)$, resp. $SF(H)$) the set of factors (resp. prefixes, resp. suffixes) of words of $H$. A subset $H$ is said to be *prefix* if no word of $H$ is a prefix of another word of $H$. The *length* of a word $w = w_1 \cdots w_n$ with $w_i$ in $A$ for $1 \leq i \leq n$ is denoted by $|w|$ and is equal to $n$. By $w^n$ one denotes the word obtained by concatenating $w$ $n$ times to itself. The set of words of length $n$ (resp. $\leq n$) of $A^*$ is denoted by $A^n$ (resp. $A^{\leq n}$).

The set of infinite sequences or infinite words on $A$ is called $A^{\mathbb{N}}$. The infinite word $www \cdots$ is denoted by $w^\omega$. Finally $F(K)$ denotes the set of finite factors or blocks of a subset $K$ of $A^{\mathbb{N}}$.

An *automaton over* $A$, $\mathcal{A} = (Q, A, E, I, T)$, is a directed graph labelled by elements of $A$; $Q$ is the set of *states*, $I \subset Q$ is the set of *initial* states, $T \subset Q$ is the set of *terminal* states and $E \subset Q \times A \times Q$ is the set of labelled *edges*. If $(p, a, q) \in E$, we write $p \xrightarrow{a} q$. The automaton is *finite* if $Q$ is finite. The automaton $\mathcal{A}$ is *deterministic* if $E$ is the graph of a (partial) function from $Q \times A$ into $Q$, and if there is a unique initial state. A subset $H$ of $A^*$ is said to be *recognizable by a finite automaton* (or *regular*) if there exists a finite automaton $\mathcal{A}$ such that $H$ is equal to the set of labels of paths starting from an initial state and ending in a terminal state.

Let $H$ be a subset of $A^*$. The *right congruence* modulo $H$ is defined on $A^*$ by

$$f \sim_H g \Leftrightarrow [\forall h \in A^*, \ fh \in H \text{ if and only if } gh \in H].$$

It is known that the set $H$ is recognizable by a finite automaton if and only if the right congruence modulo $H$ has finite index (Myhill-Nerode Theorem, see [15] or [25]). Denote by $[f]_H$ the right class of $f$ modulo $\sim_H$. Suppose that $\sim_H$ has finite index. One constructs the *minimal* deterministic automaton $\mathcal{M}$ recognizing $H$ as follows:

- the set of states of $\mathcal{M}$ is the set $\{[f]_H \mid f \in A^*\}$
- the initial state is $[\varepsilon]_H$
- the set of terminal states is equal to $\{[f]_H \mid f \in H\}$
- for every state $[f]_H$ and every $a \in A$, there is an edge $[f]_H \xrightarrow{a} [fa]_H$.

A subset $K$ of $A^{\mathbb{N}}$ is said to be *recognizable by a finite automaton* if there exists a finite automaton $\mathcal{A}$ such that $K$ is equal to the set of labels of infinite paths starting from an initial state and going infinitely often through a terminal state (Büchi acceptance condition, see [15]).

A *2-tape automaton* is an automaton over the non-free monoid $A^* \times B^*$: $\mathcal{A} = (Q, A^* \times B^*, E, I, T)$ is a directed graph the edges of which are labelled by elements of $A^* \times B^*$. Words of $A^*$ are referred to as *input words*, and words of $B^*$ are referred to as *output words*. If $(p, (f, g), q) \in E$, we write $p \xrightarrow{f/g} q$. The automaton is *finite* if the set of edges $E$ is finite (and thus $Q$ is finite). These finite 2-tape automata are also known as *transducers*. A relation $R$ of

$A^* \times B^*$ is said to be *computable by a finite 2-tape automaton* if there exists
a finite 2-tape automaton $\mathcal{A}$ such that $R$ is equal to the set of labels of paths
starting from an initial state and ending in a terminal state. A function
is computable by a finite 2-tape automaton if its graph is computable by a
finite 2-tape automaton. These definitions extend to relations and functions
of infinite words as above.

A 2-tape automaton $\mathcal{A}$ is said to be *left sequential* if edges are labelled
by elements of $A \times B^*$, if the *underlying input automaton* obtained by taking
the projection over $A$ of the label of every edge is deterministic and if every
state is terminal (see [3]). A *left subsequential* 2-tape automaton is a left
sequential automaton $\mathcal{A} = (Q, A \times B^*, E, \{i\}, \omega)$, where $\omega$ is the *terminal
function* $\omega : Q \longrightarrow B^*$, whose value is concatenated to the output word
corresponding to a computation in $\mathcal{A}$.

A 2-tape automaton $\mathcal{A}$ is said to be *letter-to-letter* if the edges are labelled
by couples of letters, that is, by elements of $A \times B$.

All the automata considered so far work implicitly from left to right, that
is to say, words are processed from left to right, but one can define similarly
*right* automata processing words from right to left.

# 7.3   Standard Representations of Numbers

We consider only non-negative real numbers.

## 7.3.1   Representation of Integers

Let $b$ be an integer $\geq 2$. The *standard b-ary representation* of an integer
$N \geq 0$ is a finite word $d_k \cdots d_0$, where digits $d_i$ belong to the alphabet
$A = \{0, \cdots, b-1\}$, and such that

$$N = \sum_{i=0}^{k} d_i b^i.$$

Such a representation is unique, with the condition that $d_k \neq 0$. We write
$\langle N \rangle_b = d_k \cdots d_0$, with the most significant digit first.

The set of all the representations of the positive integers is recognizable
by a finite automaton, and is equal to $A^*$.

**Addition of Integers.** Let $d_k \cdots d_0$ and $c_k \cdots c_0$ be two $b$-ary representations for $N$ and $M$ respectively. One can always suppose that the two
representations have the same length $k$, by padding the shortest one to the
left with enough zeroes. Let us form a new word $a_k \cdots a_0$, with $a_i = d_i + c_i$
for $0 \leq i \leq k$. Obviously, $\sum_{i=0}^{k} a_i b^i = N + M$, but $a_i$ belongs to the set
$\{0, \cdots, 2(b-1)\}$. One has to transform the word $a_k \cdots a_0$ into an equivalent
one (i.e. having the same numerical value) and belonging to $A^*$.

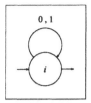

Figure 7.1: Automaton for binary representations

**Proposition 7.3.1** *Addition to the base b is a right subsequential function.*

The proof is folklore and can be found in [15]. In Figure 7.2 we show the addition automaton for base 2.

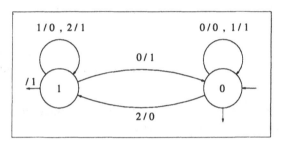

Figure 7.2: Addition base 2

**Proposition 7.3.2** *In the b-ary number system, multiplication by a fixed integer is a right subsequential function and division by a fixed integer is a left subsequential function.*

The proof is left as an exercise. Remark that multiplication is not computable by a finite 2-tape automaton (see [15]).

## 7.3.2 Representation of Real Numbers

The *standard b-ary expansion* of a positive real number $x$ is an infinite sequence $(x_i)_{k \geq i \geq -\infty}$ such that $x_i \in A = \{0, \cdots, b-1\}$ and

$$x = \sum_{-\infty}^{k} x_i b^i.$$

We note
$$\langle x \rangle_b = x_k \cdots x_0.x_{-1}x_{-2} \cdots$$
If $k < 0$, we put $x_0 = x_{-1} = \cdots = x_{k+1} = 0$.

Such an expansion is unique, with the condition that $x_k \neq 0$ and that it does not end by $(b-1)^\omega$. The set of all $b$-expansions forms the one-sided full $b$-shift.

**Proposition 7.3.3** *Addition, multiplication/division by a fixed integer of real numbers to the base $b$ are computable by a finite 2-tape automaton.*

### 7.3.3   $b$-Recognizable Sets of Integers

A set $X$ of positive integers is said to be *$b$-recognizable* if the set of $b$-representations of elements of $X$ is recognizable by a finite automaton. Two numbers $k$ and $l$ are said to be *multiplicatively independent* if there exist no positive integers $p$ and $q$ such that $k^p = l^q$. We then have the remarkable result of Cobham.

**Theorem 7.3.4** [13] *If $X$ is a set of integers which is both $k$-recognizable and $l$-recognizable in two multiplicatively independent bases $k$ and $l$, then $X$ is eventually periodic.*

The set $X$ is eventually periodic if it is a finite union of arithmetic progressions. This result implies that such a set is then $b$-recognizable for any base $b$.

There is a multidimensional version of Cobham's Theorem due to Semenov [38]. The original proofs of these two results are difficult, and several other proofs have been given, some of them using logic (see [23], [33], [32]).

## 7.4   Beta-Expansions

We now consider a generalization of number systems by taking for base a real number $\beta > 1$. Representations of real numbers with an arbitrary base $\beta > 1$, called *$\beta$-expansions*, were introduced by Rényi [36]. They arise from the orbits of a piecewise monotone transformation of the unit interval: $T_\beta : x \mapsto \beta x \pmod{1}$. Such transformations were extensively studied in ergodic theory (see [34] and the bibliography in [9]).

### 7.4.1   Definitions

Let $\beta > 1$ be a real number. A *representation in base $\beta$* (or a $\beta$-representation) of a real number $x \geq 0$ is an infinite sequence $(x_i)_{k \geq i \geq -\infty}$ such that
$$x = x_k\beta^k + x_{k-1}\beta^{k-1} + \cdots + x_1\beta + x_0 + x_{-1}\beta^{-1} + x_{-2}\beta^{-2} + \cdots$$

for a certain integer $k \geq 0$. It is denoted by

$$x = x_k x_{k-1} \cdots x_1 x_0 . x_{-1} x_{-2} \cdots$$

A particular $\beta$-representation — called the $\beta$-*expansion* — can be computed by the "greedy algorithm": denote by $[y]$ the integer part and by $\{y\}$ the fractional part of a number $y$. There exists $k \in \mathbf{Z}$ such that $\beta^k \leq x < \beta^{k+1}$. Let $x_k = [x/\beta^k]$ and $r_k = \{x/\beta^k\}$. Then for $k > i \geq -\infty$, put $x_i = [\beta r_{i+1}]$ and $r_i = \{\beta r_{i+1}\}$. We get an expansion $x = x_k \beta^k + x_{k-1}\beta^{k-1} + \cdots$. If $k < 0$ $(x < 1)$, we put $x_0 = x_{-1} = \cdots = x_{k+1} = 0$. The $\beta$-expansion of $x$ will be denoted by $\langle x \rangle_\beta$.

The digits $x_i$ obtained by this algorithm are integers from the *canonical alphabet* $A = \{0, \cdots, [\beta]\}$ if $\beta$ is not an integer, or the alphabet $A = \{0, \cdots, \beta - 1\}$ if $\beta$ is an integer. In the last case the $\beta$-expansion is just the standard expansion to the base $\beta$. We will sometimes omit the splitting point between the integer part and the fractional part of the $\beta$-expansion; then the infinite sequence is just an element of $A^{\mathbf{N}}$.

If a representation ends with infinitelymany zeros, like $v0^\omega$, it is said to be *finite*, and the ending zeros are omitted. When an expansion is of the form $vw^\omega$, the expansion is said to be *eventually periodic*.

For numbers $x < 1$, the greedy $\beta$-expansion defined above coincides with the $\beta$-expansion of Rényi [36], which can be defined by means of the $\beta$-*transformation* of the unit interval

$$T_\beta(x) = \beta x \,(\mathrm{mod}\ 1), \quad x \in [0, 1].$$

Equivalently, for $x \in [0, 1]$, let $x_1 = [\beta x]$, $r_1 = \{\beta x\}$, and for $i \geq 2$, $x_i = [\beta r_{i-1}]$, $r_i = \{\beta r_{i-1}\}$. Then $x = \sum_{i \geq 1} x_i \beta^{-i}$, with $x_i = [\beta T_\beta^{i-1}(x)]$. We note $d_\beta(x) = (x_i)_{i \geq 1}$.

However, for $x = 1$ the two algorithms differ: the greedy expansion is just $1 = 1.$, while the Rényi expansion is $d_\beta(1) = .t_1 t_2 \cdots$, with $t_k = [\beta T_\beta^{k-1}(1)]$ (the radix point is usually omitted).

Remark that the $\beta$-expansion of $x \in [0, 1]$ is finite if and only if $T_\beta^i(x) = 0$ for some $i$, and it is eventually periodic if and only if for some $i$ the orbit $\{T_\beta^i(x) : i \in \mathbf{N}\}$ is finite.

Numbers $\beta$ such that $d_\beta(1)$ is eventually periodic are called $\beta$-*numbers* and those such that $d_\beta(1)$ is finite are called *simple* $\beta$-numbers [34].

**Example 7.4.1**

1. Let $\beta$ be the Golden Ratio $(1 + \sqrt{5})/2$; then the expansion of 1 is finite, $d_\beta(1) = 11$.

2. Let $\beta = (3 + \sqrt{5})/2$; then $d_\beta(1) = 21^\omega$. The expansion is eventually periodic.

3. Let $\beta = 3/2$; then $d_\beta(1) = 101000001 \cdots$. We shall see later that it cannot be eventually periodic.

## 7.4.2   The $\beta$-Shift

Let $D_\beta$ be the set of $\beta$-expansions of numbers of $[0,1[$, and let $d_\beta :$ $[0,1[ \to D_\beta \cup \{d_\beta(1)\}$ be the function mapping $x \neq 1$ to its $\beta$-expansion, and 1 to $d_\beta(1)$. Clearly, if $x = x_k \cdots x_0.x_{-1}\cdots$ is a $\beta$-expansion, then $x/\beta^{k+1} = .x_k \cdots x_0 x_{-1}\cdots$ belongs to $D_\beta$. The set $A^{\mathbf{N}}$ is endowed with the *lexicographic ordering* (notation $<_{lex}$ ), the product topology, and the (one-sided) shift $\sigma$. The set $D_\beta$ is shift-invariant. The $\beta$-*shift* $S_\beta$ is the closure of $D_\beta$ and it is a subshift of $A^{\mathbf{N}}$. We have $d_\beta \circ T_\beta = \sigma \circ d_\beta$ on $[0,1[$, see [26] and [9].

The set $D_\beta$ is characterized as follows [34].

**Theorem 7.4.2** *Let $\beta$ be a real number greater than one, and let $d_\beta(1) = t_1 t_2 \cdots$ . Let $s$ be an infinite sequence of positive integers.*

*(i) If $d_\beta(1)$ is infinite, the condition*

$$\forall p \geq 0, \ \sigma^p(s) <_{lex} d_\beta(1)$$

*is necessary and sufficient for $s$ to belong to $D_\beta$.*

*(ii) If $d_\beta(1)$ is finite, $d_\beta(1) = t_1 \cdots t_{m-1} t_m$, then $s \in D_\beta$ if and only if*

$$\forall p \geq 0, \ \sigma^p(s) <_{lex} d_\beta^*(1) = (t_1 \cdots t_{m-1}(t_m - 1))^\omega.$$

Large inequalities in Theorem 7.4.2 give conditions for a sequence to belong to the $\beta$-shift $S_\beta$.

Let us recall some definitions on symbolic dynamical systems or subshifts (see [30]). A subshift $S$ is of *finite type* if the set $F(S)$ of its finite factors is defined by the interdiction of a finite set of words. It is said to be *sofic* if $F(S)$ is recognized by a finite automaton. And $S$ is said to be *coded* if there exists a prefix set $Y \subset A^*$ such that $S$ is the closure of the set of infinite words obtained by freely concatenating words of $Y$ (see [10] for more on coded systems, and also [30]). The name "coded" comes from the fact that a prefix set is a code. A set $C$ of finite words is called a *code* if

$$u_1 u_2 \cdots u_n = v_1 v_2 \cdots v_m$$

where $u_i$, $v_j \in C$, implies $n = m$ and $u_i = v_i$ for every $i$.

We first prove a result of A. Bertrand-Mathis (see [9]).

**Proposition 7.4.3** *The $\beta$-shift $S_\beta$ is a coded system.*

**Proof.** If $d_\beta(1) = (t_i)_{i \geq 1}$ is infinite, take $Y = \{0, \cdots, t_1 - 1, t_1 0, \cdots, t_1(t_2 - 1), \cdots\} = \{t_1 \cdots t_{n-1} a \mid a < t_n, \ n \in \mathbf{N}\}$. Clearly, $Y$ is a prefix set. We show that $D_\beta = Y^\omega$. First, let $s \in D_\beta$. By Theorem 7.4.2, $s <_{lex} d_\beta(1)$, thus it can be written as $s = t_1 \cdots t_{n_1-1} a_{n_1} s_1$, with $a_{n_1} < t_{n_1}$ and $s_1 <_{lex} d_\beta(1)$. Iterating this process, we see that $s \in Y^\omega$. Conversely, let $s = u_1 u_2 \cdots \in Y^\omega$, with $u_i = t_1 \cdots t_{n_i-1} a_{n_i}$, $a_{n_i} < t_{n_i}$. Then $s <_{lex} d_\beta(1)$. For each $p \geq 0$, $\sigma^p(s)$ starts by a word of the form $t_{j_p} t_{j_p+1} \cdots t_{j_p+r-1} b_{j_p+r}$ with $b_{j_p+r} < t_{j_p+r}$, thus $\sigma^p(s) <_{lex} t_{j_p} t_{j_p+1} \cdots t_{j_p+r-1} t_{j_p+r} t_{j_p+r+1} \cdots <_{lex} d_\beta(1)$.

If $d_\beta(1) = t_1 \cdots t_m$, let $Y = \{t_1 \cdots t_{n-1} a \mid a < t_n, \ 0 \leq n \leq m\}$. We claim that $Y^\omega = S_\beta$. First, let $s \in S_\beta$. By Theorem 7.4.2, $s \leq_{lex} d_\beta^*(1)$, thus $s = t_1 \cdots t_{n_1-1} a_{n_1} s_1$, with $n_1 \leq m$, $a_{n_1} < t_{n_1}$ and $s_1 \leq_{lex} d_\beta^*(1)$. Iterating the process we get $s \in S_\beta$. Conversely, let $s \in Y^\omega$, $s = u_1 u_2 \cdots$ $u_i = t_1 \cdots t_{n_i-1} a_{n_i}$, $n_i \leq m$. As above, one gets that, for each $p \geq 0$, $\sigma^p(s) <_{lex} d_\beta^*(1)$. ∎

Remark that $Y$ is finite if and only if $d_\beta(1)$ is finite.

The *topological entropy* of a nonempty subshift $S$ is the limit

$$h(S) = \lim_{n \to \infty} \frac{1}{n} \log s_n$$

where $s_n$ is the number of factors of length $n$ of elements of $S$.

**Proposition 7.4.4** [26] *The topological entropy of the $\beta$-shift is equal to* $\log \beta$.

Let $f_n$ be the number of words of length $n$ of $Y^*$. The *topological entropy* of $Y^*$ is defined as

$$h(Y^*) = \limsup_{n \to \infty} \frac{1}{n} \log f_n.$$

First we show that $h(S_\beta) = h(Y^*)$. Since any word of the $\beta$-shift can be uniquely written as $w = y_0$, or $w = y_1 t_1$, or $w = y_2 t_1 t_2$, etc, where $y_i \in Y^*$. Now, $s_n \leq (n+1) f_n$, hence $h(S_\beta) \leq h(Y^*)$.

The *generating series* of a set $Y$ is

$$f_Y(X) = \sum_{n \geq 0} y_n X^n$$

where $y_n$ is the number of words of length $n$ of $Y$. It is known (see [4] p. 42) that $Y$ is a code if and only if the following equality holds

$$f_{Y^*} = \frac{1}{1 - f_Y}.$$

Clearly, we have $y_0 = 0$ and for $n \geq 1$, $y_n = t_n$. Since $1 = \sum_{n \geq 1} t_n \beta^{-n}$, $\beta^{-1}$ is a simple pole for $f_{Y^*}$, and is equal to its convergence radius. We thus have $\beta = \limsup_{n \to \infty} f_n^{1/n}$, and the entropy of $Y^*$ is equal to $\log(\beta)$. ∎

The properties of the subshift $S_\beta$ are totally determined by the $\beta$-expansion of 1.

**Theorem 7.4.5** [26] *The $\beta$-shift $S_\beta$ is a system of finite type if and only if $d_\beta(1)$ is finite.*

**Theorem 7.4.6** [7] *The $\beta$-shift $S_\beta$ is a sofic system if and only if $d_\beta(1)$ is eventually periodic.*

**Proof.** Suppose that $d_\beta(1)$ is eventually periodic,
$d_\beta(1) = t_1 \cdots t_N (t_{N+1} \cdots t_{N+p})^\omega$ with $N$ and $p$ minimal and with $t_{N+jp+k} = t_{N+k}$ for $1 \leq k \leq p$, $j \geq 0$. We construct the minimal finite automaton $\mathcal{A}_\beta$ recognizing the set $F(D_\beta)$ of finite factors of $D_\beta$ as follows ([21]). The automaton $\mathcal{A}_\beta$ has $N + p$ states $q_1, \cdots, q_{N+p}$, where $q_i$, $i \geq 2$, represents the right class $[t_1 \cdots t_{i-1}]_{F(D_\beta)}$ and $q_1$ stands for $[\varepsilon]_{F(D_\beta)}$ (see Section 7.2). For each $i$, $1 \leq i \leq N + p$, there are edges labelled by $0, 1, \cdots, t_i - 1$ from $q_i$ to $q_1$, and an edge labelled $t_i$ from $q_i$ to $q_{i+1}$ if $i < N + p$. Finally, there is an edge labelled $t_{N+p}$ from $q_{N+p}$ to $q_{N+1}$. Let $q_1$ be the only initial state, and all states be terminal. That $F(D_\beta)$ is precisely the set recognized by the automaton $\mathcal{A}_\beta$ follows from Theorem 7.4.2. Remark that, when the $\beta$-expansion of 1 happens to be finite, say $d_\beta(1) = t_1 \cdots t_m$, with $t_n = 0$ for $n > m$, the same construction applies with $N = m, p = 0$ and all edges from $q_m$ (labelled $0, 1, \cdots, t_m - 1$) leading to $q_1$. In that case, the automaton is local (see [1]).

Suppose now that $d_\beta(1) = (t_i)_{i \geq 1}$ is neither finite nor eventually periodic. There exists an infinite sequence of indexes $i_1 < i_2 < i_3 < \cdots$ such that the sequences $t_{i_k+1} t_{i_k+2} t_{i_k+3} \cdots$ be all different for all $k \geq 1$. Thus there exists $p \geq 0$ such that $w = t_{i_1+1} \cdots t_{i_1+p-1} = t_{i_2+1} \cdots t_{i_2+p-1}$, and for instance, $t_{i_1+p} < t_{i_2+p}$. One has $t_1 \cdots t_{i_1} w t_{i_1+p} \in F(D_\beta)$, $t_1 \cdots t_{i_2} w t_{i_2+p} \in F(D_\beta)$, $t_1 \cdots t_{i_2} w t_{i_1+p} \in F(D_\beta)$, but $t_1 \cdots t_{i_1} w t_{i_2+p}$ does not belong to $F(D_\beta)$. Hence $t_1 \cdots t_{i_1}$ and $t_1 \cdots t_{i_2}$ are not right congruent modulo $F(D_\beta)$. The number of right congruence classes is thus infinite, and $F(D_\beta)$ is not recognizable by a finite automaton. ∎

## 7.4.3   Classes of Numbers

Recall that an *algebraic integer* is a root of a monic polynomial with integral coefficients. An algebraic integer $\beta > 1$ is called a *Pisot number* if all its Galois conjugates have modulus less than one; it is a *Salem number* if all its conjugates are less than or equal to one in modulus and at least one conjugate has modulus one; and it is a *Perron number* if all its conjugates have modulus less than $\beta$.

**Example 7.4.7**

1. Every integer is a Pisot number. The golden ratio $(1 + \sqrt{5})/2$ and $(3 + \sqrt{5})/2$ are Pisot numbers, with minimal polynomial $X^2 - X - 1$ and $X^2 - 3X + 1$ respectively.

2. A non-integral rational number cannot be an algebraic integer.

3. $(5 + \sqrt{5})/2$ is a Perron number; it is neither Pisot nor Salem.

**Proposition 7.4.8** [6] *If $\beta$ is a Pisot number then the $\beta$-shift $S_\beta$ is sofic.*

In fact, more is known, owing to the following results.

**Proposition 7.4.9** [6], [37] *If $\beta$ is a Pisot number then every number of $\mathbb{Q}(\beta)$ has an eventually periodic $\beta$-expansion.*

**Proposition 7.4.10** [37] *If every rational number of $[0, 1]$ has an eventually periodic $\beta$-expansion, then $\beta$ must be Pisot or Salem.*

On the other hand there is a gap between Pisot and Perron numbers (see [29]).

**Proposition 7.4.11** *If $S_\beta$ is sofic then $\beta$ is a Perron number.*

Thus when $\beta$ is a non integral rational number (for instance 3/2), $S_\beta$ cannot be sofic.
We have also this property on conjugates of $\beta$-numbers.

**Proposition 7.4.12** [40], [17] *If $\beta$ is a $\beta$-number, that is if $S_\beta$ is sofic, then the Galois conjugates of $\beta$ have modulus less than the golden ratio.*

**Remark 7.4.13** If $\beta$ is a Perron number with a real conjugate $> 1$, then $d_\beta(1)$ cannot be eventually periodic.

**Example 7.4.14** Take $\beta = (5 + \sqrt{5})/2$. Since it has a real conjugate $> 1$, then $S_\beta$ is not sofic.

**Example 7.4.15** There are Perron numbers which are neither Pisot nor Salem such that $S_\beta$ is of finite type: for instance the root $\beta > 1$ of $X^4 - 3X^3 - 2X^2 - 3$, with $d_\beta(1) = 3203$.

For Salem numbers, very few facts are known.

**Proposition 7.4.16** [11] *If $\beta$ is a Salem number of degree 4 then $d_\beta(1)$ is eventually periodic.*

### 7.4.4    Normalization in Base $\beta$

We have seen above that any number from $[0, 1]$ has a $\beta$-expansion by the
Rényi algorithm. But such a number can also have several $\beta$-representations.
The $\beta$-expansion is the greatest for the lexicographic ordering.

**Example 7.4.17** Let $\beta = (1 + \sqrt{5})/2$, and $x = (3 - \sqrt{5})/2 = 1/\beta^2$. The $\beta$-expansion of $x$ is $\langle x \rangle_\beta = 01$ but $0011$ and $001(01)^\omega$ are different $\beta$-representations of $x$.

Denote by $\pi((x_i)_{i \geq 1})$ the real number $\sum_{i \geq 1} x_i \beta^{-i}$. To add two $\beta$-expansions, we proceed as in an integral base. Let $x = (x_i)_{i \geq 1}$ and $y = (y_i)_{i \geq 1}$, with $x_i$ and $y_i$ belonging to the canonical alphabet $A = \{0, \cdots, [\beta]\}$. Then $x + y = (x_i + y_i)_{i \geq 1}$, where $x_i + y_i$ belongs to the alphabet $\{0, \cdots, 2[\beta]\}$. What we have to do now is to transform this $\beta$-representation into the $\beta$-expansion of $x + y$.

More generally, let $C$ be a finite alphabet of integers. The *normalization function* on $C$

$$\nu_C : C^{\mathbf{N}} \longrightarrow A^{\mathbf{N}}$$

is the partial function which maps a sequence $s = (s_i)_{i \geq 1}$, $s_i \in C$, such that $0 \leq \pi(s) \leq 1$, to the $\beta$-expansion of $\pi(s)$.

First we get the following.

**Proposition 7.4.18** [19] *Let* $C = \{0, \cdots, c\}$, *where $c$ is an integer* $\geq 1$. *Normalization* $\nu_C : C^{\mathbf{N}} \longrightarrow A^{\mathbf{N}}$ *is a function computable by a finite letter-to-letter 2-tape automaton if and only if the set*

$$Z(\beta, c) = \{ s = (s_i)_{i \geq 0} \mid s_i \in \mathbf{Z}, \ |s_i| \leq c, \ \sum_{i \geq 0} s_i \beta^{-i} = 0 \}$$

*is recognizable by a finite automaton.*

**Example 7.4.19** Figure 7.3 represents the automaton recognizing the set $Z(\beta, 1)$ of infinite words on the alphabet $\{-1, 0, 1\}$ having numerical value $0$ to the base $\beta = (1 + \sqrt{5})/2$. Note that, since $\pi((10)^\omega) = 1$, every state is terminal. The notation $\bar{1}$ stands for the signed digit $-1$.

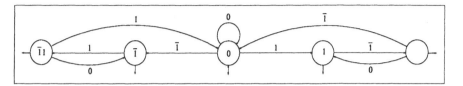

Figure 7.3: Automaton recognizing the set of infinite words on the alphabet $\{-1, 0, 1\}$ having numerical value 0 in base $(1 + \sqrt{5})/2$

The fact that normalization can be realized by a *letter-to-letter* 2-tape automaton is important because it means that computations are done without unbounded delay. This property permits to show that the alphabet $A' = \{0, \cdots, [\beta], [\beta] + 1\}$ plays a special role.

**Proposition 7.4.20** [20] *Normalization* $\nu_C : C^{\mathbb{N}} \longrightarrow A^{\mathbb{N}}$ *is a function computable by a finite letter-to-letter 2-tape automaton for every alphabet $C$ of (possibly negative) integers if and only if* $\nu_{A'} : A'^{\mathbb{N}} \longrightarrow A^{\mathbb{N}}$ *is a function computable by a finite letter-to-letter 2-tape automaton.*

**Theorem 7.4.21** [2] *The following conditions are equivalent:*

(i) *the normalization* $\nu_C : C^{\mathbb{N}} \longrightarrow A^{\mathbb{N}}$ *is a function computable by a finite letter-to-letter 2-tape automaton on any alphabet $C$ of integers,*

(ii) $\nu_{A'} : A'^{\mathbb{N}} \longrightarrow A^{\mathbb{N}}$ *is a function computable by a finite letter-to-letter 2-tape automaton,*

(iii) $\beta$ *is a Pisot number.*

**Corollary 7.4.22** *If $\beta$ is a Pisot number, then addition, subtraction and multiplication/division by a fixed integer are computable by a finite 2-tape automaton.*

# 7.5 $U$-Representations

We now consider another generalization of the notion of number system, which applies only to the representation of integers.

## 7.5.1 Definitions

Let $U = (u_n)_{n \geq 0}$ be a strictly increasing sequence of integers with $u_0 = 1$. A *representation in the system $U$* — or a *$U$-representation* — of a nonnegative integer $N$ is a finite sequence of integers $(d_i)_{k \geq i \geq 0}$ such that

$$N = \sum_{i=0}^{k} d_i u_i.$$

Such a representation will be written $d_k \cdots d_0$, most significant digit first.

We say that a word $d = d_k \cdots d_0$ is *lexicographically greater* than a word $f = f_k \cdots f_0$, and this will be denoted by $d >_{lex} f$, if there exists an index $k \geq i \geq 0$ such that $d_k = f_k, \ldots, d_{i-1} = f_{i-1}$ and $d_i > f_i$. Among all possible $U$-representations $d_k \cdots d_0$ of a given nonnegative integer $N$ one is distinguished and called the *normal U-representation* of $N$: it is the greatest in the lexicographic ordering. It is sometimes called the *greedy* representation, since it can be obtained by the following greedy algorithm (see [18]): given integers $m$ and $p$ denote by $q(m, p)$ and $r(m, p)$ the quotient and the remainder of the Euclidean division of $m$ by $p$. Let $k \geq 0$ such that $u_k \leq N < u_{k+1}$ and let $d_k = q(N, u_k)$ and $r_k = r(N, u_k)$, and, for $i = k - 1, \cdots, 0$, $d_i = q(r_{i+1}, u_i)$ and $r_i = r(r_{i+1}, u_i)$. Then $N = d_k u_k + \cdots + d_0 u_0$.

By convention the normal representation of $0$ is the empty word $\varepsilon$. Under the hypothesis that the ratio $u_{n+1}/u_n$ is bounded by a constant as $n$ tends to infinity, the integers of the normal $U$-representation of any integer $N$ are bounded and contained in a *canonical* finite alphabet $A$ associated with $U$. The set of normal $U$-representations of all nonnegative integers is denoted by $L(U)$.

**Example 7.5.1** Let $U = (u_n)_{n \geq 0}$ be the sequence of Fibonacci numbers

$$u_n = u_{n-1} + u_{n-2}, \quad u_0 = 1, \quad u_1 = 2.$$

Then the canonical alphabet is equal to $A = \{0, 1\}$ and $L(U)$ is the set of words whithout the factor $11$, and not beginning by a $0$,

$$L(U) = 1\{0, 1\}^* \setminus \{0, 1\}^* 11\{0, 1\}^*.$$

## 7.5.2   The Set $L(U)$

The first result is due to Shallit.

**Proposition 7.5.2** [39] *If the set $L(U)$ is recognizable by a finite automaton, then $U$ must be a linear recurrence with integral coefficients.*

**Proof.** We follow the proof given by Loraud [31]. Recall that a formal series with coefficients in $\mathbb{N}$ is said to be $\mathbb{N}$-*rational* if it belongs to the smallest class containing polynomial with coefficients in $\mathbb{N}$, and closed under addition, multiplication and star operation, where $F^*$ is the series $1 + F + F^2 + F^n + \cdots =$

$1/(1-F)$, $F$ being a series such that $F(0) = 0$. A $\mathbb{N}$-rational series is necessarily $\mathbb{Z}$-rational, and can be written $P(X)/Q(X)$, with $P(X)$ and $Q(X)$ in $\mathbb{Z}[X]$, and $Q(0) = 1$. Thus the sequence of coefficients of a $\mathbb{N}$-rational series satisfies a linear recurrent relation with coefficients in $\mathbb{Z}$.

It is a classical fact that, if $L$ is recognizable by a finite automaton, then the series $f_L(X) = \sum_{n \geq 0} l_n X^n$, where $l_n$ denotes the number of words of length $n$ in $L$, is $\mathbb{N}$-rational (see [5]).

We now prove that, if $L(U)$ is recognizable by a finite automaton, then the series $U(X) = \sum_{n \geq 0} u_n X^n$ is $\mathbb{N}$-rational. Let $l_n$ be the number of words of length $n$ in $L(U)$. The series $f_{L(U)}(X) = \sum_{n \geq 0} l_n X^n$ is $\mathbb{N}$-rational. We have $u_n = l_n + \cdots + l_0$, because the number of words of length $\leq n$ in $L(U)$ is equal to the number of integers smaller than $u_n$, the normal representation of which has length $n+1$. Thus $U(X) = f_{L(U)}(X)/(1-X)$, and it is $\mathbb{N}$-rational. ∎

When the sequence $U$ is linearly recurrent, we say that $U$ defines a *linear numeration system*.

In the sequel we expose the nice results of Hollander [24] giving conditions on $U$ such that the set $L(U)$ is recognizable by a finite automaton. Let $U$ be a linear recurrent sequence of integers. If $\lim_{n \to \infty}(u_n/u_{n-1}) = \beta$ for a real $\beta > 1$ then $U$ is said to satisfy the *dominant root condition* and $\beta$ is called the dominant root of the recurrence.

In the case that $d_\beta(1)$ is finite or eventually periodic, there is a polynomial which naturally arises from the pattern of the digits. If $d_\beta(1)$ is eventually periodic, $d_\beta(1) = t_1 \cdots t_N(t_{N+1} \cdots t_{N+p})^\omega$, $\beta$ satisfies the polynomial

$$B(X) = X^{N+p} - \sum_{i=1}^{N+p} t_i X^{N+p-i} - X^N + \sum_{i=1}^{N} t_i X^{N-i}.$$

Similarly, if $d_\beta(1)$ is finite, $d_\beta(1) = t_1 \cdots t_m$, we let

$$B(X) = X^m - \sum_{i=1}^{m} t_i X^{m-i}.$$

Note that $B(X)$ depends on the choice of $N$ and $p$ (or $m$). We refer to any such polynomial as an *extended beta polynomial* for $\beta$. If $N$ and $p$ (or $m$) are chosen to be minimal, the polynomial we get is referred to as the *beta polynomial* for $\beta$.

**Theorem 7.5.3** [24] *Assume that the linearly recurrent sequence $U$ has dominant root $\beta > 1$. Then*

*(i) The set $L(U)$ is recognizable by a finite automaton only if $d_\beta(1)$ is finite or eventually periodic.*

222                                                                    *C. Frougny*

*(ii) If $d_\beta(1)$ is eventually periodic, then $L(U)$ is recognizable by a finite automaton if and only if $U$ satisfies an extended beta polynomial for $\beta$.*

*(iii) If $d_\beta(1)$ is finite, then*
   * *if $U$ satisfies an extended beta polynomial for $\beta$ then $L(U)$ is recognizable by a finite automaton,*
   * *if $L(U)$ is recognizable by a finite automaton then $U$ satisfies a polynomial of the form $(X^m - 1)B(X)$ where $B(X)$ is an extended polynomial for $\beta$ and $m$ is the length of $d_\beta(1)$.*

Note that, in the case where $d_\beta(1)$ is finite, initial conditions of the recurrence play a crucial role.

**Example 7.5.4** Let the characteristic polynomial of the recurrence $U$ be $B(X) = (X - 1)(X - 3)$. With initial conditions $u_0 = 1$ and $u_2 = 4$, $L(U)$ is recognizable by a finite automaton. With initial conditions $u_0 = 1$ and $u_2 = 2$, $L(U)$ is not recognizable by a finite automaton [24].

### 7.5.3   Normalization in the Linear Numeration System $U$

The *numerical value* in the system $U$ of a representation $w = d_k \cdots d_0$ is equal to $\pi(w) = \sum_{i=0}^k d_i u_i$. Let $C$ be a finite alphabet of integers. The *normalization* in the system $U$ on the alphabet $C$ is the partial function

$$\nu_C : C^* \longrightarrow A^*$$

that maps a word $w$ of $C^*$ such that $\pi(w)$ is non negative to the normal $U$-representation of $\pi(w)$.

In the sequel, we assume that $U$ is defined by the linear recurrence relation

$$u_n = a_1 u_{n-1} + a_2 u_{n-2} + \cdots + a_m u_{n-m}, \quad a_i \in \mathbb{Z}, \quad a_m \neq 0, \quad n \geq m. \quad (7.1)$$

In that case, the canonical alphabet $A$ associated with $U$ is $A = \{0, \cdots, K\}$ where $K < \max(u_{i+1}/u_i)$ is bounded. The polynomial $P(X) = X^m - a_1 X^{m-1} - \cdots - a_m$ will be called the *characteristic polynomial* of the recurrence relation (7.1).
Let

$$Z(U, c) = \{d_k \cdots d_0 \mid d_i \in \mathbb{Z}, \ |d_i| \leq c, \ \sum_{i=0}^k d_i u_i = 0\}$$

be the set of words on $\{-c, \cdots, c\}$ having numerical value 0 in the system $U$. Then the following holds.

**Proposition 7.5.5** [19] *Let* $U$ *be a linear recurrent sequence of integers satisfying (7.1).*
**A.** *If* $L(U)$ *and* $Z(U,c)$ *are recognizable by a finite automaton then normalization on* $C = \{0, \cdots, c\}$

$$\nu_C : C^* \longrightarrow A^*$$

*is a function computable by a finite 2-tape automaton.*
**B.** *Assume that the characteristic polynomial* $P$ *has a dominant root* $> 1$. *If* $\nu_C$ *is computable by a finite 2-tape automaton then this automaton can be chosen letter-to-letter, and* $L(U)$ *and* $Z(U,c)$ *are recognizable by a finite automaton.*

**Example 7.5.6** Below is the automaton recognizing the set of words on the alphabet $\{-1,0,1\}$ having numerical value 0 in the Fibonacci numeration system. The only terminal state is 0.

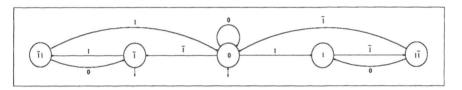

Figure 7.4: Automaton recognizing the set of words on $\{-1,0,1\}$ having value 0 in the Fibonacci numeration system

Recall that $A = \{0, \ldots, K\}$ is the canonical alphabet. As for $\beta$-expansions, there is a critical alphabet.

**Proposition 7.5.7** [20] *Under the same hypotheses as in Part B of Proposition 7.5.5 above, let* $A' = \{0, \cdots, K, K+1\}$. *Then normalization* $\nu_C : C^* \longrightarrow A^*$ *is a function computable by a finite letter-to-letter 2-tape automaton for every alphabet* $C$ *of (possibly negative) integers if and only if* $\nu_{A'} : A'^* \longrightarrow A^*$ *is a function computable by a finite letter-to-letter 2-tape automaton.*

The next results are analogous to those on normalization to the base $\beta$, but more complicated.

**Theorem 7.5.8** [21] *Let* $U$ *be a linear recurrent sequence of integers satisfying (7.1). If the characteristic polynomial* $P$ *of* $U$ *is the minimal polynomial of a Pisot number* $\beta > 1$, *then for any alphabet* $C$ *of (possibly negative) integers, normalization* $\nu_C$ *is a function computable by a finite 2-tape automaton.*

**Example 7.5.9** Figure 7.5 represents a letter-to-letter finite 2-tape automaton realizing the normalization on $\{0,1\}$ in the Fibonacci numeration system. For simplicity, we assume that input and output words have the same length.

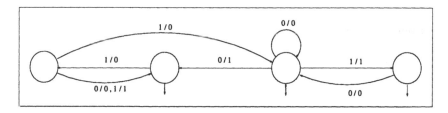

Figure 7.5: Normalization on $\{0,1\}$ in the Fibonacci numeration system

**Corollary 7.5.10** *If $U$ is a linear recurrent sequence such that the characteristic polynomial is the minimal polynomial of a Pisot number, then addition, subtraction, multiplication/division by a fixed integer are functions computable by a finite 2-tape automaton.*

We have a kind of converse of Theorem 7.5.8.

**Theorem 7.5.11** [21] *If the characteristic polynomial $P$ of $U$ is the minimal polynomial of a Perron number $\beta > 1$ which is not Pisot, then for every integer $c \geq K + 1$, normalization $\nu_C$ on $C = \{0, \cdots, c\}$ is not computable by a finite 2-tape automaton.*

**Example 7.5.12** Let $u_{n+4} = 3u_{n+3} + 2u_{n+2} + 3u_n$, with (for instance) $u_0 = 1$, $u_1 = 4$, $u_2 = 15$, $u_3 = 54$. In that case, the canonical alphabet is $A = \{0, \cdots, 3\}$. The characteristic polynomial has a dominant root $\beta$ which is Perron but not Pisot. We have $d(1, \beta) = 3203$, so the $\beta$-shift $S_\beta$ is a system of finite type.

From Theorem 7.5.11 we know that for any alphabet $C \supseteq \{0, \cdots, 4\}$, normalization on $C$ is not computable by a finite 2-tape automaton. Nevertheless, it can be shown with an *adhoc* proof that normalization on $A$ is computable by a finite letter-to-letter 2-tape automaton.

### 7.5.4 $U$-Recognizable Sets of Integers

The definition of $b$-recognizable sets of integers can be generalized as follows. A set $X$ of positive integers is $U$-recognizable if the set of normal $U$-representations of numbers of $X$ is recognizable by a finite automaton.

There are many works on generalizations of the Cobham and Semenov theorems (see [16, 12, 35]).

We first give a result on linear numeration systems. Let $n \geq 1$. A subset $X \subset \mathbb{N}^n$ is said to be *U-definable* if $X$ is first-order definable in the structure $\langle \mathbb{N}; +, V_U \rangle$, where the function $V_U : \mathbb{N} \longrightarrow \mathbb{N}$ is defined as follows: $V_U(0) = u_0 = 1$, and for every positive integer $m$, if the normal $U$-representation of $m$ is $a_k \cdots a_j 0^j$, with $a_j \neq 0$, then $V_U(m) = u_j$.

**Proposition 7.5.13** [12] *Let $U$ be a linear numeration system such that its characteristic polynomial is the minimal polynomial of a Pisot number. Then for every $n$, a set $X \subset \mathbb{N}^n$ is $U$-recognizable if and only if it is $U$-definable.*

**Theorem 7.5.14** [8] *Let $\beta$ and $\beta'$ be two multiplicatively independent Pisot numbers, and let $U$ and $U'$ be two linear numeration systems whose characteristic polynomial are the minimal polynomial of $\beta$ and $\beta'$ respectively. For every $n \geq 1$, if $X \subset \mathbb{N}^n$ is $U$- and $U'$-recognizable then $X$ is definable in $\langle \mathbb{N}, + \rangle$.*

The proof uses Theorem 7.5.8. When $n = 1$, the result says that $X$ is eventually periodic.

In the context of substitutions, a more general result in the one-dimensional case has been recently proved by F. Durand.

**Theorem 7.5.15** [14] *If a non-periodic sequence $X$ on a finite alphabet is the image by a morphism of a fixed point of both a primitive substitution $\sigma$ and a primitive substitution $\tau$, then the dominant eigenvalues of $\sigma$ and $\tau$ are multiplicatively dependent.*

# Bibliography

[1] M.P. Béal and D. Perrin, *Symbolic Dynamics and Finite Automata*, Handbook of Formal Languages, ed. D. Rozenberg and A. Salomaa, **2**, pp. 463–503 (1997).

[2] D. Berend and Ch. Frougny, *Computability by Finite Automata and Pisot Bases*, Math. Systems Theory **27**, pp. 274–282 (1994).

[3] J. Berstel, *Transductions and Context-Free Languages*, Teubner (1979).

[4] J. Berstel and D. Perrin, *Theory of Codes*, Academic Press (1985).

[5] J. Berstel and C. Reutenauer, *Rational Series and Their Languages*, Springer-Verlag (1988).

[6]  A. Bertrand, *Développements en Base de Pisot et Répartition Modulo 1*, C. R. Acad. Sc., Paris **285**, pp. 419–421 (1977).

[7]  A. Bertrand-Mathis, *Développement en Base $\theta$, Répartition Modulo un de la Suite $(x\theta^n)_{n \geq 0}$, Langages Codés et $\theta$-Shift*, Bull. Soc. Math. France **114**, pp. 271–323 (1986).

[8]  A. Bès, *An Extension of the Cobham-Semenov Theorem*, J. of Symbolic Logic, to appear.

[9]  F. Blanchard, *$\beta$-Expansions and Symbolic Dynamics*, Theor. Comp. Sci. **65**, pp. 131–141 (1989).

[10]  F. Blanchard and G. Hansel, *Systèmes Codés*, Theor. Comp. Sci. **44**, pp. 17–49 (1986).

[11]  D. Boyd, *Salem Numbers of Degree four have Periodic Expansions*, Number Theory, ed. J.-H. de Coninck and C. Levesque, Walter de Gruyter, pp. 57–64 (1989).

[12]  V. Bruyère and G. Hansel, *Bertrand Numeration Systems and Recognizability*, Theor. Comp. Sci. **181**, pp. 17–43 (1997).

[13]  A. Cobham, *On the Base-Dependence of Sets of Numbers Recognizable by Finite Automata*, Math. Systems Theory **3**, pp. 186–192 (1969).

[14]  F. Durand, *A Generalization of Cobham's Theorem*, Theory of Computing Systems **32**, pp. 169–185 (1998).

[15]  S. Eilenberg, *Automata, Languages and Machines*, vol. A, Academic Press (1974).

[16]  S. Fabre, *Une Généralisation du Théorème de Cobham*, Acta Arithmetica **67/3**, pp. 197–208 (1994).

[17]  L. Flatto, J.C. Lagarias and B. Poonen, *The Zeta Function of the Beta Transformation*, Ergodic Theory Dynamical Systems **14**, pp. 237–26 (1994).

[18]  A.S. Fraenkel, *Systems of Numeration*, Amer. Math. Monthly **92(2)**, pp. 105–114 (1985).

[19]  Ch. Frougny, *Representation of Numbers and Finite Automata*, Math. Systems Theory **25**, pp. 37–60 (1992).

[20]  Ch. Frougny and J. Sakarovitch, *Automatic Conversion from Fibonacci Representation to Representation in Base $\phi$, and a Generalization*, Internat. J. Algebra Comput., to appear.

[21] Ch. Frougny and B. Solomyak, *On Representation of Integers in Linear Numeration Systems*, in Ergodic Theory of $Z^d$-Actions, ed. M. Pollicott and K. Schmidt, London Mathematical Society Lecture Note Series **228**, pp. 345–368 (1996).

[22] W. Gilbert, *Radix Representations of Quadratic Fields*, J. Math. Anal. Appl. **83**, pp. 264–274 (1981).

[23] G. Hansel, *A Propos d'un Théorème de Cobham*, Actes de la Fête des Mots, Rouen (1982).

[24] M. Hollander, *Greedy Numeration Systems and Recognizability*, Theory of Computing Systems **31**, pp. 111-133 (1998).

[25] J. E. Hopcroft and J. D. Ullman, *Introduction to Automata Theory, Languages, and Computation*, Addison-Wesley (1979).

[26] S. Ito and Y. Takahashi, *Markov Subshifts and Realization of β-Expansions*, J. Math. Soc. Japan **26**, pp. 33–55 (1974).

[27] I. Kátai and J. Szabó, *Canonical Number Systems*, Acta Sci. Math. **37**, pp. 255–280 (1975).

[28] D.E. Knuth, *The Art of Computer Programming*, vol. 2, Seminumerical Algorithms, 2nd ed., Addison-Wesley (1988).

[29] D. Lind, *The Entropies of Topological Markov Shifts and a Related Class of Algebraic Integers*, Ergod. Th. & Dynam. Sys. **4**, pp. 283–300 (1984).

[30] D. Lind and B. Marcus, *Symbolic Dynamics and Coding*, Cambridge University Press (1995).

[31] N. Loraud, *β-Shift, Systèmes de Numération et Automates*, Journal de Théorie de Nombres de Bordeaux 7, pp. 473–498 (1995).

[32] C. Michaux and R. Villemaire, *Presburger Arithmetic and Recognizability of Sets of Natural Numbers by Automata: new Proofs of Cobham's and Semenov's Theorems*, Annals of Pure and Applied Logic **17**, pp. 251–277 (1996).

[33] A. Muchnik, *Definable Criterion for Definability in Presburger Arithmetic and its Application*, unpublished manuscript, Institute of New Technologies, Moscow (1991).

[34] W. Parry, *On the β-Expansions of Real Numbers*, Acta Math. Acad. Sci. Hungar. **11**, pp. 401–416 (1960).

C. Frougny

[35] F. Point and V. Bruyère, *On the Cobham-Semenov Theorem*, Theory of Computing Systems **30**, pp. 197–220 (1997).

[36] A. Rényi, *Representations for Real Numbers and their Ergodic Properties*, Acta Math. Acad. Sci. Hungar. **8**, pp. 477–493 (1957).

[37] K. Schmidt, *On Periodic Expansions of Pisot Numbers and Salem Numbers*, Bull. London Math. Soc. **12**, pp. 269–278 (1980).

[38] A.L. Semenov, *The Presburger Nature of Predicates that are Regular in Two Number Systems*, Siberian Math. J. **18(2)**, pp. 289–299 (1977).

[39] J. Shallit, *Numeration Systems, Linear Recurrences, and Regular Sets*, I.C.A.L.P. 92, Wien, Lecture Notes in Computer Science **623**, pp. 89–100 (1992).

[40] B. Solomyak, *Conjugates of Beta-Numbers and the Zero-Free Domain for a Class of Analytic Functions*, Proc. London Math. Soc. **(3)68**, pp. 477–498 (1994).

[41] W. Thurston, *Groups, Tilings, and Finite State Automata*, AMS Colloquium Lecture Notes, Boulder (1989).

# Chapter 8

# A NOTE ON THE TOPOLOGICAL CLASSIFICATION OF LORENZ MAPS ON THE INTERVAL

*Rafael LABARCA*
*Departamento de Matemática y Ciencias de la Computación*
*Universidad de Santiago de Chile*
*Casilla 307, Correo 2, Santiago*
*Chile*

In this note we characterize the itineraries associated to Lorenz maps and the conjugacy classes for this class of maps.

## 8.1 Introduction

In this note we will discuss the topological classification problem for discontinuous piecewise monotone maps on the interval.

**Definition 8.1.1** Let $f : X \to X$ and $g : Y \to Y$ be two continuous maps of topological spaces $X$ and $Y$ respectively. A continuous and onto map $\pi : X \to Y$ such that $g \circ \pi = \pi \circ f$ is called a factor map and $f$ is said to be an extension of $g$. We also say that $g$ and $f$ are semi conjugated. If $\pi$ is an homeomorphism we say that $f$ and $g$ are topologically conjugate or topologically equivalent.

Assume that $X = Y$. In this situation we can ask for necessary and sufficient conditions for the maps $f$ and $g$ to be topologically conjugated maps. This was one of the central topics in dynamics in the last thirty years. See for instance [3], [5], [7], [4], [13] and the references therein.

In this notes we discuss these questions for the special class of Lorenz maps of the interval.

**Definition 8.1.2**

a) Let $J \subseteq \mathbb{R}$. A map $f : J \to J$ is strictly increasing if for any $x, y \in J$, $x < y$, we have $f(x) < f(y)$.

b) Let $I = [0,1]$ and $c_0 = 0 < c_1 < c_2 < \ldots < c_{n-1} < c_n = 1$ be any sequence of $I$. A Lorenz map is a function

$$f : I \backslash \{c_1, c_2, \ldots, c_{n-1}\} \to I$$

which satisfies that $f|_{[0,c_1[}, f|_{]c_1,c_2[}, \ldots, f|_{]c_{n-1},1]}$ are continuous and strictly increasing maps (for $A \subseteq I$, $f|_A$ is the restriction of $f$ to $A$).

Let us denote this class of maps by $U(c_1, c_2, \ldots, c_{n-1})$.

Under additional conditions on their expanding properties, this class of maps has been studied by several authors in the past, due principally to the fact that they arise in the study of singular cycles (see [2], [8], [10], [11], [14], [15]).

For the class of expanding Lorenz maps the topological classification problem has been mainly solved by Hubbard and Sparrow in [9], and for continuous (strictly) piecewise monotone maps by S. Baldwin in [1].

In his contribution, S. Baldwin, provides a complete set of invariants for the conjugacy classes of piecewise monotone maps. Moreover, he provides a criterion to decide whether or not a potential invariant is actually realized by some piecewise monotone continuous function.

In this work we go on in this direction and we provide a version of these results for Lorenz maps. We make explicit a condition imposed by Hubbard and Sparrow in [9].

We have to point out that Glendinning and Sparrow have made (see [6]) a complete and interesting study of the dynamics and the kneading theory of topologically expansive Lorenz maps.

In a forthcoming paper (together with C. Moreira, see [12]) we study the bifurcation theory for generic two parameter families of Lorenz maps.

## 8.2    Statements of the Results

In paragraphs two and three we work with $n = 2$, $c_1 = c$ and with the additional conditions $\lim_{x \to c^-} f(x) = f(c^-) = 1$, $\lim_{x \to c^+} f(x) = f(c^+) = 0$, while in paragraph four we deal with the general case. We denote for simplicity $U(c) = U_c$.

Let $f \in U_c$. The first step is to construct a map $F$ associated to $f$ with the following property: the dynamics of the map $F$ will "essentially" be the dynamics of the map $f$.

To begin we introduce a small interval $I_\varepsilon(c)$ whose length is $\varepsilon$, instead of $c \in [0,1]$. For $n \geq 1$ and $x \in f^{-n}(\{c\})$ we introduce an interval, $I_{\frac{\varepsilon}{3^n}}(x)$, whose length is $\dfrac{\varepsilon}{3^n}$, instead of $x \in [0,1]$. Let $\widetilde{I}$ be the resulting interval (see Figure 8.1).

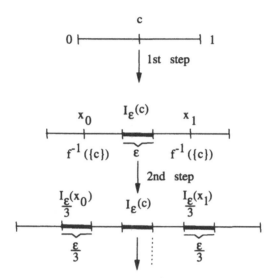

Figure 8.1:

We now modify $f$ in such a way that if $x \in f^{-n}(\{c\})$, $n \geq 1$, $x \neq 0$, $x \neq 1$, then $F\left(I_{\frac{\varepsilon}{3^n}}(x)\right) = I_{\frac{\varepsilon}{3^{n-1}}}(f(x))$ in a strictly monotone way (see Figure 8.2).

In the case $0 \in f^{-n}(\{c\})$ (or $1 \in f^{-n}(\{c\})$) we modify the map $f$ so that $F^n\left(I_{\frac{\varepsilon}{3^n}}(0)\right) = [c_1, c^+]$ (resp. $F^n\left(I_{\frac{\varepsilon}{3^n}}(1)\right) = [c^-, c_2]$) in a strictly monotone way. Here we use the notation $I_\varepsilon(c) = [c^-, c^+]$ and the points $c_1$ and $c_2$ satisfy

$$c^- < c_2 < \frac{c^- + c^+}{2} < c_1 < c^+.$$

To define the values $F(c^+)$ and $F(c^-)$ we have two possibilities:

(i) $0 \in f^{-n}(\{c\})$ or $1 \in f^{-m}(\{c\})$ for some $(n, m) \in \mathbb{N} \times \mathbb{N}$ or

(ii) $0 \notin f^{-n}(\{c\})$ and $1 \notin f^{-n}(\{c\})$ for any $n \geq 1$.

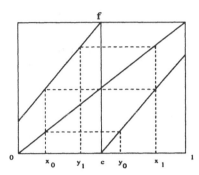 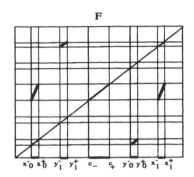

Figure 8.2: Construction of the map $F$. The first and the second step.

In the first case let $I_{\frac{5}{3^n}}(0) = ]0^-, 0^+]$ and $I_{\frac{5}{3^m}}(1) = [1^-, 1^+[$. Under these conditions we define $F(c^+) = 0^+$, $F(c^-) = 1^-$. In the second case we denote $\widetilde{I} = [0^+, 1^-]$ and we define $F(c^+) = 0^+$, $F(c^-) = 1^-$.

Let $C = \cup_{n \geq 0} f^{-n}(\{c\})$, and let $\mathcal{G} = \cup_{x \in C} \text{int}(I(x))$ be the union of the interior of the introduced intervals. Let $D = \widetilde{I} \backslash \mathcal{G}$.

We consider $I_0 = \widetilde{I} \backslash ]c^-, 1^+[$ and $I_1 = \widetilde{I} \backslash ]0^-, c^+[$. Now for any $x \in D$ and for any $n \geq 0$, we have $F^n(x) \in I_0 \cup I_1$, hence we can define a sequence $\Theta_F(x) : \mathbb{N} \cup \{0\} \to \{0, 1\}$ by $\Theta_F(x)(i) = j$ if and only if $F^i(x) \in I_j$. Thus, we have defined a map $\Theta_F : D \to \Sigma_2$ which satisfies $\sigma \circ \Theta_F = \Theta_F \circ F$. Here $\Sigma_2 = \{\Theta : \mathbb{N} \cup \{0\} \to \{0, 1\}\}$ is the set of one sided sequences with symbols 0 and 1 (a full shift in two symbols), and $\sigma : \Sigma_2 \to \Sigma_2$ is the usual shift map, $\sigma(\Theta)(i) = \Theta(i + 1)$.

**Remark 8.2.1**

1) In the set of sequences $\Sigma_2$ we will use the topology induced by the distance

$$d(\alpha, \beta) = \sum_{i=0}^{\infty} \frac{\bar{d}(\alpha(i), \beta(i))}{2^i},$$

where $\bar{d}(i, j) = \left\{ \begin{array}{ll} 1 & i \neq j \\ 0 & i = j \end{array} \right.$

2) In the set $\Sigma_2$ we have the lexicographical order defined as follows: given $\alpha, \beta \in \Sigma_2$ such that $\alpha(i) = \beta(i)$ for $0 \leq i < n$ and $\alpha(n) \neq \beta(n)$, if $\alpha(n) > \beta(n)$ then $\beta \leq \alpha$. When $\alpha \neq \beta$ and $\beta \leq \alpha$ we will use the notation $\beta < \alpha$.

It is obvious that $\Theta_F : D \rightarrow \Sigma_2$ is a continuous map which is onto its image $I_f = \Theta_F(D)$. Let $a = \Theta_F(0^+)$ and $b = \Theta_F(1^-)$ be the kneading sequences associated to $f$. It is clear that $a \leq \sigma^j(a) \leq b$ and $a \leq \sigma^j(b) \leq b$ for any $j \in \mathbb{N}$.

Initially we state the following result.

**Lemma 8.2.2** *Let $\Theta \in \Sigma_2$ be a sequence such that $a \leq \sigma^j(\Theta) \leq b$ for any $j \geq 0$, then there exists $x \in D$ such that $\Theta_F(x) = \Theta$.*

Now for $\Theta \in \Sigma_2$ such that $a \leq \sigma^j(\Theta) \leq b$ for any $j \geq 0$ we denote $I_\Theta = \{x \in D : \Theta_F(x) = \Theta\}$.

**Lemma 8.2.3** *$I_\Theta$ is a point or an interval.*

Let $I_f = I_f^1 \cup I_f^2$ be a disjoint decomposition of the set $I_f$, given by $I_f^1 = \{\Theta \in \Sigma_2 \colon a \leq \sigma^j(\Theta) \leq b$ for any $j \geq 0$ and there exists a unique $x \in D$ such that $\Theta_F(x) = \Theta\}$. That is, $I_f^1$ is the set of itineraries corresponding to images of points under $\Theta_F$ and $I_f^2 = I_f \backslash I_f^1$. It is obvious that $I_f^2$ is a countable set of sequences.

For any $\Theta \in I_f^2$, let $I_\Theta = [x_\Theta^-, x_\Theta^+]$, with $x_\Theta^- < x_\Theta^+$, be the interval which satisfies $\Theta_F(I_\Theta) = \{\Theta\}$. It is clear that $F : I_\Theta \rightarrow F(I_\Theta)$ is a homeomorphism and $F(I_\Theta) = I_{\sigma(\Theta)}$. We observe that $x < x_\Theta^-$, $x \in D$, implies $\Theta_F(x) \leq \Theta$ and $x_\Theta^+ < x$, $x \in D$, implies $\Theta_F(x) \geq \Theta$. Under these conditions we have the following result.

**Lemma 8.2.4** *Let $\Theta \in I_f^2$. If $F^i(I_\Theta) \cap I_\Theta \neq \emptyset$ for some $i \geq 1$, then $F^i(I_\Theta) = I_{\sigma^i(\Theta)} \subseteq I_\Theta$.*

**Corollary 8.2.5** *Let $\Theta \in I_f^2$. Assume there are positive integers $j, p$ such that $F^{j+p}(I_\Theta) \cap F^j(I_\Theta) \neq \emptyset$. Then $I_{\sigma^{j+p}(\Theta)} = F^{j+p}(I_\Theta) \subseteq F^j(I_\Theta) = I_{\sigma^j(\Theta)}$.*

**Remark 8.2.6** Under conditions of Lemma 8.2.4 we have that $\Theta$ is a periodic point for the shift map. Under conditions of Corollary 8.2.5 we have that $\Theta$ is a eventually periodic point for the shift map.

Let $DP_f$ be the set $\{\Theta \in I_f^2 : \Theta$ is a periodic point for the shift map$\}$ and $DEP_f$ be the set $\{\Theta \in I_f^2 : \Theta$ is a eventually periodic point for the shift map$\}$. Put $DE_f = I_f^2 \backslash (DP_f \cup DEP_f)$. For any $\Theta \in DE_f$ we must have $F^i(I_\Theta) \cap F^j(I_\Theta) = \emptyset$ for any $i, j \geq 0$, $i \neq j$. That is, $I_\Theta$ is a wandering interval for the dynamics of the map $F$.

We remark the following: let $\Theta_1, \Theta_2 \in DE_f$ and $J_i = \cup_{j \geq 0} F^j(I_{\Theta_i})$, $i = 1, 2$. Then the map $F|_{J_1}$ is topologically equivalent to $F|_{J_2}$. In fact, let $h : I_{\Theta_1} \rightarrow I_{\Theta_2}$ be the linear homeomorphism. For $x \in F^i(I_{\Theta_1})$ we define $h(x) = F^i(h(\bar{x}))$, where $\bar{x} \in I_{\Theta_1}$ is the point which satisfies $F^i(\bar{x}) = x$.

Clearly, $F^i \circ h(\bar{x}) = h \circ F^i(\bar{x})$. We note that this map does not necessarily extend to all the interval $\tilde{I}$.

Let $K$ be the set

$$K = \{g : [0,1] \to [0,1] : g \text{ is an increasing homeomorphism onto its image}\}.$$

Let $M = K/_\sim$ be the space of equivalence classes, where $\sim$ means "topological equivalence". For any $\Theta \in DP_f$, it is clear that there is a homeomorphism $\varphi : I_\Theta \to [0,1]$ such that $\varphi \circ F^i \circ \varphi^{-1} \in K$, where $i \geq 1$ is the first positive integer defined by $F^i(I) \subseteq I$. Let $R_i(f,\Theta)$ denote the equivalence class $[\varphi \circ F^i \circ \varphi^{-1}]_\sim \in M$. This equivalence class represents the dynamics of the map $F^i|_{I_\Theta}$.

In the same way, for any $\Theta \in DEP_f$, we have that there are integers $j, p \in \mathbb{N}$ and a homeomorphism $\varphi : F^j(I_\Theta) \to [0,1]$ such that $\varphi \circ F^p \circ \varphi^{-1} \in K$, where $p \geq 1$ is the integer defined by $F^p(F^j(I_\Theta)) \subseteq F^j(I_\Theta)$. We denote by $R_{j+p}(f,\Theta)$ the equivalence class $[\varphi \circ F^p \circ \varphi^{-1}]_\sim \in M$. This equivalence class represents the dynamics of the map $F^p|_{F^j(I_\Theta)}$.

**Lemma 8.2.7** *Let $f$, $g \in U_c$ be two Lorenz maps. If $f \sim g$ then,*

a) $I_f = I_g$,

b) $DP_f = DP_g$, $DEP_f = DEP_g$, $DE_f = DE_g$,

c) *for any $\Theta \in DP_f$ let $i = i_\Theta(f) \in \mathbb{N}$ be the integer such that $F^i(I_\Theta) = I_\Theta$ and $R_i(f,\Theta)$ be the corresponding equivalence class. We have that $\Theta \in DP_g$, $i_\Theta(f) = i_\Theta(g) = i$ and $R_i(f,\Theta) = R_i(g,\Theta)$,*

d) *for any $\Theta \in DEP_f$ let $j = j_\Theta(f) \in \mathbb{N}, p = p_\Theta(f) \in \mathbb{N}$ be the integers such that $F^{j+p}(I_\Theta) \subseteq F^j(I_\Theta)$. We have that $\Theta \in DEP_g$, $j_\Theta(g) = j_\Theta(f) = j$, $p_\Theta(g) = p_\Theta(f) = p$ and $R_{j+p}(f,\Theta) = R_{j+p}(g,\Theta)$.*

The converse of this result is also true.

**Lemma 8.2.8** *Assume that $f$, $g \in U_c$ are maps which satisfy:*

1. $I_f^1 = I_g^1$, $DP_f = DP_g$, $DEP_f = DEP_g$, $DE_f = DE_g$,

2. *for any $\Theta \in DP_f$ with period $i = i_\Theta$, we have $R_i(f,\Theta) = R_i(g,\Theta)$ and*

3. *for any $\Theta \in DEP_f$ we have $j = j_f(\Theta) = j_g(\Theta)$, $p = p_\Theta(f) = p_\Theta(g)$ and $R_{j+p}(f,\Theta) = R_{j+p}(g,\Theta)$.*

*Then $f$ is topologically equivalent to $g$.*

This lemma completes the characterization of the conjugacy class of a map $f \in U_c$. At this point we have to observe that this result is, essentially, a "folklore result" for this theory. We use the occasion to provide a proof of it.

We now state an interesting consequence of these results for expansive maps. Let us recall first the definition of expansive map.

**Definition 8.2.9** We say that $f \in U_c$ is an expansive map if there is $\varepsilon > 0$ with the following property: for any $x, y \in I \setminus \{c\}$, $x \neq y$, we can find $i \in \mathbb{N}$ such that $|f^i(x) - f^i(y)| > \varepsilon$.

Under these conditions, it is clear that $I_f = I_f^1 = \{\Theta \in \Sigma_2 : a_f \leq \sigma^j(\Theta) \leq b_f,$ for any $j \geq 0\}$, and we have

**Corollary 8.2.10** ([9]) *For expansive maps $f$, $g \in U_c$ we have $f \sim g$ if and only if $I_f = I_g$.*

**Remark 8.2.11** We remark that in the expansive case $I_f$ is a Cantor set.

Let us point out that for any $f \in U_c$ we have constructed sequences $a_f$, $b_f \in \Sigma_2$ that satisfy $a_f \leq \sigma^j(a_f) \leq b_f$, $a_f \leq \sigma^j(b_f) \leq b_f$ for any $j \geq 0$.

**Theorem 8.2.12** *Let $a$, $b \in \Sigma_2$ be two sequences that satisfy*

$$a \leq \sigma^j(a) \leq b, \quad a \leq \sigma^j(b) \leq b, \text{ for any } j \geq 0. \tag{8.1}$$

*There is a map $f \in U_c$ such that $a_f = a$, $b_f = b$.*

The last two results were proved by J. H. Hubbard and C. T. Sparrow in [9] for the special class of expansive maps. In this situation the sequences $a_f$ and $b_f$ satisfy:

$$a_f \leq \sigma^j(a_f) < b_f, \ a_f < \sigma^j(b_f) \leq b_f, \text{ for any } j \geq 0. \tag{8.2}$$

An interesting question, addressed in [9], is to characterize the sequences $(a, b) \in \Sigma_2 \times \Sigma_2$ that satisfy (8.2). The next results provide an answer to this question for sequences that satisfy (8.1). Essentially, we explain what (8.2) means in terms of 0's and 1's.

We now consider sequences $a$, $b \in \Sigma_2$ such that $a = (0, \dots)$ and $b = (1, \dots)$. Let $\Lambda_{a,b}$ denote the set

$$\Lambda_{a,b} = \{\theta \in \Sigma_2 : a \leq \sigma^n(\theta) \leq b, \text{ for all } n \in \mathbb{N} \}.$$

In the sequel we give necessary and sufficient conditions to have $a \in \Lambda_{a,b}$ or $b \in \Lambda_{a,b}$. First assume that $a = (0, 0, \dots) = \mathbb{O}$. Clearly $a \in \Lambda_{a,b}$. Assume that $b = (b_0, b_1, \ldots, b_k, \mathbb{O})$. Associated with $b$ we have a finite bisequence $A_b$ given by

$$A_b = \begin{bmatrix} 1 & 0 & 1 & \ldots & 1 \\ M_1(b) & N_1(b) & M_2(b) & \ldots & M_r(b) \end{bmatrix}$$

constructed in the following way: $M_1(b)$ is the number of consecutive one's starting from $b_0$; then we count the number of zero's and we denote it by $N_1(b)$ and so on.

Under these conditions we have the following result.

**Lemma 8.2.13** $b \in \Lambda_{a,b}$ *if and only if* $M_i(b) \le M_1(b)$ *for any* $i = 1, 2, \ldots, r$, *and each time that* $M_i(b) = M_1(b)$ *we have* $N_1(b) \le N_i(b)$.

Now, let us assume that $\sigma^j(b) \ne \mathbb{O}$ for all $j \in \mathbb{N}$. If $\sigma^j(b) = (1, 1, \ldots) = \mathbb{1}$ for some $j \in \mathbb{N}$ then $b \notin \Lambda_{a,b}$ unless $b = \mathbb{1}$. Hence, without loss of generality we can assume that $\sigma^j(b) \ne \mathbb{1}$.

As before, we can associate to $b$ an infinite bisequence:

$$A_b = \begin{bmatrix} 1 & 0 & 1 & 0 & \ldots \\ M_1(b) & N_1(b) & M_2(b) & N_2(b) & \ldots \end{bmatrix}$$

In the same way as in the previous case we have the result.

**Lemma 8.2.14** $b \in \Lambda_{a,b}$ *if and only if* $M_i(b) \le M_1(b)$ *for any* $i > 1$, *and each time that* $M_i(b) = M_1(b)$ *then* $N_1(b) \le N_i(b)$.

Let us assume that $a \ne \mathbb{O}$, $b \ne \mathbb{1}$ and $\sigma^i(a) \ne \mathbb{O}$, $\sigma^i(a) \ne \mathbb{1}$, $\sigma^i(b) \ne \mathbb{O}$, $\sigma^i(b) \ne \mathbb{1}$ for all $i \in \mathbb{N}$. As in the previous cases we can associate to $a$ and $b$ the infinite bisequences $A_a$, $A_b$.

**Lemma 8.2.15** $b \in \Lambda_{a,b}$ *if and only if*

*(i)* $M_i(b) \le M_1(b)$ *and* $N_i(b) \le N_1(1a)$ *for any* $i \in \mathbb{N}$,

*(ii) if* $N_1(1a) = N_i(b)$ *then* $M_{i+1}(b) \ge M_1(1a)$,

*(iii) if* $M_i(b) = M_1(b)$ *then* $N_1(b) \le N_i(b)$.

We point out that a similar characterization is true for an element $\theta \in \Lambda_{a,b}$, $\theta = (1, \ldots)$. This result provides necessary and sufficient conditions for sequences $1a$ and $b$ to belong to the set $\Lambda_{a,b}$. Now we will assume that this is not the case and we will compute the extreme points in $\Lambda_{a,b}$.

**Remark 8.2.16** Assume $b \in \Sigma_2$, $b \ne \mathbb{1}$, $b = (1, \ldots)$, is a sequence that satisfies

$$\sup\{\sigma^n(b) : n \ge 1\} = \mathbb{1}.$$

In this situation for any $a \in \Sigma_2$, $a = (0, \ldots)$, we must have $b \notin \Lambda_{a,b}$. This set of sequences includes the preimages of $\mathbb{1}$ and all the sequences with dense orbit in $\Sigma_2$.

Now, let us assume that $a = \mathbb{O}$ and $b \notin \Lambda_{a,b}$. In this situation we can find $k_b \in \mathbb{N}$ such that $\sigma^{k_b+1}(b) > b$ and $\sigma^i(b) \leq b$, $i = 0, 1, \ldots k_b$. Let $b = (b_0, b_1, \ldots, b_{k_b}, b_{k_b+1}, \ldots)$ and $\beta$ be the periodic sequence whose period is $(b_0, b_1, \ldots, b_{k_b})$.

**Theorem 8.2.17** $\beta = \sup(\Lambda_{a,b})$.

Assume that $a \neq \mathbb{O}$ and $a \notin \Lambda_{a,b}$. As before let $k_a \in \mathbb{N}$ be the first integer such that: $\sigma^{k_a+1}(a) \notin [a, b] = \{\theta \in \Sigma_2 : a \leq \theta \leq b\}$. Let $a = (a_0, a_1, \ldots, a_{k_a}, a_{k_a+1}, \ldots)$ and $\alpha$ be the periodic sequence whose period is $(a_0, a_1, \ldots, a_{k_a})$.

**Theorem 8.2.18** $\alpha = \inf(\Lambda_{a,b})$.

Let us now state a result which is a generalization of Lemma 8.2.15. To do so, let $a_0^+ = (0, \ldots), a_1^- = (0, \ldots)$, $a_1^+ = (1, \ldots)$, and $a_2^- = (1, \ldots)$ be sequences that satisfy $a_0^+ \leq a_1^- < a_1^+ \leq a_2^-$.

Let $\Lambda(a_0^+, a_1^-, a_1^+, a_2^-)$ be the set

$$\Lambda(a_0^+, a_1^-, a_1^+, a_2^-) = \{\theta \in \Sigma_2 : \sigma^i(\theta) \in [a_0^+, a_1^-] \cup [a_1^+, a_2^-], \text{ for any } i \in \mathbb{N}\}.$$

Let $a = \sigma(a_1^+)$ and $b = \sigma(a_1^-)$. We assume (without loss of generality) that $a_0^+ \leq a$ and $b \leq a_2^-$. In this situation for $\theta_0 = 0$ or 1 we define

$$\theta_0^r \Lambda_{a,b} = \{\underbrace{(\theta_0, \ldots, \theta_0}_{r \text{ times}}, \theta) : \theta \in \Lambda_{a,b}\}.$$

We have,

**Theorem 8.2.19**

$$\Lambda(a_0^+, a_1^-, a_1^+, a_2^-) = \bigcup_{r \geq 0} \{0^r \Lambda(a, b) \cup 1^r \Lambda(a, b)\} \cap [a_0^+, a_2^-].$$

**Remark 8.2.20** To finish let us observe that we have used the fact that $f(c^-) = 1$ and $f(c^+) = 0$. For instance, if we have $f([0, c^-[) \subseteq [0, c^-[$ and $f([c^+, 1[) \subseteq ]c^+, 1]$ then $a_f = \mathbb{O}$ and $b_f = \mathbb{1}$, and Lemma 8.2.2 is not true.

In section 8.4 we state some of these results for the general case.

# 8.3    Proof of the Results

**Proof of Lemma 8.2.2.** A simple proof of this fact is the following: modify the map F in such a way that the new modification, say G, is an extension of the map F and is semiconjugated to the fullshift $\Sigma_2$. For the map G (whose domain is the interval $J$) and for any $\Theta \in \Sigma_2$ we have that the set

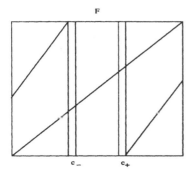 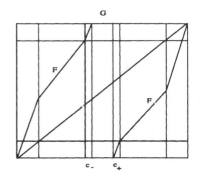

Figure 8.3: Extension of the map $F$.

$I_\Theta = \{x \in I : \Theta_G(x) = \Theta\}$ is nonempty (see Figure 8.3). Now, since the map $G$ is an extension of the map $F$ we have, for any $\Theta \in \Sigma_2$ satisfying the hypothesis of the lemma, that the respective set is included in the domain of definition of the map F. Now the lemma follows. ∎

**Proof of Lemma 8.2.3.** This is an easy consequence of the fact that $x \leq y$ implies $\Theta_F(x) \leq \Theta_F(y)$ for $x, y \in D$. ∎

**Proof of Lemma 8.2.4.** In fact, since $\Theta_F$ is order-preserving and $I_\Theta = [x_\Theta^-, x_\Theta^+]$, we have that $x < x_\Theta^-$ implies $\Theta_f(x) < \Theta_f(x_\Theta^-) = \Theta$ and that $x_\Theta^+ < x$ implies $\Theta = \Theta_F(x_\Theta^+) < \Theta_F(x)$.

For $x \in I_\Theta \cap F^i(I_\Theta)$ we have $x = F^i(y)$ for some $y \in I_\Theta$. Since $\Theta = \Theta_F(x) = \Theta_F(F^i(y)) = \sigma^i \circ \Theta_F(y) = \sigma^i(\Theta)$ we conclude that $\Theta$ is a periodic orbit for $\sigma^i$. Let $z \in F^i(I_\Theta)$, then there is a point $y \in I_\Theta$ such that $F^i(y) = z$. Under these conditions $\Theta_F(z) = \Theta_F(F^i(y)) = \sigma^i(\Theta_F(y)) = \sigma^i(\Theta) = \Theta$, that is $z \in I_\Theta$. Therefore, we have $F^i(I_\Theta) \subseteq I_\Theta$. ∎

The proof of Corollary 8.2.5 follows similarly.

The proof of the Lemma 8.2.7 follows from the following fact: the existence of a topological conjugacy between $f$ and $g$ implies that the combinatorial type of the map $f$ and $g$ are the same.

Let us now give a proof of Lemma 8.2.8.

**Proof of Lemma 8.2.8.** We construct a homeomorphism $h : [0,1] \to [0,1]$ such that $h(c) = c$ and $h \circ f = g \circ h$.

Since $DP_f = DP_g$ and for any $\Theta \in DP_f$ we have $R_{i_\Theta}(f, \Theta) = R_{i_\Theta}(g, \Theta)$, then we can construct a homeomorphism

$h : \cup_{\Theta \in DP_f} \cup_{i=0}^{i_\Theta - 1} F^i(I_\Theta) \to \cup_{\Theta \in DP_f} \cup_{i=0}^{i_\Theta - 1} G^i(I_\Theta)$ which satisfies $h \circ F = G \circ h$.

In the same way, we construct a homeomorphism

$$h : \bigcup_{\Theta \in DEP_f} \left[ \bigcup_{i=0}^{j_\Theta - 1} F^i(I_\Theta) \cup \bigcup_{i=0}^{p_\Theta - 1} F^i(F^{j_\Theta}(I_\Theta)) \right]$$

$$\rightarrow \bigcup_{\Theta \in DEP_g} \left[ \bigcup_{i=0}^{j_\Theta - 1} G^i(I_\Theta) \cup \bigcup_{i=0}^{p_\Theta - 1} G^i(G^{j_\Theta}(I_\Theta)) \right]$$

which satisfies $h \circ F = G \circ h$.

Now, since the dynamics of $F : D_f \rightarrow D_f$ and $G : D_g \rightarrow D_g$ are semi-conjugated to $\sigma : I_f \rightarrow I_f$ we can use these semiconjugacies to extend our previous $h$ to all of $D_f$. We thus construct $h : D_f \rightarrow D_g$ such that $h \circ F = G \circ h$.

To obtain $h : [0,1] \rightarrow [0,1]$ such that $h(c) = c$ and $h \circ f = g \circ h$ we only need to collapse all the intervals that we have introduced instead of the elements of the preimages $f^{-n}(\{c\})$, $n \geq 1$. Since this map is order-preserving and bijective then it is continuous. We construct its inverse map analogously. ∎

Let us now give a proof of Theorem 8.2.12.

**Proof of Theorem 8.2.12.** If $a = (0,0,0,\ldots) = \mathbb{0}$ and $b = (1,1,\ldots,1) = \mathbb{1}$, the linear map, $L$, given by Figure 8.4 satisfies the conditions.

Assume that one of the sequences $a$ or $b$ satisfies $a \neq (0,0,\ldots)$ or $b \neq (1,1,\ldots)$. Let $x,y \in ([0,1] \setminus \{c\})$ be the points which satisfy $\Theta_L(x) = a$, $\Theta_L(y) = b$ for the previous linear map. Let $I_0 \subseteq ]x,y[$ be the set $\{z \in [x,y] : L(z) \notin ]x,y[\}$ (see Figure 8.5). We identify all the points in $I_0$ with the point $c$. We proceed in a similar way with all the preimages $L^{-i}(I_0) \subseteq [x,y]$.

Now we reparametrize the set $[\bar{x},\bar{y}]$ (if it is necessary we introduce intervals in the place of points) to obtain a map that satisfies the conclusion of the lemma. ∎

Let us now give a proof of Theorem 8.2.17.

**Proof of Theorem 8.2.17.** In the sequel we will use $I_0 = [\mathbb{0},\mathbb{01}] \subseteq \Sigma_2$ and $I_1 = [\mathbb{10},\mathbb{1}] \subseteq \Sigma_2$.

Let us now assume that $a = \mathbb{0}$ and $b \notin \Lambda_{a,b}$. In this situation we can find $k = k_b \in \mathbb{N}$ such that $\sigma^{k+1}(b) > b$, that is, $\sigma^k(b) \in ]0b,0\mathbb{1}[$ and $\sigma^i(b) \in [\mathbb{0},b]$, $i = 0,1,\ldots k$. Let $b = (b_0, b_1, \ldots, b_k, b_{k+1}, \ldots)$ and $\beta$ be the periodic sequence whose period is $(b_0, b_1, \ldots, b_k)$. First we prove that $\beta \in \Lambda_{a,b}$.

Since $\sigma^{k+1}(b) > b$ we have

$$(b_{k+1}, b_{k+2}, \ldots, b_{2k+1}, b_{2k+2}, \ldots) > (b_0, b_1, \ldots).$$

If $b_{k+1} = b_0$, $b_{k+2} = b_1, \ldots, b_{k+j+1} = b_j$, for some $j \leq k-1$, and $b_{k+j+2} > b_{j+1}$ then obviously $b > \beta$. Otherwise, the same is true for some sequence

$$(b_{k+1}, b_{k+2}, \ldots, b_{2(k+1)}, b_{2(k+1)+1}, \ldots, b_{3(k+1)}, b_{3(k+1)+1} \ldots)$$

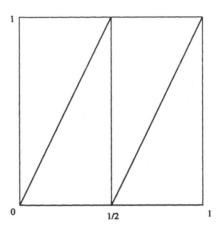

Figure 8.4: The map $L$

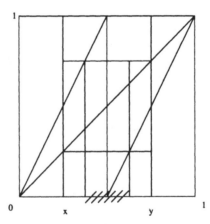

Figure 8.5: Let $g : [x,y]|_\sim = [\overline{x},\overline{y}] \to [\overline{x},\overline{y}]$ be the resulting map. This map $g$ satisfies $\Theta_g(\overline{x}) = a$, $\Theta_g(\overline{y}) = b$.

and we conclude that $\beta < b$. This implies $\beta \in [0, b]$.

Let now consider

$$I_{b_0}, I_{b_0,b_1} = I_{b_0} \cap \sigma^{-1}(I_{b_1}), \ldots, I_{b_0,b_1,\ldots,b_k} = I_{b_0} \cap \sigma^{-1}(I_{b_1}) \cap \ldots \cap \sigma^{-k}(I_{b_k}).$$

Clearly

$$\sigma(I_{b_0,b_1}) = I_{b_1}, \sigma^2(I_{b_0,b_1,b_2}) = I_{b_2}, \ldots, \sigma^k(I_{b_0,b_1,\ldots,b_k}) = I_{b_k}.$$

Since $[\beta, b] \subseteq I_{b_0,b_1,\ldots,b_k}$ we conclude that $\sigma([\beta, b]) = [\sigma(\beta), \sigma(b)] \subseteq I_{b_1,\ldots,b_k}$, $\sigma^2([\beta, b]) = [\sigma^2(\beta), \sigma^2(b)] \subseteq I_{b_2,\ldots,b_k}, \ldots, \sigma^k([\beta, b]) = [\sigma^k(\beta), \sigma^k(b)] \subseteq I_{b_k}$, therefore $\sigma^j(\beta) \leq \sigma^j(b)$ for $j = 0, \ldots, k$. Since $\sigma^j(b) \leq b$ for $j = 0, \ldots, k$, we get $\beta \in \Lambda_{a,b}$.

Let us now prove that $\beta = \sup(\Lambda_{a,b})$. First we observe, since the shift map is expansive, that:

$$I_{\Theta(0)} \cap \sigma^{-1}(I_{\Theta(1)}) \cap \sigma^{-2}(I_{\Theta(2)}) \ldots = \cap_{i \geq 0} \sigma^{-i}(I_{\Theta(i)}) = \Theta$$

for any $\Theta \in \Sigma_2$.

Let $\alpha \in \Sigma_2$ be a sequence that satisfies $\beta < \alpha < b$. We have $\sigma^i(\alpha) \in [\sigma^i(\beta), \sigma^i(b)]$, $i = 0, 1, \ldots, k$ and $\sigma^{k+1}(\alpha) \in [\sigma^{k+1}(\beta), \sigma^{k+1}(b)] = [\beta, \sigma^{k+1}(b)]$. If $\sigma^{k+1}(\alpha) > b$ then $\alpha \notin \Lambda_{a,b}$. Otherwise $\sigma^{k+1+j}(\alpha) \in [\sigma^j(\beta), \sigma^j(b)]$ and $\sigma^{2(k+1)}(\alpha) \in [\beta, \sigma^{k+1}(b)]$. Since, we have $d(\beta, \sigma^{2(k+1)}(\alpha)) > d(\beta, \sigma^{k+1}(\alpha))$, then $\sigma^{2(k+1)}(\alpha) > b$ or $\sigma^{2(k+1)}(\alpha) \in ]\sigma^{k+1}(b), b[$ and so on. Hence, we conclude, since the shift map is expansive, that there is $j_0 \in \mathbb{N}$ such that $\sigma^{j_0(k+1)}(\alpha) > b$. This implies $\alpha \notin \Lambda_{a,b}$. This fact, serves to conclude the proof of Theorem 8.2.17. ∎

In the same way we prove Theorem 8.2.18. The results in Lemma's 8.2.13, 8.2.14, 8.2.15 and Theorem 8.2.19 are easy and we left them to the reader.

# 8.4   The General Case

Now we deal with the general case for Lorenz maps. We have to consider the set $C = \cup_{i=1}^{n-1} \cup_{j \geq 0} f^{-j}(\{c_i\})$. Since the map $f \in U(c_1, \ldots, c_{n-1})$ is discontinuous in the set $\{c_1, c_2, \ldots, c_{n-1}\}$ we have that the itinerary $\Theta_f(x)$ is not defined for $x \in C$. In the complement of this set, which we will call $D$, the map $x \to \Theta_f(x)$ is continuous. We observe that the right and left itineraries are well defined for any point in $[0, 1]$. Let $\Theta_f(c_i^+) = \lim_{x \downarrow c_i} \Theta_f(x)$ where the limit is taken over all $x \in D$ located to the right hand side of the point $c_i$. In a similar way we define $\Theta_f(c_i^-)$. We use the right hand and left hand definitions with $c_0$ and $c_n$ respectively. In the same way we proceed to define the itineraries for $x \in C$. Now the itineraries are points in the full shift of $n$ symbols $\Sigma_n = \{\Theta : \mathbb{N} \cup \{0\} \to \{0, 1, \ldots, n-1\}\}$. Let

$I_f = \{\Theta_f(x^+)|x \in I\} \cup \{\Theta_f(x^-)|x \in I\}$ be the set of itineraries associated to $f$.

Let us denote by

$$A_f = [\Theta_f(0^+), \Theta_f(c_1^-)] \cup [\Theta_f(c_1^+), \Theta_f(c_2^-)] \ldots [\Theta_f(c_{n-1}^+), \Theta_f(1^-)].$$

We have the following result.

**Theorem 8.4.1** $I_f = \cap_{n \geq 0} \sigma^{-n}(A_f)$.

**Proof.** It is clear, for any $x \in D$, that $\Theta_f(x) \in A_f$. As a consequence we obtain $\{\Theta_f(x^+), \Theta_f(x^-)\} \subseteq A_f$. Since $f(D) \subseteq D$ and $\Theta_f \circ f = \sigma \circ \Theta_f$ on the set $D$, then we must have $\Theta_f(D) \subseteq \cap_{n \geq 0} \sigma^{-n}(A_f)$.

Assume $f(c_i^+) = \lim_{x \to c_i^+}(f(x)) \in D$. In this situation, $\sigma(\Theta_f(c_i^+)) = \lim_{x \to c_i^+}(\sigma \circ \Theta_f(x)) = \lim_{x \to c_i^+}(\Theta_f \circ f(x)) = \Theta_f \circ f(c_i^+) \in D$, that is, $\theta(c_i^+) \in \sigma^{-1}(A_f)$. Since $f^n(f(c_i^+)) \in D$ we must have $\theta_f(c_i^+) \in \cap_{n \geq 0} \sigma^{-n}(A_f)$.

Now assume $f(c_i^+) = c_j$. In this case, $\sigma(\Theta_f(c_i^+)) = \lim_{x \to c_i^+}(\sigma \circ \Theta_f(x)) = \lim_{x \to c_i^+}(\Theta_f \circ f(x)) = \Theta_f(\lim_{x \to c_i^+} f(x)) = \Theta_f \circ f(c_j^+)$. That is, $\theta_f(c_i^+) \in \sigma^{-1}(A_f)$.

If $f(c_j^+) \in D$ we will have $\theta_f(c_i^+) \in \sigma^{-n}(A_f)$ and we are done. If $f(c_j^+) = c_k$ then we will have $\theta_f(c_i^+) \in \sigma^{-2}(A_f)$ and we can obtain the result inductively. Hence, we conclude $I_f \subseteq \cap_{n \geq 0} \sigma^{-n}(A_f)$.

To obtain the converse we will make an additional construction. Let us assume that $f(0) \in ]0, 1[$ and $f(1) \in ]0, 1[$. For any $i = 1, 2, \ldots, n - 1$ let $I_i = [c_i^-, c_i^+]$ be an interval whose length is $\frac{\varepsilon}{n}$. We put it in the place of $c_i$. For any $x \in f^{-1}(\{c_i\})$ let us put an interval $I_i^1(x) = [x_{i,1}^-, x_{i,1}^+]$ whose length is $\frac{\varepsilon}{3n^2}$. Now we define a map $F_1$ extending $f$ with the property $F_1(I_i^1(x)) = I_i$ in an increasing way. Inductively, in any point $x \in f^{-j}(\{c_i\})$ we introduce an interval $I_i^j(x) = [x_{i,j}^-, x_{i,j}^+]$, whose length is $\frac{\varepsilon}{3^j n^{j+1}}$ and we define $F_j$ extending $F_{j-1}$ with the property $F_j(I_i^j(x)) = I_i^{j-1}(f(x))$ in an increasing way.

Denote by $\tilde{I}$ the resulting interval and by $F_0 : \tilde{I} \setminus \cup_{i=1}^{n-1}]c_i^-, c_i^+[\to \tilde{I}$ the resulting map. Let $J_i = [d_i^-, d_i^+] \subseteq ]c_i^-, c_i^+[$ be a closed interval. Now, we extend the map $F_0$ to the set $\tilde{I} \setminus \cup_{i=1}^{n-1}]d_i^-, d_i^+[$ in the following way. Let $\tilde{I} = [\tilde{a}, \tilde{b}]$:

1. if $F_0(\tilde{a}) = \tilde{a}$ and $F_0(\tilde{b}) = \tilde{b}$, then $F(c_i^+) = f(c_i^+)$, $F(c_i^-) = f(c_i^-)$, $F(d_i^-) = \tilde{b}, F(d_i^+) = \tilde{a}$, and the restrictions $F|_{]c_i^-, d_i^-[}$ and $F|_{]d_i^+, c_i^+[}$ are linear maps,

2. if $F_c(\tilde{a}) \in ]\tilde{a}, \tilde{b}[$ and $F_0(\tilde{b}) \in ]\tilde{a}, \tilde{b}[$ then we add intervals $[\tilde{a} - \varepsilon, \tilde{a}]$ and $[\tilde{b}, \tilde{b} + \varepsilon]$, and we proceed to extend the map $F$ to all of $\tilde{J} = [\tilde{a} - \varepsilon, \tilde{b} + \varepsilon]$, as in the previous case, in such a way that $F(\tilde{a} - \varepsilon) = \tilde{a} - \varepsilon$ and $F(\tilde{b} + \varepsilon) = \tilde{b} + \varepsilon$.

The resulting map satisfies the following property. Let

$$\Lambda_F = \{x \in \tilde{I} : F^n(x) \text{ is defined for any } n \geq 0\},$$

then the restriction map $F|_{\Lambda_F} : \Lambda_F \to \Lambda_F$ is semiconjugated to the shift map $\sigma : \Sigma_n \to \Sigma_n$. In particular the semiconjugacy $h : \Lambda_F \to \Sigma_n$ is a surjective map and, therefore, for any $\Theta \in \Sigma_n$ we have that the set $I_\Theta = \{x \in \tilde{I} : h(x) = \Theta\}$ is non-empty. This map, h, is exactly the itinerary function and it is order preserving between $\tilde{I}$ and $\Sigma_{n-1}$ with the lexicographical order. Hence the set $I_\Theta$ is a point or an interval.

Since $\Theta_F(c_i^+) = \Theta_f(c_i^+)$ and $\Theta_F(c_i^-) = \Theta_f(c_i^-)$ we have that

$$I_f = h(\cap_{n \geq 0} F^{-n} \left([0^+, c_1^-] \cup [c_1^+, c_2^-] \ldots [c_{n-1}^+, 1^-]\right)) \supset \cap_{n \geq 0} \sigma^{-n}(A_f).$$

In this way we get the result. ∎

Now, completed the proof of this theorem, we obtain, in the same way as before, the results in Lemma 8.2.2 - Lemma 8.2.8 for maps in $U(c_1, c_2, \ldots, c_{n-1})$.

We point out that given $\Theta \in \Sigma_2$ such that $\sigma^j(\Theta) \in A_f$, $j \geq 0$, then, since $I_f = \cap_{j \geq 0} \sigma^{-j}(A_f)$, there is $x \in I$ such that $\Theta = \Theta_f(x_i^+)$ or $\Theta = \Theta_f(x_i^-)$. Hence we get a result similar to Lemma 8.2.2.

In the same way we get the result in Lemma 8.2.4 and Corollary 8.2.5.

**Remark 8.4.2** Now we have to observe the following. The main idea behind the results in Section 8.2 and 8.3 is the fact that the full shift $\sigma : \Sigma_2 \to \Sigma_2$ is topologically equivalent to the dynamics of the map $L_2$ given in Figure 8.6.

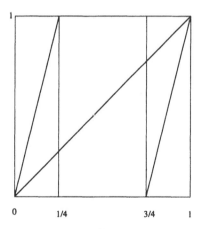

Figure 8.6: The map $L_2$.

For any Lorenz map $f : [0,1] \setminus \{c\} \to [0,1]$ let $x_f^- \in [0, \frac{1}{4}]$ and $x_f^+ \in [\frac{3}{4}, 1]$ be the points given by $\Theta_{L_2}(x_f^-) = a_f$, $\Theta_{L_2}(x_f^+) = b_f$.

Let $x_f^- < x_f^{+,-} < x_f^{-,+} < x_f^+$ be the points defined by $f_\infty(x_f^{+,-}) = x_f^+$, $f(x_f^{-,+}) = x_f^-$. We have that $I_f = \cap_{j=0}^\infty \sigma^{-j}([\Theta_{L_2}(x_f^-), \Theta_{L_2}(x_f^{+,-})] \cup [\Theta_{L_2}(x_f^{-,+}), \Theta_{L_2}(x_f^+)]$. That is essentially (that is up to semi conjugacy) the dynamics of the map $f$ is a factor of the dynamics of the map $L_2$.

For the general case we will have a similar result for the map $L_n$ and, therefore, we can get the same kind of result.

For instance, in this situation Theorem 8.2.12 has the following form: let

$$\Theta(0^+) \leq \Theta_1^- < \Theta_1^+ \leq \Theta_2^- < \cdots \leq \Theta_{n-1}^- < \Theta_{n-1}^+ \leq \Theta(1^-)$$

be sequences in $\Sigma_n$ such that, if

$$A = [\Theta(0^+), \Theta_1^+] \cup \cdots \cup [\Theta_{n-1}^+, \Theta(1^-)]$$

then $\sigma^n(\Theta) \in A$ for any $\Theta \in \{\Theta(0^+), \Theta_1^-, \dots, \Theta(1^-)\}$ and $n \in \mathbb{N}$.

**Theorem 8.4.3** *There is a map $f \in U(c_1, \dots, c_{n-1})$ such that*

$$\Theta_F(0^+) = \Theta(0^-), \dots, \Theta_F(1^-) = \Theta(1^-).$$

**Proof.** If all the sequences are different then it is enough to consider a linear map, $L$, topologically equivalent to the full shift $\sigma : \Sigma_n \to \Sigma_n$. We distinguish the points $x(0^+) < x_1^- < \cdots < x_{n-1}^+ < x(1^-)$ in $I$ which correspond, under the conjugacy, to the sequences $\Theta(0^+), \Theta_1^-, \dots, \Theta_{n-1}^+, \Theta(1^-)$. Then we identify the intervals $J_1 = ]x_1^-, x_1^+[, \dots, J_{n-1} = ]x_{n-1}^-, x_{n-1}^+[$ with points $x_1 \dots x_{n-1}$ and the intervals $[0, x(0^+)]$ with $x(0^+)$ and $[x(1^-), 1]$ with $x(1^-)$. In the same way we proceed with the connected components of the preimages $L^{-i}(J_s)$ which are included in $[x(0^+), x(1^-)]$. Let $\tilde{I}$ be the resulting set and $\tilde{L} : \tilde{I} \to \tilde{I}$ be the resulting map. Now we reparametrize the set $\tilde{I}$ (if necessary introducing intervals in the place of points) to obtain a new domain, which is an interval $I$, and a new map $f : I \setminus \{c_0, \dots, c_{n-1}\} \to I$ which satisfies the conclusion of the theorem.

Now, let us assume that two of the given sequences are equal, for instance $\Theta(0^+) = \Theta_1^-$. Let us consider $\{L^n(x(0^+)) : n \geq 0\}$. In any of these points we introduce an interval $I_n$ whose length is $\frac{\varepsilon}{3^n}$. We modify the map $L$ in such a way that the modification sends the interval $I_n$ into the interval $I_{n+1}, n \geq 0$. For the resulting map we proceed as before identifying the right hand limit of the interval $I_0$ with the left hand limit of $[\Theta_1^+, \Theta_2^-]$ and so on. ∎

**Acknowledgment.** Partially supported by Fondecyt grants 1941080, 1961212, 1970720 and Dicyt- Usach.

The author wishes to thank IMPA - Brazil and ICTP - Trieste - Italy for their support and hospitality while preparing part of these notes.

# Bibliography

[1] S. Baldwin, *A Complete Classification of the Piecewise Monotone Functions on the Interval*, Trans. Am. Math. Soc. **319(1)**, pp. 155–178 (1990).

[2] R. Bamón, R. Labarca, R. Mañé, M.J. Pacifico, *The Explosion of Singular Cycles*, M.J. Publ. Math. I.H.E.S. (1993).

[3] S.S. Chern, S. Smale, *Global Analysis. Proceedings of Symposia in Pure Mathematics*, Vol XIV, AMS (1970).

[4] W. de Melo, J. Palis, *Geometric Theory of Dynamical Systems*, Springer Verlag (1982).

[5] W. de Melo, S. Van Strien, *One Dimensional Dynamics*, Springer Verlag (1993).

[6] P. Glendinning, C. T. Sparrow, *Prime and Renormalizable Kneading Invariants and the Dynamics of Expanding Lorenz Maps*, Physica D **62**, pp. 22–50 (1993).

[7] J. Guckenheimer, P. Holmes, *Nonlinear Oscillations, Dynamical Systems and Bifurcations of Vector Fields*, Springer Verlag (1983).

[8] J. Guckenheimer, R. F. Williams, *Structural Stability of Lorenz Attractors*, Publ. Math. I.H.E.S. **50**, pp. 59–72 (1979).

[9] J. H. Hubbard, C. T. Sparrow, *The Classification of Topologically Expansive Lorenz Maps*, Comm. on Pure and App. Math. **43(4)**, pp. 431–444 (1990).

[10] R. Labarca, *Bifurcations of Contracting Singular Cycles*, Ann. Scient. de L'école Norm Sup. **T28(4)** (1995).

[11] R. Labarca, *Unfolding Singular Cycles*, preprint (1996).

[12] R. Labarca, C. Moreira, *On the Problem of the Existence of Universal Two Parameter Families for Discontinuous Maps on the Interval and the Bifurcation Scenario for Lorenz-like Flows*, preprint (1999).

[13] C. Robinson, *Dynamical Systems, Stability, Symbolic Dynamics and Chaos*, CRS Press (1995).

[14] Ch. Tresser, R. F. Williams, *Splitting Words and Lorenz Braids*, Physica D **62**, pp. 15–21 (1993).

[15] R. F. Williams, *The Structure of the Lorenz Attractors*, Publ. Math. I.H.E.S. **50**, pp. 94-112 (1979).

Printed in the United States
By Bookmasters